The Chloroplast:
From Molecular Biology to Biotechnology

NATO Science Series

A Series presenting the results of activities sponsored by the NATO Science Committee. The Series is published by IOS Press and Kluwer Academic Publishers, in conjunction with the NATO Scientific Affairs Division.

General Sub-Series

A. Life Sciences	IOS Press
B. Physics	Kluwer Academic Publishers
C. Mathematical and Physical Sciences	Kluwer Academic Publishers
D. Behavioural and Social Sciences	Kluwer Academic Publishers
E. Applied Sciences	Kluwer Academic Publishers
F. Computer and Systems Sciences	IOS Press

Partnership Sub-Series

1. Disarmament Technologies	Kluwer Academic Publishers
2. Environmental Security	Kluwer Academic Publishers
3. High Technology	Kluwer Academic Publishers
4. Science and Technology Policy	IOS Press
5. Computer Networking	IOS Press

The Partnership Sub-Series incorporates activities undertaken in collaboration with NATO's Partners in the Euro-Atlantic Partnership Council – countries of the CIS and Central and Eastern Europe – in Priority Areas of concern to those countries.

NATO-PCO-DATA BASE

The NATO Science Series continues the series of books published formerly in the NATO ASI Series. An electronic index to the NATO ASI Series provides full bibliographical references (with keywords and/or abstracts) to more than 50000 contributions from international scientists published in all sections of the NATO ASI Series.
Access to the NATO-PCO-DATA BASE is possible via CD-ROM "NATO-PCO-DATA BASE" with user-friendly retrieval software in English, French and German (© WTV GmbH and DATAWARE Technologies Inc. 1989).

The CD-ROM of the NATO ASI Series can be ordered from: PCO, Overijse, Belgium.

3. High Technology – Vol. 64

The Chloroplast: From Molecular Biology to Biotechnology

edited by

Joan H. Argyroudi-Akoyunoglou
Institute of Biology,
NCSR "Demokritos",
Athens, Greece

and

Horst Senger
Fachbereich Biologie/Botanik,
Philipps Universität,
Marburg, Germany

Kluwer Academic Publishers

Dordrecht / Boston / London

Published in cooperation with NATO Scientific Affairs Division

Proceedings of the NATO Advanced Research Workshop on
The Chloroplast: From Molecular Biology to Biotechnology
Kolymbari-Chania, Crete, Greece
10–15 August 1998

A C.I.P. Catalogue record for this book is available from the Library of Congress

ISBN 0-7923-5576-8

Published by Kluwer Academic Publishers,
P.O. Box 17, 3300 AA Dordrecht, The Netherlands.

Sold and distributed in North, Central and South America
by Kluwer Academic Publishers,
101 Philip Drive, Norwell, MA 02061, U.S.A.

In all other countries, sold and distributed
by Kluwer Academic Publishers,
P.O. Box 322, 3300 AH Dordrecht, The Netherlands.

Printed on acid-free paper

Printed in the Netherlands

This meeting is dedicated to the memory of all colleagues and friends who are no longer with us

George Akoyunoglou
Warren Butler
Alma Gal
Gottfried Galling
Shimon Klein
Jerry Schiff
Hans Schweiger
Philip Thornber

CONTENTS

Chloroplast Envelope Membrane

Pigment Biosynthesis

Preface

From August 10 to August 15, 1998, an international Advanced Research Workshop-Lecture Course on **The chloroplast: from Molecular Biology to Biotechnology** was held at the Orthodox Academy of Crete, Kolymbari-Chania, on the island of Crete, Greece.

After five previous meetings on the chloroplast topic in Marburg (1975), Spetses (1978), Rhodos (1985), Aghia Pelaghia, Crete (1991) and Marburg (1995) this conference proved again that chloroplast research is continuously in the focus of intensive research interest.

The meeting, sponsored by NATO and supported by the Federation of the European Societies for Plant Physiology (FESPP) and the Greek Ministry of Development (General Secretariat of Research and Development), was held under the auspices of the International Society for Chloroplast Development and the National Center for Scientific Research "Demokritos", Athens-Greece.

Aim of the workshop was to bring together experts and students from different disciplines, coming from various countries around the world, studying chloroplast biogenesis from different perspectives in an effort to propose biotechnological approaches, *via* genetic manipulation of the organelle, applicable in solving problems of economic importance. Ninety scientists (including observers) coming from 19 countries actively participated and discussed recent advances in the field. During the meeting it became clear that as a result of the progress made in molecular biology, including genomic sequence, and in biochemistry over recent years, this exciting field of chloroplast development is continuously promoted by renewed interest in as yet unsolved but very important questions. Most important, it has become clear that genetic manipulation of the organelle is a powerfull tool that can now be used to solve problems of economic importance.

The current volume contains close to 60 papers presented as lectures or posters on topics of most intense interest in the field, namely: the three-dimensional structure of the photosynthetic units and their components; the regulation of gene expression; the role of the chloroplast envelope membrane in chloroplast biogenesis and import of nucleus-encoded chloroplast proteins into the plastid; the regulatory mechanisms in pigment biosynthesis, and in the synthesis, assembly and turnover of the photosynthetic units; the genetic manipulation of the organelle as a tool for biotechnological applications.

We want to express our thanks to all those who contributed to the meeting and to this volume. Our sincere thanks are due to the members of the organizing committee, Professors Laurie Bogorad, Harvard University, Cambridge, MA, USA; Navassard Karapetyan, Moscow University, Moscow, Russia, Itzhak Ohad, Hebrew University of Jerusalem, Jerusalem, Israel, who helped in every way to make this meeting stimulating and scientifically productive.

Our thanks are also due to the staff of the Orthodox Academy of Crete for their warm hospitality, to Mr. Leonidas Pastelakos for his excellent technical assistance and Mrs Lily Fotidas for her efficient and never expiring assistance.

October 1998

Joan Argyroudi-Akoyunoglou
Horst Senger

STRUCTURE OF PHOTOSYNTHETIC MEMBRANE PROTEINS

DINOFLAGELLATE LIGHT-HARVESTING PROTEINS: GENES, STRUCTURE AND RECONSTITUTION

R.G. HILLER*, M.J. BROUGHTON*, P.M. WRENCH*,
F.P. SHARPLES*, D.J. MILLER# AND J. CATMULL#

*School of Biological Sciences, Macquarie University, N. Ryde, NSW Australia 2109 and #Dept of Biochemistry and Molecular Biology, James Cook University, Townsville, QLD Australia 4811

KEYWORDS/ABSTRACT: *Peridinin / chlorophyll a / transit peptide / heterologous expression / amino acid sequence / peridinin-chlorophyll a-protein*

Background information on both the intrinsic light-harvesting complex (LHC) and peridinin-chlorophyll a-protein (PCP) is given. Amino acid sequences and introns of both the mature proteins and the chloroplast transit peptides have been analysed and a different route to the chloroplast is postulated. Two distinct forms of PCP are sufficiently dissimilar that they may not be homologous and no ancestor for either can be deduced. Heterologous expression of apoPCP and its reconstitution to functional PCP is reported.

1. **Introduction**

It is now widely accepted that several groups of eukaryotic algae are the result of a serial symbiosis between two eukaryotes and this is reflected in an increase from two to three or four membranes surrounding the chloroplast. The complexities of combining several genomes via this symbiotic route may be further confused by lateral gene transfer from unrelated organisms. Dinoflagellates have three membranes surrounding their chloroplasts and genes for two photosynthetic proteins, RUBISCO and PCP may originate in lateral gene transfer. Dinoflagellates are primarily heterotrophs; some collect chloroplasts from other algae for temporary use, while others have real chloroplasts. Amongst the latter are dinoflagellates, having the unique carotenoid, peridinin, as a major photosynthetic pigment. They along with the Cryptophytes, are unusual in that they have acquired a dual component light-harvesting system as a means of extending absorbance into the green spectral region. The components in Dinoflagellates are an intrinsic thylakoid complex (LHC) whose apoproteins have homologous counterparts in all algal groups [1] and a unique water soluble peridinin-chl a-protein (PCP) [2]. This paper reviews our understanding of the genes and structure of both systems with emphasis on work published within the last five years together with recent new data from the authors' laboratories.

J. H. Argyroudi-Akoyunoglou and H. Senger (eds.),
The Chloroplast: From Molecular Biology to Biotechnology, 3–10.

2. The Intrinsic Chl a-c-peridinin Complex (LHC)

2.1 HISTORICAL PERSPECTIVE

Methods for the separation of chl-protein complexes from higher plant chloroplasts were developed and refined between 1966 and 1985. They were based on SDS polyacrylamide gels in conjunction with other detergents, notably Deriphat and were applied to different algal groups with varying success. In general the methods resulted in unstable complexes from which pigments were differentially stripped and in the case of the Dinoflagellates comigration of peptides from the abundant soluble PCP. Independently in 1993 two groups [3,4] solubilised well washed thylakoids in n-decyl maltoside and separated the light-harvesting proteins by centrifugation through sucrose gradients. A deep brown band at the top of the gradient retained excellent energy transfer from carotenoids (mostly peridinin) to chl a, was enriched in chl *c* and contained only one or two peptides of ~19kDa. This functional light-harvesting protein was used in a range of basic spectroscopic studies as well as limited protein sequencing [3]. The latter demonstrated that despite equivocal antibody cross reactivity, the intrinsic dinoflagellate LHC was part of the superfamily of chl-binding proteins characterised by three transmembrane helices and which binds a great part of the chl in all algal groups as well as in higher plants.

2.2 PROTEIN AND GENE SEQUENCES

The LHC material from the sucrose gradients was sufficiently pure to allow direct protein sequencing, although several slightly different mature proteins were present. Additional complete sequences have been obtained by RT-PCR [4] and genomic sequencing and all are presented in fig 1. They show high identity in regions that can be tentatively assigned to transmembrane helices 1 and 3 by comparison with the higher plant LHC model [6]. The RT-PCR experiments [5] showed that the mature LHC peptides must start off as a polyprotein since one LHC peptide followed another and each was cleaved following the motif SPLR to give new C and N termini. Northern blotting established that the polyprotein might contain as many as 10 mature protein units since the mRNA was ~6kb. As the mRNA was so large and efforts to obtain the 5' end proved difficult, it was decided to screen a genomic library to gain insight into the chloroplast transit sequence as well as the gene structure. Using one of the RT-PCR products as a probe a clone was obtained and 3386 bp has been sequenced (Embl accession no AJ 009670) The 5' end is complete with a putative leader sequence of 95 amino acids, 4 complete and one partial mature protein units and six introns. Of the mature protein units, that immediately following the transit peptide closely resembles (95% identity) the consensus sequence obtained by protein sequencing and all others show greater differences. Sequence 4 is the most diverse and the region between residues 101 and 130 has been assigned on the assumption that the number of amino acids is constant and that this region of DNA is interrupted by an intron with conventional 5'and 3' borders.

```
  1                                                               60
a XFENERGVQD PVGFFDPLGF TADGSVENFK RLAQTEIKHG RVAMLATMGY ITPEITGKLP
b a**s*l***a *t*******l ss****d*** *rras***** ********** m*****a*f*
c a****l**aa *t**w****l ak***mka** *rras***** *i******** ****l***f*
1 a********* ********** ********** *rr******* ********** **********
2 a**s*****p ********** *k**n****t *rra****** *I******** ****l#frfd
3 a********* ********** ********** *rr******* ********** *********k
4 a**s*l***p ****w****s *k**draa*q *rra**f*** *I******** *****a**f*
  61                                                             I20
a GYLSPSTGVK YDDIPNGLGA ISKVPAAGWG QMIAYAAFSE LSQDQSAGTP AAEGDFGFKV
b ****y*qss* fa*v****a* m****vl**a *va**g*vcd v********* g*a*******
c ******m*l* *e******** *******v*a *il***fyc* q*****e*sa gea*******
1 *****d**** ********** ********** ********c* ********** **********
2 ****r*a*** *v******** f****fv**v **v**cllv* I*****q*a* sq#*******
3 ****r***** ********** ********** ********c* ********** **********
4 *e**#*e#l* ft*v*t**a* lvg**f***l *i*l*cyavh alass*nia* cqspplfsgg
  121                                                170
a LTSSDPAELE KKLSAELANG RLAMMAIIGM FFQDGLTGSA WGDWASYTAS PL
b I**e*eet*k r**ns***** *********l ********g* y****n**d* **r
c ***k*eeg** r**ns*I*** ********** ********** y****nf*** **r
1 ********** ******I*** ********** ********** ********** **r
2 ****n***k* ***r****** ********** *********p *****n**d* **r
3 ********** ******s*** ********** ********** *****n**d* **r
4 wvgk**tlrk ****vd**** ********** *********p *****n**d* **r
```

Figure 1. Alignment of LHC mature protein sequences. a, direct protein sequencing; b and c, RT-PCR derived sequences; 1-4, genomic derived sequences in order from transit sequence; *, identity; #, a space.

Permitting the intron borders to depart from the eukaryotic consensus did not improve the match and the possibility that the clone contains a pseudogene must be considered. Introns have also been noted in the proteobacterial unique typeII Rubisco from *Symbiodinium* [7] and are compared with those from the *Amphidinium* LHC gene in table 1. There is a mixture of typical eukaryotic 5' and 3' borders with 11 out of 12 3' borders being AG.

Table1. Introns in nuclear genes coding for dinoflagellate photosynthetic proteins.

	Amphidinium LHC			*Symbiodinium* RUBISCO		
Intron	*5'Join*	*3'Join*	*No of Bases*	*5' Join*	*3' Join*	*No of bases*
1	CAGGT	AGGCT	189	CAGGA	AGGAA	211
2	CAGGT	AGGCT	70	AGGGC	AGGTA	163
3	CAGGC	AGGCT	117	CAGGA	AGGAA	212
4	GCAGT	AGCCC	119	AGGGC	AGGTA	163
5	AGTGT	AGCCA	105	ATGGC	AGGCA	123
6	CAGGC	AGGCT	180	GGTGA	GGCCT	455

A striking feature is that each of the LHC mature protein coding regions is split by an intron (introns 1,2,3 and 6) at the conserved Gly at position 158. Alternative splicing of the primary transcript could generate different mixes of mature peptides in response to different environmental conditions. Some evidence for this is shown in Fig 2 where at low and medium light intensities the mRNA is 6kb and 3kb but only the 3kb species is observed at high light [8]. Earlier work also reported minor hybridisation to a 1.5kb message. A consequence of such a splicing mechanism would be that mature protein species adjacent to the chloroplast transit peptide would always be produced.

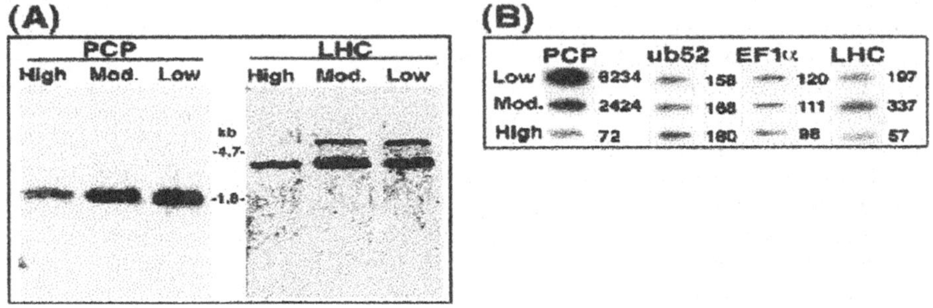

Figure 2. PCP and LHC mRNA levels in *Amphidinium* grown under different light regimes. A, Northern analysis; B, Quantitative slot blots with housekeeping genes ub52 and EF1a for comparison [9].

Upstream of the coding region no typical eukaryotic promoters can be identified as is the case for PCP in both *Amphidinium* and *Gonyaulax* [8], although it has not been established that active genes were sequenced. Establishing the nature of the promoters of dinoflagellate light-harvesting may prove difficult in view of the high cell DNA content and many gene copies. Transcription of the LHC gene or genes is promoted by growth at low light intensities although the effect is not as great as for PCP [9] as shown in Fig 2.

```
o o+ o    o    o     o        -   o+o o+ oo- o oooo o oo
MSLSKPTVALATLAAAASPAAAFHASPNVMHPAANEGMQTRTQTRVSSEMSASSSSFTLSS
  o   o        o+   +   +o        - -+
VAVSGVATLAAAAVLMANSRHVAKPQAKTVVM ⇑ AFENER ..
```

Figure 3. The transit peptide of *Amphidinium* LHC. The arrow marks the known final cleavage point. Positively charged residues, +; negatively charged residues, -; hydrophylic residues, o; are indicated above the sequence. Hydrophobic regions are underlined.

The chloroplast transit peptide shows a high degree of complexity. It is much longer at 95 amino acids compared to that of PCP, which has ~35 residues after removing the lumen-directing domain and it is possible that LHC and PCP take different routes to their final thylakoid locations. The LHC transit sequence has some similarity with those [10,11] of *Euglena* LHC and Rubisco ssu, in that a hydrophobic N-terminal region is followed by an extended hydrophylic domain and a second hydrophobic domain of ~17 residues. In *Euglena* pre LHCP is synthesised as a polyprotein on membrane bound ribosomes and is transported to the chloroplast via the the Golgi apparatus [12], although the roles of the different transit peptide domains apart from the first have not been elucidated.

2.3 STRUCTURE

As noted above most models of algal LHC are based on the structure of Pea LHC [4]. This modeling can be justified on the degree of identity in the transmembrane helix regions together with conserved chl *a*-binding residues, especially in helices 1 and 3. The loops which join the helices are much shorter as is the N-terminal region. Despite the shorter loops, LHCs, which contain fucoxanthin or peridinin, bind many more carotenoids (5-10) per mature protein than does the higher plant LHC. In *Amphidinium* the visible CD spectra in the carotenoid region resembles that of PCP whose structure is known at atomic dimensions and where the peridinins are in van der Waals contact and arranged in pairs crossing at ~55°. Again by analogy with Pea LHC, one pair of peridinins or possibly diadinoxanthin would be arranged in the manner of lutein. A feature of dinoflagellate LHCs is the splitting of the chl c Qy transitions into 630nm and 645nm forms [3,4], which as shown by LD spectra, are preferentially oriented in the plane of the membrane as are long wavelength transitions of chl *a*. In contrast, there is little LD in the region of the peridinin absorbance, so these molecules are randomly oriented or have their transitions at the magic angle (55° to the membrane normal).

3 The Soluble Peridinin Chlorophyll a-Proteins (PCPs)

3.1 HISTORICAL PERSPECTIVE

PCP was purified in 1976 from *Amphidinium*, *Gonyaulax* and *Heterocapsa* (formerly *Glenodinium*) [2]. It was established that the apoproteins are either 32kDa or 15 kDa although the holoproteins are of similar mol mass and spectroscopic properties. Antibodies raised against either form recognise PCP, even across genera. The consensus pigment composition was 8 peridinins:2chl *a* in a functional protein of ~ 40 kDa and arranged as two pairs of peridinin molecules at right angles around each chl [13]. On isoelectric focusing gels, native PCP separates into a number of isoforms, which are species characteristic, suggesting that the variation in pI is largely genetically determined. In vitro translation showed that the PCP genes are nuclear encoded and that the preprotein is about 6kDa larger than the mature protein. In 1994 Norris and Miller [14] published the first full length PCP cDNA sequence and this confirmed the original hypothesis that the 32 kDa PCP apoprotein probably arose from a fusion of two smaller PCP genes.

3.2 PROTEIN AND GENE SEQUENCES

Full length cDNA sequences are available for the most abundant *Amphidinium* [15], *Gonyaulax* [16] and Symbiodinium [14] PCPs. The sequences of the mature protein from all three are ~85% identical although the transit peptide sequences are more variable. The mature protein sequences are also 70% identical to a composite sequence [15] for the dimeric PCP from *Heterocapsa.* A second distinct PCP ("high salt form") was isolated during the purification of large amounts of PCP for crystallisation. This contains only 6 peridinins:2 chl a compared to 8 peridinins:2chl a and its derived mature protein sequence is only 31% identical to other PCPs. One of these genes has

```
  1 MVRSGKKAVVLAAVAFCATSVVQKSHGFVPSPLRQRAAAAGAAAASAATM- 50
    |.| |:| :.::|:|:...|   ...:|||:|  .|... :|.:| :| :
  1 MAR-GRKLLTVGALALAVAS---WRPAFVPGPRTARSNVMAASPALPALLA 47
 51 FAPAAFADEIGDAAKKLGDASYAFAKEVDWNN----GIFLQAPGKLQPLE 96
    .|||||||:|:||.|:::||.|::|...||..    : ::..::  .| :
 48 AAPAAFADGIADASKRFSDATYPIAEKFDWGGSSAIAKYIADASASNPRQ 97
 97 ALKAIDKMIVMGAAADPKLLKAAAEAHHKAIGSV-SGPNGVTSRADWDNV 145
    |  |::|:: :| . ||||::||.||| ||::|. ..:. :.|:.|:..|
 98 AALAVEKLLEVGLTMDPKLVRAAVEAHSKALDSAKTNAKLMASKEDFAAV 147
146 NAALGRVIASVPENMVMDVYDSVSKITDPKVPAYMKSLVSGADAEKAYEG 195
    |.||:|:|||...... .: .....  ...:.: : :  .: :|||||::
148 NEALARMIASADKQKFAALRTAFPE--SRELQGKLFGGNNAFEAEKAYDS 195
196 FLAFKDVVKKSQVTSAAGPATV---------PSGDKIGVAAQQLSEASYPF 237
    | |:...|:.. :.:| :|...         .::::.:| ||..:|||.||:
196 FKALTSAVRDASINGAKAPVIAEAARAERYVGDGPVGRAAKKFSEATYPI 245
238 LKEIDWLSDV----YMKPLPGVSAQQSLKAIDKMIVMGAQADGNALKAAA 283
    :...:|| ...    |:.. .: .:.   .:||| : ::   :.||:..|.
246 MDKLDWGKSPEISKYIETASAKNPKMMADGIDKTLEVALTMNQNAINDAV 295
284 EAHHKAI.GSIDATG.VTSAADYAAVNAALGRVIASVPKSTVMDVYNAMA 331
     || :|| |.::..| |.. .|:| || ||:::||...........:..|::
296 FAHVRAIKGALNTPGLVAERDDFARVNLALAKMIATAHPAKFKALLTAFP 345
332 GATDTSIPLNMFSKVNPLDANAAAKAFYTFKDVVQAAR 368
    |..|  :.:.:|.  || :|.|| ..| .:...|...
346 GNAD--LQMALFAANNPEQAKAAYETFVALTSAVASS  380
```

Figure 4. Alignment of two forms of PCP from *Amphidinium* using the programme 'Bestfit'. **Bold,** a typical abundant PCP; ordinary face, 'high salt' PCP. |, identity; : and ., similarity; -, a space.

been sequenced (Embl no AJ006994) and the translated sequence is aligned with that of the main PCP form from *Amphidinium* in Fig 4. Identity of 31% is only reached with the insertion of a significant number of gaps and there are extensive regions lacking any similarity. We speculate that the two forms of PCP may not be homologous and that the degree of similarity is the result of functional constraints in the location of the pigments. At low light intensities there is increased translation and transcription of all PCP forms and there is an additional blue-light effect which results in an increased representation of low pI species. No promoters have yet been identified although they may be remote and the up and downstream sequences have no characteristic eukaryotic features.

3.3 STRUCTURE

The structure of one form of PCP has been solved at 2.0A by X-ray crystallography [16]. The monomers show a pseudo twofold symmetry and the predominant secondary structure is alpha helical (Fig 5). The helices form a hydrophobic cage which encloses the pigments. There are two chromophore clusters of 4 peridinins and 1chl *a*, with a cluster in each domain. Efficient energy transfer from peridinin to chl *a* is achieved by the approach of the pigments at van der Waals distance. The monomers form a non-crystallographic trimer which ensures energy equilibration over all six chl *a* molecules before it is transferred to the inner antennae by a Förster mechanism.

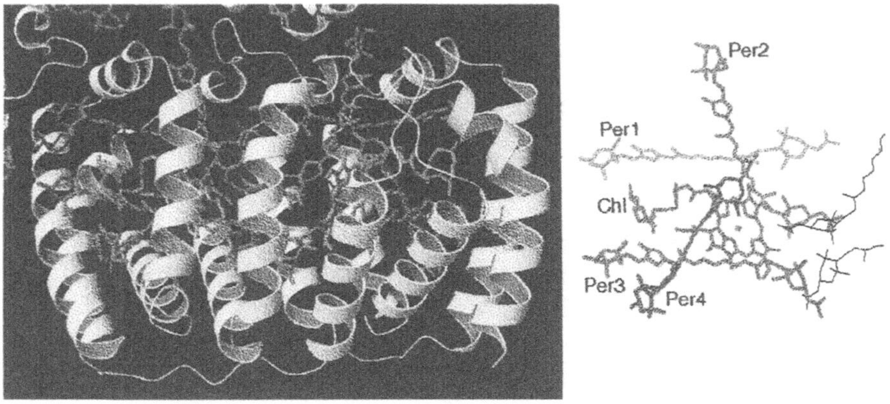

Figure 5 A monomer of PCP (left panel): the pseudo two fold axis symmetry between N- and C-terminal domains is vertical through the middle with the N-terminus bottom right . The right panel shows the C-terminal domain pigment cluster together with part of 2 peridinins from the adjacent pigment cluster.

3.4 HETEROLOGOUS EXPRESSION AND RECONSTITUTION

Complete understanding of the spectroscopy and function of PCP can only be achieved when structures of several naturally occurring forms are known or the existing protein can be manipulated in predictable ways eg by mutagenesis. To achieve the latter, reconstitution of PCP from isolated pigments and heterologously expressed apoPCP is essential. This has been achieved and our procedure is summarised in Fig 6. Reconstituted PCP is identical to native PCP as judged by absorbance, visible CD and fluorescence. Fortuitously, the expressed N-terminal domain alone (a 'half' PCP!) assembles better than the full length construct but forms a dimer as judged by size exclusion chromatography. Consequently, the effect on the chromophores of a site directed mutagenesis resulting in amino acid substitution will be duplicated in the holoprotein. Our first targets will be conserved hydrophobic residues close to the peridinins eg Trp 23 and residues involved in chl binding eg His 66. The reconstitution system will also be used to generate larger quantities of the 'High Salt'PCP so that its structure may also be solved.

Isolate pigments from native PCP ⇒ Dry ⇒ Redissolve in Ethanol
⇓
Mix pigments ⇒ Size exclusion
protein (1:1 mol) at 4° Chromatography
⇑ PCP ~25%Yield
Clone PCP coding ⇒ Heat ⇒ apoPCP into ⇒ Extract protein and
region into pNDE shock inclusion bodies purify by ion exchange

Figure 6 Flow sheet for heterologous expression and reconstitution of PCP

4 ACKNOWLEDGEMENTS

This work has been supported by the Australian Research Council (Grant No A19600918) and Macquarie and James Cook Universities. We thank Eckhard Hoffman (Konstanz University) for Figure 5.

5 REFERENCES

1) Green BR and Pichersky E (1994) Photosynth Res 39: 149-162
2) Prézelin BB (1987) Photosynthetic physiology of dinoflagellates. In Taylor FJR (ed) The biology of dinoflagellates, pp174-223 Blackwell Oxford
3) Hiller RG, Wrench PM, Gooley AP, Shoebridge G and Breton G. (1993) Photochem Photobiol 57: 125-131
4) Iglesias-Prieto R, Govind NS and Trench RK, (1993) Philos Trans Roy Soc Lond Ser B 340: 381-392
5) Hiller RG, Wrench PM and Sharples FP (1995) FEBS Lett 363: 175-178
6) Kuhlbrandt W, Wang DN and Fujiyoshi Y (1994) Nature 367: 614-621
7) Rowan R, Whitney SM, Fowler AM and Yellowlees D (1996) Plant Cell 8: 539-553
8) Le QH, Markovic P, Hastings JW, Jovine RV and Morse D. (1997) Mol Gen Genet 255: 595-604
9) ten Lohuis MR and Miller DJ (1998) Plant Physiol 117: 189-197
10) Kishore R, Muchhal US and Schwartzbach SD (1993) Proc Acad Sci USA 90: 1845-11849
11) Chan RL, Keller M, Canaday J, Weil J-H and Imbault P (1990) EMBO J 9: 333-338
12) Muchal US and Schwartzbach SD (1992) Plant Mol Biol 18: 287--299
13) Song PS, Koka P, Prézelin BB and Haxo FT (1976) Biochemistry 15:4422-4427
14) Norris BJ and Miller DJ (1994) Plant Mol Biol 24: 673-677
15) Sharples FP, Wrench PM and Hiller RG (1996) Biochim Biophys Acta 1276: 117-123.
16) Hofmann E, Wrench PM, Sharples FP, Hiller RG, Welte W and Diedereichs K (1996) Science 272: 1788-1791

ORGANISATION OF THE PHOTOSYSTEM TWO LIGHT HARVESTING SYSTEM

J. BARBER
Wolfson Laboratories, Biochemistry Department
Imperial College of Science, Technology & Medicine
London SW7 2AY
UK

1. Introduction

Like other types of photosystems, Photosystem II (PSII) consists of a reaction center served by a light harvesting system. Higher plants, algae and cyanobacteria all contain an inner PSII antenna system composed of CP43 and CP47 which are chlorophyll *a* (chl*a*) binding proteins encoded by the psbB and psbC genes, respectively. They bind about 15 chls each (1) and have 6 transmembrane helices (2). Six chl*a* molecules are bound to the PSII reaction center complex composed of the D1 and D2 proteins (3) thus giving a total of about 36 chl*a* molecules associated with the PSII core complex. In higher plants and green algae, the outer PSII antenna is made up of chl*a*/chl*b* binding proteins encoded by the nuclear located cab genes (4). The dominant forms (Lhcb1, Lhcb2 and Lhcb3) make up the light harvesting complex, LHCII, which exists as a trimer (5), binding about 42 chl each (24 chl*a* and 18 chl*b*). In addition, three minor forms of Cab proteins are found in PSII; Lhcb4, also known as CP29, Lhcb5 (CP26) and Lhcb6 (CP24). Some Cab proteins are more tightly held to the PSII core while others are more easily dissociated, leading to the concept of an inner and outer pool of LHCII. On average the total antenna size corresponds to about 250 chl per reaction center. In red algae and cyanobacteria the secondary antenna is contained within the phycobilisomes.

In this brief paper I outline how recent structural studies are starting to reveal the organisation of the light harvesting systems in PSII of higher plants.

2. Photosystem II - multisubunit complex

Taken as a whole, PSII is a macromolecular complex consisting of over 30 different subunits with an estimated molecular mass of 600 kDa or more (6). Many of the subunits bind chlorophyll but many do not. Except for extrinsic proteins associated with the Oxygen Evolving Complex (OEC) most of the non-chlorophyll binding proteins have molecular masses less than 10 kDa and have a single transmembrane helix (7). As far as I know a single PSII complex containing a full complement of proteins and chlorophylls has never been isolated. Nevertheless, biochemical methods have been used to obtain isolated forms of PSII with differing degrees of subunit and

J. H. Argyroudi-Akoyunoglou and H. Senger (eds.),
The Chloroplast: From Molecular Biology to Biotechnology, 11–18.

pigment composition. It is these types of preparations that we are using to elucidate the structural organisation of PSII.

2.1. LHCII-PSII SUPERCOMPLEX

Using β-dodecyl maltoside and sucrose density centrifugation we have developed procedures to isolate from spinach a stable and highly active form of PSII which not only contains the oxygen evolving core but also a complement of Cab proteins (8). When care is taken to stabilize the OEC proteins by the inclusion of glycine betaine in the preparation media and sucrose gradients, the resulting complex has functional activities similar to that of PSII enriched membrane fragments of the BBY type (9,10). Fig 1.1 shows a top view of this supercomplex obtained by electron microscopy which is dimeric having total molecular mass of about 725 kDa (11). It binds about 100 chl (75 chl*a* and 25 chl*b*) per reaction center (200 chl per dimeric complex) and immunoblotting shows the presence of Lhcb1, Lhcb2, Lhcb4 (CP29) and Lhcb5 (CP26), as well as CP47 and CP43 (8). The pigment level is consistent with one LHCII trimer and a single copy of CP29, CP26, CP43 and CP47 per reaction center. The supercomplex measures about 26.8 x 12.3 nm and is consistent with the size of dimeric ESs particles observed by Seibert *et al* (12) in intact granal membranes by freeze-etching. However, an average PSII usually has about 250 chl per reaction center suggesting that *in vivo* at least another three LHCII trimers can be associated with the supercomplex. Presumably these trimers represent the outer pool of LHCII. Recently, two groups (13,14) have reported the preparation of PSII complexes larger than the supercomplex which contains additional LHCII and also contains Lhcb6 (CP24). Electron microscopy confirms that two additional LHCII trimers can be bound to the supercomplex (i.e. one extra per reaction center).

2.2. PSII CORES

As Figs. 1.3 and 1.4 show, LHCII and the other Cab proteins of supercores located at the peripheral of the supercomplex can easily be dislodged to yield a PSII core complex with approximately 35 chl per reaction center and dimensions of 17.2 x 9.7 nm (8,11). With sufficient care (sucrose density gradients plus glycine betaine), these cores retain the OEC proteins and have high functional properties (15).

As shown in Fig. 1.2, the cores are derived from the central region of the supercomplex and therefore remain dimeric. Excess detergent treatment or treatment with phospholipase A2 causes them to monomerize (Hankamer, Kruse and Barber unpublished). Although the monomeric cores of PSII are able to evolve oxygen, they tend to be more unstable than their dimeric counterparts (8). The PSII cores can be washed free of extrinsic proteins to yield non-oxygen evolving cores which have recently been used to make 2D crystals and analyzed by electron microscopy to yield the 3D structure shown in Fig. 2 (ref. 16).

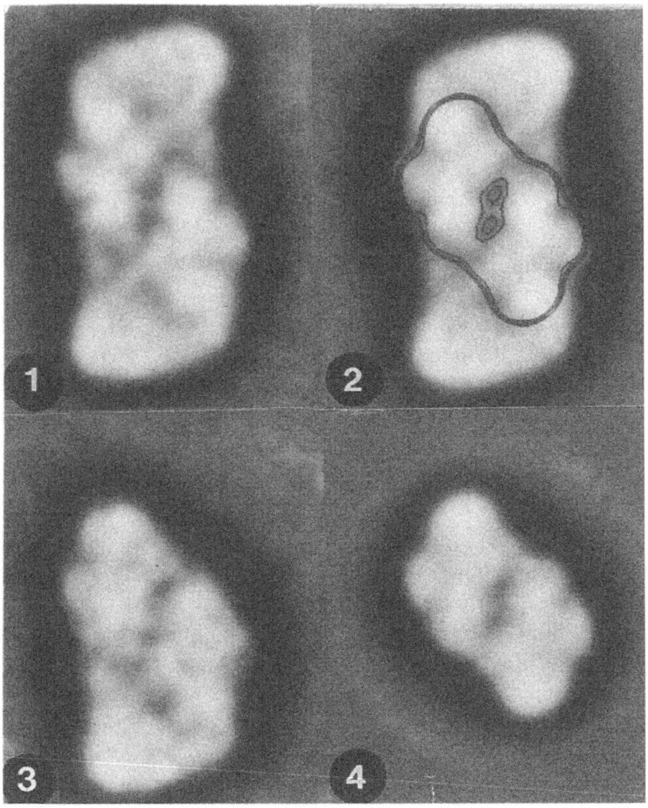

Fig. 1 Averaged top views of negatively stained PSII complexes isolated from spinach viewed by electron microscopy, (1) supercore complex, (2) supercore complex showing central position of the PSII complex minus one set of LHCII/Cab proteins and revealing the edge of the centrally located core dimer, (4) PSII core dimer (from ref.11).

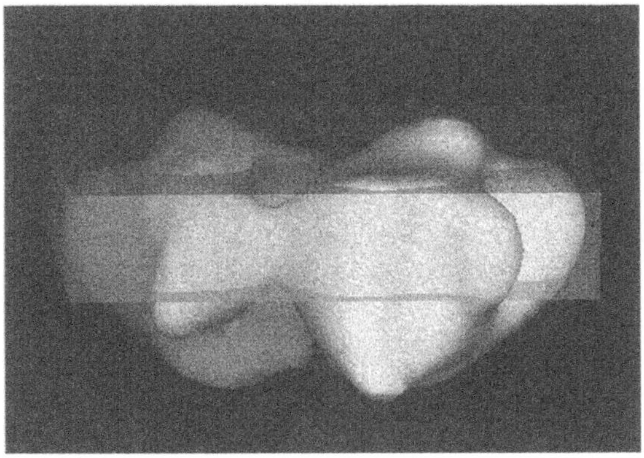

Fig. 2 3D structure of the non-oxygen evolving PSII core dimer of spinach determined by electron crystallography after negative staining viewed from the side with the lumenal surface downwards (16). The proposed position of the membrane is shown as a semi-transparent gray box, 3.5 nm in depth.

2.3. CP47-RC COMPLEX

Using detergents such as heptylthioglycopyranoside (HTG), it is possible to remove CP43 from PSII cores to produce a CP47-RC complex. Using sucrose density gradients, we obtained a dimeric form of this complex which contained 21 chl*a* and consisted of the D1, D2 and CP47 proteins, together with the products of the psbE, F, I, L, K, T and W genes (17). The dimeric CP47-RC also bound a plastoquinone and showed Q_A activity (18). When monomerized, the CP47-RC complex lost the plastoquinone, the PsbL and K proteins and a small amount of chlorophyll (17).

Recently we have grown highly ordered 2D crystals of the CP47-RC complex and a projection map at 8Å determined (19). We have now extended this electron crystallographic work to obtain a 3D structure which is in the process of being published (20). At 8Å resolution we have been able to identify the transmembrane helices and assign densities to chlorophylls bound within CP47 and the D1/D2 heterodimer. The conclusions of this work are:

i) The CP47-RC complex contains 23 transmembrane helices of which 10 can clearly be identified as those belonging to the D1 and D2 proteins and 6 to CP47.
ii) The D1 and D2 helices are arranged in a manner similar to the L and M subunits of the purple bacterial RC, but with some interesting differences, e.g. transmembrane helix III seems to be straighter at the lumenal ends than in the bacterial counterparts.
iii) The CP47 helices are organised as a circle of 3x pairs of helices, an arrangement almost identical to that of the 6 N-terminal helices of the PSI RC proteins, PsaA and B.
iv) The remaining 7 transmembrane helices probably belong to the proteins PsbE, PsbF, PsbI, PsbL, PsbK, PsbT and PsbW (17) but as yet their assignment to a particular density has not been made.
v) The arrangement of the core porphorins within the D1/D2 heterodimer involved in primary electron transfer is similar to that found in the purple bacterial reaction center except that there is no 'special pair'. The chlorophylls corresponding to the 'special pair' are spaced further apart, in line with the multimer model of Durrant *et al.* (21).
vi) A 'special pair' is not an absolute requirement for rapid primary charge separation in photosynthesis.
vii) Chlorophyll ligated to D1 His198 seems to be the closest porphorin to Tyr_Z161 located at the lumenal end of helix three.
viii) Densities, corresponding probably to 14 chlorophyll molecules, can be identified in the CP47 helical bundle. They tend to be arranged in two layers towards the membrane surface.
ix) The organisation of the CP47-RC complex in the 2D crystal is consistent with the dimeric nature of the PSII complex.

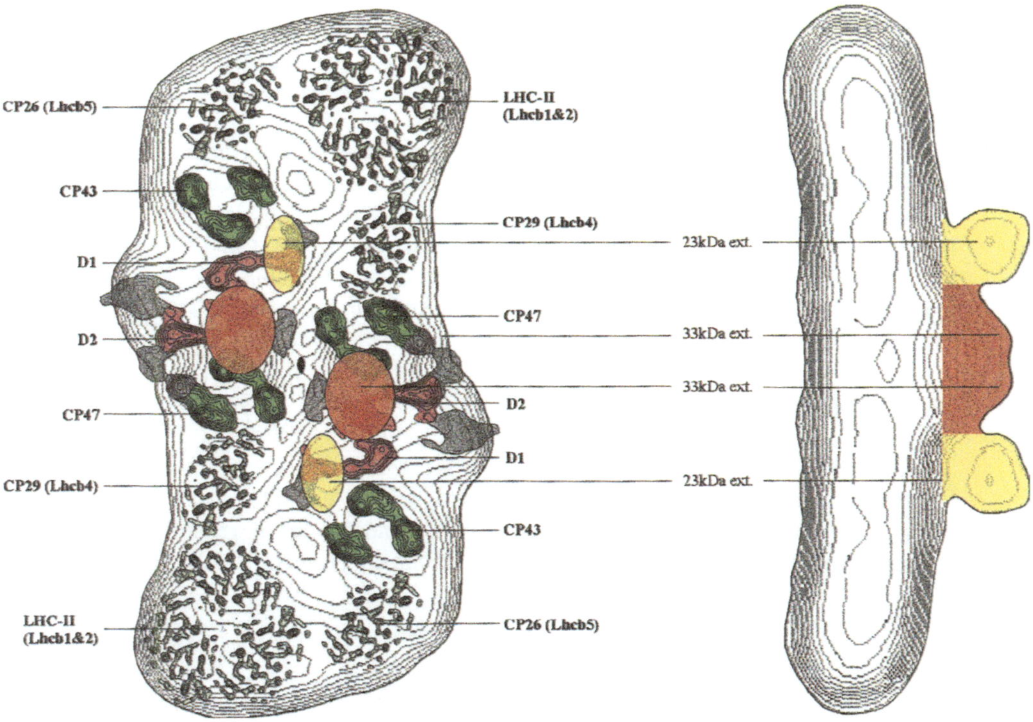

Fig. 3 Proposed position of the major subunits of PSII within the PSII-LHCII supercomplex. The mapping is based on biochemical and cross linking data, as well as recent structural information obtained from electron crystallography (ref. 22).

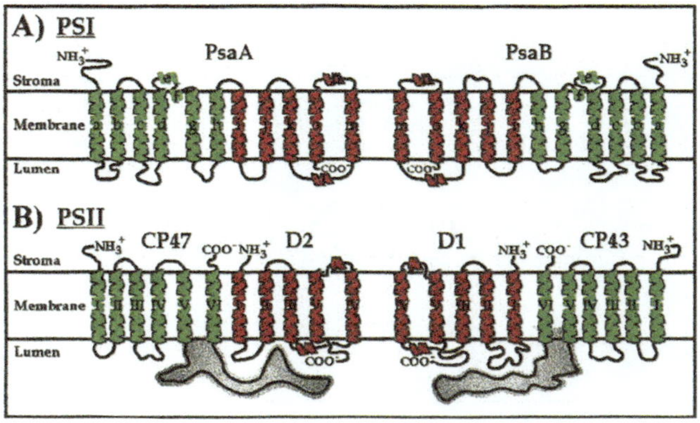

Fig. 4 Comparison of the transmembrane segments of the PSII and PSI reaction center core proteins emphasizing their structural similarities.

3. Mapping subunit positions

Given the electron microscopy data derived from our single particle and 2D crystallization studies we can now give a model for the organisation of the light harvesting systems of PSII using the supercomplex as a framework as shown in Fig. 3.

Note that Lhcb4 (CP29) and Lhcb5 (CP26) act as conduits for transferring energy from the LHCII trimer to the PSII core. Also of importance is that, in contrast to a previous model (23), CP43 and CP47 are shown to be on opposite sides of the D1/D2 heterodimer. Given the similarity between the helix arrangement of CP47 and the N-terminus of the photosystem I (PSI) reaction center proteins (24), it has been reasonable to postulate that PSII and PSI are structurally related (7,25) as shown diagramatically in Fig. 4.

4. Position of transmembrane helices

To prove, absolutely, that CP43 and CP47 are located on opposite sides of the D1 and D2 heterodimer requires a more detailed analyses of the 2D crystals of the core complex containing all four proteins. To this end we have recently conducted cryoelectron microscopy on such crystals and obtained a projection map (26).

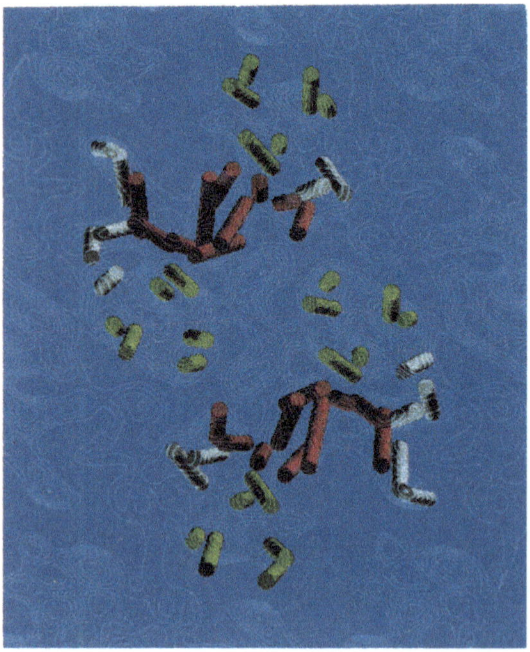

Fig. 5 Projection map of PSII cores and suggested localization of the transmembrane helices of the D1 and D2 reaction centre proteins, CP47 and CP43 within it. The resolution of the map is about 10Å and helical positions are in part taken from ref. 20. The map is from Hankamer, Morris and Barber (unpublished).

The PSII core dimer crystals similar in appearance and size to those reported in ref.16, were grown by reconstituting purified core dimers with thylakoid lipids. Judged from SDS-PAGE they bound the 33 kDa extrinsic protein and presumably maintained oxygen evolving properties. Low dose images of the frozen, unstained crystals were recorded in a Philips CM200FEG electron microscope using a Gatan cryo-specimen holder on Agfa Scientia film. After digitizing, the images were corrected for long range disorder and averaged projection maps calculated by Fourier methods.

Fig. 5 is an example of a projection map obtained. Although the resolution of the map is less than that for the CP47-RC complex (~10Å verses 8Å), it is sufficient to identify the transmembrane helices that are common to both crystals. In this way we are able to identify the densities that correspond to the transmembrane helices of CP43 (see Fig. 5). The lower resolution map which was reported in ref. 26 already indicated that the CP43 helical cluster was on the opposite side of the D1/D2 heterodimer to that of CP47. The new high resolution data will be published in full in due course but it unequivocally supports the earlier conclusion and shows that the six transmembrane helices of CP43 are related to those of CP47 by the same approximate pseudo two fold symmetry axis to which the D1 and D2 proteins are related.

5. Acknowledgments

The work and conclusions presented in this article would not have been possible without the input of colleagues who are named in the primary references. I also acknowledge the BBSRC for financial support.

6. References

1. Barbato, R., Race, H.L., Friso, G. and Barber, J. (1991) Chlorophyll levels in pigment binding proteins of PSII: A study based on the chlorophyll to cytochrome ratio in different PSII preparations. *FEBS Lett.* **286**, 86-90.
2. Bricker, T.M. (1990) The structure and function of CPa-1 and Cpa-2 in photosystem II. *Photosynth. Res.* **24**, 1-113.
3. Zheleva, D., Hankamer, B. and Barber, J. (1996) Heterogeneity and pigment composition of isolated PSII reaction centers. *Biochemistry* **35**, 15074-15079.
4. Jansson, S. (1994) The light harvesting a/b binding proteins. *Biochim. Biophys. Acta* **1184**, 1-19.
5. Kühlbrandt, W., Wang, D.N. and Fujiyoshi, Y. (1994) Atomic model of plant light-harvesting complex by electron crystallography. *Nature* **367**, 614-621.
6. Hankamer, B., Barber, J. and Boekema, E.J. (1997) Structure and membrane organisation of PSII in gren plants. *Annu. Rev. Plant Phys. Mol. Biol.* **48**, 641-671.
7. Barber, J., Nield, J., Morris, E.P., Zheleva, D. and Hankamer, B. (1997) The structure, function and dynamics of photosystem II. *Physiol. Plant.* **100**, 817-827.
8. Hankamer, B., Nield, J., Zheleva, D., Boekema, E.J., Jansson, S. and Barber, J. (1997) Isolation and biochemical characterisation of monomeric and dimeric PSII complexes from spinach and their relevance to the organisation of photosystem II in vivo. *Eur. J. Biochem.* **243**, 422-429.
9. Boekema, E.J., Nield, J., Hankamer, B. and Barber, J. (1998) Localisation of the 23 kDa subunit of the oxygen evolving complex of PSII by electron microscopy. *Eur. J. Biochem.* **252**, 268-276.

10. Eshaghi, S., Andersson, B. and Barber, J. (1998) Isolation of a highly active PSII-LHCII supercomplex from thylakoid membranes by a direct method. Submitted for publication

11. Boekema, E.J. Hankamer, B., Bald, D., Kruip, J., Nield, J., Boonstra, A.F., Barber, J. and Rögner, M. (1995) Supramolecular structure of the photosynthetic complex from green plants and cyanobacteria. *Proc. Natl. Acad. Sci. USA* **92**, 175-179.

12. Seibert, M., DeWitt, M. and Staehelin, L.A. 91987) Structural localisation of the oxygen evolving apparatus to multimeric (tetrameric) particles on the lumenal surface of freeze-etched photosynthetic membranes. *J. Cell. Biol.* **105**, 2257-2265.

13. Boekema, E.J., van Roon, H. and Dekker, J.P. (1998) Specific association of PSII and light-harvesting complex II in partially solubilised photosystem II membranes. *FEBS Lett.* **424**, 95-99.

14. Harrer, R., Bassi, R., Testi, M.G. and Schaefer, C. (1998) Nearest-neighbour analysis of a PSII complex from *Marchantia polymorpha* L (liverwort) which contains reaction center and antenna proteins. *Eur. J. Biochemistry* in press.

15. Catucci, L., Dörner, W., Hankamer, B., Vass, I. and Barber, J. (1998) Isolation and characterisation of oxygen evolving PSII complexes of spinach in the presence of glycine betaine. in Proc. XI Inter. Photosynth. Congress, Budapest. Kluwer Academic Publ. Dordrecht, The Netherlands in press

16. Morris, E.P., Hankamer, B., Zheleva, D., Friso, G. and Barber, J. (1997) The 3-D structure of a photosystem II core complex determined by electron crystallography. *Structure* **5**, 837-849.

17. Zheleva, D., Sharma, J., Panico, M., Morris, H.R. and Barber, J. (1998) Isolation and characterisation of monomeric and dimeric CP47-RC PSII complexes. *J. Biol. Chem.* **273**, 16122-16127.

18. Bianchetti, M., Zheleva, D., Deak, Z., Zharmuhamedov, S., Klimov, V.V, Nugent, J.H.A., Vass, I. and Barber, J. (1998) Comparison of the functional properties of the monomeric and dimeric forms of the isolated CP47-RC complex. *J. Biol. Chem.* **273**, 16128-16133.

19. Rhee, K.-H., Morris, E.P., Zheleva, D., Hankamer, B., Kühlbrandt, W. and Barber, J. (1997) Two-dimensional structure of plant photosystem II at 8Å resolution. *Nature 389*, 522-526.

20. Rhee, K.-H., Morris, E.P., Barber, J. and Kühlbrandt, W. (1998) 3-D structure of photosystem II reaction center at 8Å resolution. *Nature* in press

21. Durrant, J.R., Klug, D.R., Kwa, S.L.A., van Grondelle, R., Porter, G. and Dekker, J.P. (1995) A multimer model for P680, the primary electron donor of photosystem II. *Proc. Natl. Acad. Sci. USA* **92**, 4798-4802.

22. Barber, J. (1998) Photosystem Two. *Biochim. Biophys. Acta* **1365**, 269-277.

23. Rögner, M., Boekema, E.J. and Barber, J. (1996) How does PSII split water? The structural basis of efficient energy conversion. *Trends Biochem. Sci.* **21**, 44-49.

24. Fromme, P., Witt, H.T., Schubert, W.-D., Klukas, O., Saenger, W. and Krauss, N. (1996) Structure of photosystem I at 4.5Å resolution: a short review including evolutionary aspects. *Biochim. Biophys. Acta* **1275**, 76-83.

25. Schubert, W.-D., Klukas, O., Saenger, W., Witt, H.T. and Fromme, P. (1998) A common ancestor for oxygenic and anoxygenic photosynthetic systems - a comparison based on the structural model of photosystem I. *J. Mol. Biol.* **280**, 297-314.

26. Hankamer, B., Morris, E.P. and Barber, J. (1998) Cryoelectron mciroscoy of photosystem II shows that CP43 and CP47 are located on opposite sides of the D1/D2 reaction centre proteins. in Proc. XI Inter. Photosynth. Congress, Budapest. Kluwer Academic Publ. Dordrecht, The Netherlands in press

THREE-DIMENSIONAL STRUCTURE OF PHOTOSYSTEM I

Present State of the X-Ray Crystallographic Studies

N. KRAUSS
Institut für Kristallographie, Freie Universität Berlin
Takustr. 6, D-14195 Berlin, Germany

1. Introduction

In water oxidising photosynthesis, the primary processes responsible for the conversion of the energy of visible light into chemical energy take place at two multi-subunit protein pigment complexes, the photosystems I and II (PSI and PSII), which are embedded in the thylakoid membrane. PSI belongs to the larger group of type-I photosynthetic reaction centres (RC), characterised by iron sulphur clusters as intrinsic terminal electron acceptors, whereas PSII is a representative of the type-II RC, which possess quinone cofactors as their terminal acceptors [1]. The structures of the type-II RC from the purple bacteria *Rhodopseudomonas viridis* [2,3] and *Rhodobacter sphaeroides* [4,5,6,7,8] are well investigated at resolutions of 3 Å or better. Because of the known similarities between these reaction centres (PbRC) and PSII their structures were supposed to be of relevance for the PSII structure [9]. In contrast, the crystallographic analysis of PSI isolated from the cyanobacterium *Synechococcus elongatus* at 4 Å resolution [10,11] provides the most detailed structural information about a type-I reaction centre up to now.

Cyanobacterial PSI consists of 11 protein subunits, of which two large, homologous subunits denoted PsaA and PsaB and subunits PsaF, PsaI, PsaJ, PsaK, PsaL and PsaM contain hydrophobic stretches which should span the membrane. PsaC, PsaD and PsaE are extrinsic subunits located at the stromal side of the PSI complex (see [12,13] for reviews). A PSI monomer binds approximately 100 chlorophyll *a* (Chl*a*) molecules belonging to the core antenna system and the electron transfer chain, 15 to 25 carotenoid and two phylloquinone molecules and three Fe_4S_4 clusters. The following cofactors of the electron chain have been identified spectroscopically: A primary electron donor P700 which consists of a Chl*a* dimer, a primary acceptor A_0 identical with a Chl*a* monomer, a phylloquinone as the secondary acceptor A_1, an iron-sulphur cluster F_X as an intermediate and the iron sulphur clusters F_A and F_B as terminal acceptors.

2. Quality of the structural model of photosystem I

The first structural model of PSI which could be obtained from the X-ray crystallographic analysis of single crystals of trimeric PSI [14] was derived from an electron density map at 6 Å resolution [15]. It consisted of the positions of the three iron sulphur clusters of PSI as well as numerous α-helical fragments and Chl*a* cofactors. Subsequently, further improvement of the crystal quality [16] could be translated into electron density maps at 4.5 Å [17] and 4.0 Å resolution [10,11]. Compared to the model at 6 Å resolution, the present structural model of PSI (Fig. 1) has improved as regards the larger number of α-helices and Chl*a* cofactors and the identification of some connections between α-helices and of the position of a single phylloquinone cofactor. Because of the relatively low resolution of the electron density maps complete tracing of

J. H. Argyroudi-Akoyunoglou and H. Senger (eds.),
The Chloroplast: From Molecular Biology to Biotechnology, 19–26.

the polypeptide backbones of the protein subunits of PSI as well as the identification of amino acid side chains were not feasible. As a consequence, the resulting structural model is not complete and accurate enough to be suitable for crystallographic refinement, a procedure which is used to improve the structural model and the electron density maps and in addition serves as a proof for the quality of the model. The validity of the present structure of PSI relies much more on its compatibility with the biochemical and biophysical data available for this complex than it would be the case for a crystallographically refined model at higher resolution.

Fig. 1. The PSI trimer represented by most of its constituting structural elements: α-helices, Chl*a* a and iron-sulphur clusters, view from the stromal side. Figure modified from [11].

3. Assignment of structural elements to functional building blocks of PSI

As the primary structures of the 11 protein subunits constituting PSI of *Synechococcus elongatus* are known [18], an unambiguous assignment of respective parts of the structural model to these subunits would be straightforward if the identification of the amino acid side chains and of the complete polypeptide backbones from the electron density map would be possible. An assignment of spectroscopically identified cofactors to the structure would then be possible if the interactions with individual side chain residues are known, i.e. from investigations on PSI mutants. Although this type of procedure cannot be applied so far, the huge knowledge about structural details of PSI provided by various techniques and combinations of them, e.g. electron microscopy, spectroscopic methods, chemical crosslinking studies and mutagenesis experiments could be used to gain insight into the spatial architecture of PSI on the basis of the pure arrangement of structural elements.

The identification of the membrane intrinsic part of the structure and of the stromal and lumenal sides of the PSI complex was already possible at 6 Å resolution, because the majority of α-helices is arranged similarly to the packing of transmembrane α-helices known from the structures of other membrane integral proteins, and because the three Fe_4S_4 clusters, of which the two terminal electron acceptors F_A and F_B are known to be bound to subunit PsaC located at the stromal side of the complex, could unambiguously be located outside the membrane intrinsic region. The electron density map at 4 Å resolution reveals eleven pairs of transmembrane α-helices and four pairs of lumenal and stromal surface α-helices to be related by a pseudo-twofold rotation axis [11] (Fig. 2). The region occupied by these α-helices was assigned to the heterodimer consisting of the homologous subunits PsaA and PsaB, the number of transmembrane helices per each subunit being in agreement with secondary structure predictions based on hydrophobicity analyses. The regions defined by the remaining transmembrane helices belong to subunits PsaF, PsaI, PsaJ, PsaK, PsaL and PsaM which contain hydrophobic stretches as derived from their primary structures. Several alternative assignments of these six subunits to individual transmembrane α-helices are still possible.

The pseudo-symmetry axis is oriented perpendicular to the membrane plane and passes through the centre of the Fe_4S_4 cluster located closest to the membrane intrinsic region, a position well compatible with the cluster F_X coordinated symmetrically by two cystein residues each of a sequence motif strictly conserved in both PsaA and PsaB. The remaining iron-sulphur clusters are identical with F_A and F_B. The inner membrane space close to F_X, i.e. around the pseudo-twofold rotation axis, is delimited by a palisade-like structure formed by a total of ten transmembrane α-helices of PsaA and PsaB and harbours the Chl*a* and phylloquinone cofactors of the electron transfer chain, which are thereby spatially separated from the cofactors constituting the core antenna system of PSI.

In case of the three extrinsic sububits PsaC, PsaD and PsaE known to be located on the stromal side of the PSI complex, structural models of relevance for PsaC and PsaE are available. A first step towards the location of PsaC was the identification of the iron-sulphur clusters F_A and F_B in the electron density map which are bound to this subunit. The supposed similarity between the structures of $2Fe_4S_4$ ferredoxins, e.g. from *Peptococcus asaccharolyticus* (PaFd, [19]) and PsaC [20,21] was not only confirmed by the identical centre to centre distances of 12 Å between the Fe_4S_4 clusters in these ferredoxins and PsaC, but also by the good agreement of large portions of the PaFd structure with the electron density map at 4 Å resolution in the region surrounding both Fe_4S_4 clusters. Assuming that the orientations of the iron-sulphur clusters within PsaC is similar as in PaFd, their orientations within the PSI complex inferred from the PaFd model fitted into the electron density map are consistent with EPR experiments on the PSI crystals from *Synechococcus elongatus* [22,23]. For a structural model of PsaC derived from the PaFd structure, two alternative orientations within the PSI complex are still possible because of a twofold pseudo-symmetry of the corresponding electron density region. Thus, although the Cystein residues of PsaC which coordinate F_A and F_B have been identified [24,25], an unambiguous assignment of these clusters to the corresponding two positions located in the X-ray structure on the basis of purely structural arguments is not possible at present. The structure of PsaE from *Synechococcus* sp. PCC 7002 has been determined by NMR-studies in solution [26]. Surprisingly, an electron density region corresponding to this structural model, which after minor alterations should be suitable as a starting model for PsaE from *Synechococcus elongatus*, could not be detected in the crystal structure of PSI.

The detection of some loops connecting α-helices, especially those core α-helices surrounding the electron transfer chain, in connection with the possible coordination of F_X and the P700 Chl*a* within the structural model and the known binding sequence for

F_X [27] led to the assignment of the five C-terminal hydrophobic stretches of PsaA and PsaB to the five pairs of transmembrane core α-helices. Consequently, PsaA and PsaB were interpreted to be divided into a reaction centre domain, consisting of the five C-terminal hydrophobic stretches, and a core antenna domain, which is formed by the six N-terminal hydrophobic stretches of these subunits.

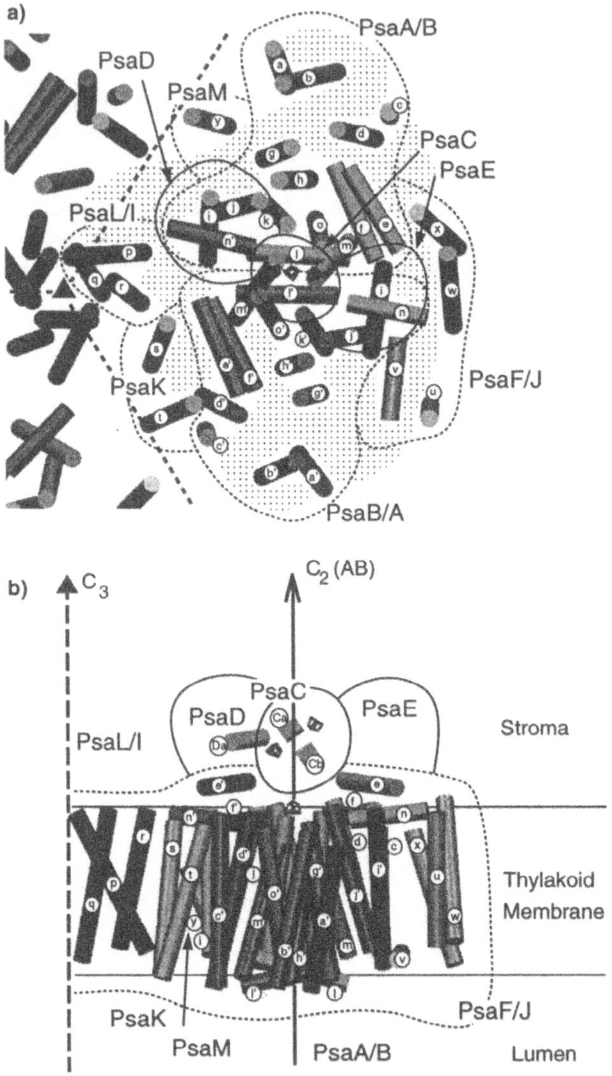

Fig. 2.Arrangement of the α-helices of one PSI monomer and tentative assigment of PSI subunits. *(a)* View from the stroma side, the region occupied by the Chl*a* constituting the core antenna system is indicated by grey shading. *(b)* View along the mebrane plane. Figure modified from [11].

4. Results of general relevance for the research on PSI

Despite the relatively low resolution of the crystallographic structure analysis of PSI, some new insights into the relationship between structure and function of this complex could be obtained from the present structural model.

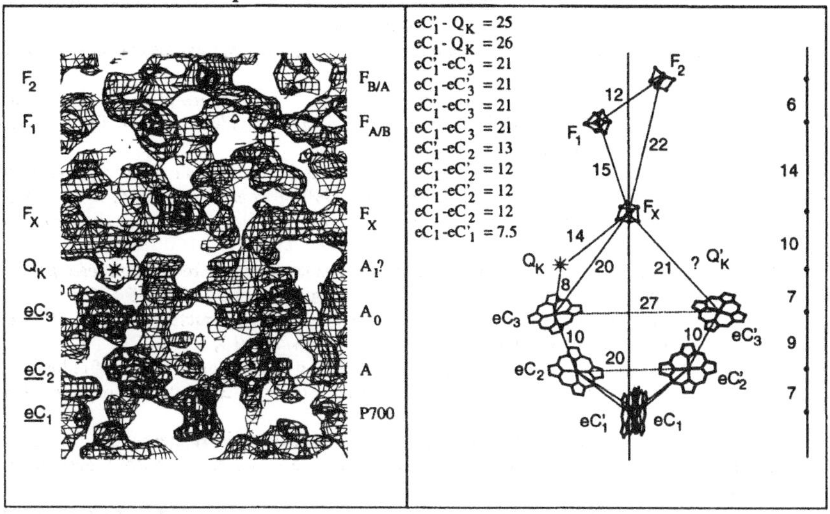

Fig. 3. (left) Region of the electron density map containing the cofactors of the electron transfer chain of PSI, assignment to the spectroscipacally identified cofactors indicated on the right side. View along the membrane plane. *(right)* Arrangement of cofactors (same view direction) with centre to centre distances indicated.

For the membrane intrinsic cofactors of the electron transfer chain, two pseudo-symmetrical branches of cofactors are observed, with the exception of the phylloquinone denoted $\mathbf{Q_K}$ (Fig. 3), possibly indentical with A_1, which lacks a symmetrically equivalent position, presumably because of larger local errors in the electron density of the corresponding region. Intermediate between the pair of Chl*a* $\mathbf{eC_1}$, $\mathbf{eC_1'}$, assigned to P700, and the Chl*a* pair $\mathbf{eC_3}$, $\mathbf{eC_3'}$, one or both of which could be identical with A_0, a pair of Chl*a* monomers $\mathbf{eC_2}$, $\mathbf{eC_2'}$ is located (Fig. 3), for which no indication had been reported prior to the structure analysis at 6 Å resolution. Taking into account the structural model, the fast kinetics of primary charge separation could only be explained if these Chl*a*, denoted accessory Chl*a* in analogy to the accessory bacteriochlorophyll *a* of PbRC, are intermediate electron acceptors between P700 and A_0 [28,11].

The asymmetric arrangement of the iron-sulphur clusters with different distances from the clusters bound to PsaC (denoted $\mathbf{F_1}$ and $\mathbf{F_2}$ in Fig. 3 because of the ambiguity in assigning them to F_A and F_B) to F_X (15 Å and 22 Å) suggested a sequential route of electron transfer to ferredoxin or flavodoxin.

PSI from *Synechococcus elongatus* is the first representative of a photosynthetic system possessing a reaction centre and a core antenna system for which a three-dimensional structural model exists. Out of the 89 Chl*a* identified in the electron density map at 4 Å resolution, 83 belong to the core antenna system. Within a PSI monomer, the Chl*a* form a continuous net with inter-Chl*a* distances (centre to centre) of ≤ 16 Å. This net is linked to the electron transfer chain only via two Chl*a* which are related by the pseudo-symmetry axis of the PsaA/PsaB heterodimer, suggesting a special route of excitation energy transfer to the primary donor via these Chl*a*, although it is unlikely that energy transfer is exclusively occuring via an individual route [29].

5. Comparison of the known structural models of photosynthetic reaction centres

Despite little sequence homology, superposition of the RC-domain of PSI (PSI-RC) and PbRC reveals striking similarities between the structures formed by subunits L and M of PbRC including the electron transfer chain and PSI-RC. This supports the hypothesis that both photosynthetic systems share a common evolutionary origin. As sequence homology had been reported for the antenna binding protein CP47 of PSII and the six N-terminal hydrophobic stretches of subunit PshA of the heliobacterial RC (PSH) [30] and subsequently a more general sequence homology had been shown for the six hydrophobic stretches of CP47 and CP43 of PSII and the six N-terminal hydrophobic spans of PsaA, PsaB and PshA [31], structural models were suggested for the cores of PSII, PSH and the RC from green sulphur bacteria (PSC) [32]. These models were based on the known structures of PSI and PbRC. From the sequence homology of PshA (PSH), PscA (PSC), PsaA and PsaB (PSI) structural similarties were inferred for the RC- and core antenna domains of PSH, PSC and PSI. The PSII model is described by a PbRC-like RC-core connected to a core antenna domain consisting of CP43 and CP47 on opposite sides of this core. The six transmembrane α-helices of CP43 and CP47 are arranged as trimers of dimers resembling the structures formed by the six N-terminal hydrophobic spans of PsaA and PsaB. Electron crystallographic studies on PSII sub-core complexes resulting in a projection map at 8 Å resolution [33] seem to be compatible with this model.

6. Approaches to improve the present structural model of PSI

Recently, an improved native X-ray diffraction data set of 3.5 Å resolution together with data of a mercury derivative obtained by co-crystallisation have been included into calculations of new phases using the method of multiple isomorphous replacement as implemented in the program SHARP [34]. These phases, subsequently improved by density modification techniques (SOLOMON, [35]) resulted in an improved electron density map at 4 Å resolution [36]. The most important new features in the model derived from this map are a second phylloquinone cofactor at a position related by the pseudo-twofold symmetry axis of the PsaA/PsaB heterodimer to the previsously identified phylloquinone [37], and a first model of PsaE within the PSI complex [36]. Only minor alterations of the NMR structural model [26] were necessary to fit it into the electron density map. According to this model, the N- and C-termini of PsaE are directed towards the stromal surface of the PSI complex.

Although the PSI crystals have reached a quality where X-ray diffraction can be detected up to about 3 Å resolution if synchrotron radiation sources are used [16], an electron density map of this resolution is not avaialable for two reasons. First, because of the large mosaicity (≥ 0.7°) of the crystals coupled to the large unit cell constants it is difficult to record separated reflection profiles at the desired high resolution limit of 3 Å. Second, because the crystals, especially of the heavy atom derivatives, are sensitive to radiation damage if the experiment is carried out at a synchrotron source at moderate temperatures (e.g. 277 K), complete data sets collected under these conditions have always been of lower maximum resolution. Recently, this problem has been overcome by establishing a procedure to carry out the experiments under cryogenic conditions, at a temperature of 100 K, without loosing the initial diffraction quality of the PSI crystals after freezing. As expected, radiation damage of the crystals could be drastically reduced under these conditions even when the highly intense X-ray beam at beam line ID2 (EMBL outstation at ESRF, Grenoble) was used. These experiments resulted in the first complete data sets of the PSI crystals of about 3 Å resolution, possibly forming the basis for a significant improvement in the structure analysis of PSI in a near future.

7. Acknowledgments

This work was supported by the Deutsche Forschungsgemeinschaft

(Sonderforschungsbereich 312) and the Bundesministerium für Bildung und Forschung. Synchrotron beam-time allocation at ESRF (EMBL oustation, Grenoble), EMBL oustation and MPG ASMB (DESY, Hamburg), SRS (Daresbury), LURE (Orsay) and Photon Factory (Tsukuba) is gratefully acknowledged.

References

1. Blankenship, R.E. (1992) Origin and early evolution of photosynthesis, *Photosynth. Res.* **33**, 91-111.
2. Deisenhofer, J., Epp, O., Miki, K., Huber, R. and Michel, H. (1985) Structure of the protein subunits in the photosynthetic reaction centre from *Rhodopseudomonas viridis* at 3 Å resolution, *Nature* **318**, 618-624.
3. Deisenhofer, J., Epp, O., Sinning, I. and Michel, H. (1995) Crystallographic refinement at 2.3 Å resolution and refined model of the photosynthetic reaction centre from *Rhodopseudomonas viridis*, *J. Mol. Biol.* **246**, 429-457.
4. Allen, J.P., Feher, G., Yeates, T.O., Rees, D.C., Deisenhofer, J., Michel, H. and Huber, R. (1986) Structural homology of reaction centres from *Rhodobacter sphaeroides* and *Rhodopseudomonas viridis* as determined by X-ray diffraction, *Proc. Natl. Acad. Sci. USA* **83**, 8589-8593.
5. Chang, C.H., el-Kabbani, O., Tiede, D. Norris, J. and Schiffer, M. (1991) Structure of the membrane bound protein photosynthetic reaction from *Rhodobacter sphaeroides*, *Biochemistry* **30**, 5352-5360.
6. Ermler, U., Fritsch, G., Buchanan, S.K. and Michel, H. (1994) Structure of the photosynthetic reaction centre from *Rhodobacter sphaeroides* at 2.65 Å resolution: cofactors and protein-cofactor interactions, *Structure* **2**, 925-936.
7. Arnoux, B., Ducruix, A., Reiss-Husson, F., Lutz, M., Norris, J., Schiffer, M. and Chang, C.H. (1989) Structure of spheroidene in the photosynthetic reaction centre from *Rhodobacter sphaeroides* Y, *FEBS Letters* **258**, 47-50.
8. Arnoux, B., Gaucher, J.-F., Ducruix, A. and Reiss-Husson, F. (1995) Structure of the photochemical centre of a spheroidene-containing purple bacterium, *Rhodobacter sphaeroides* Y, at 3 Å resolution. *Acta Crystallogr.* **D51**, 368-379.
9. Michel, H. and Deisenhofer, J. (1988) Relevance of the photosynthetic reaction center from purple bacteria for the structure of photosystem II, *Biochemistry* **27**, 1-7.
10. Krauß, N., Schubert, W.-D., Klukas, O., Fromme, P., Witt, H.T. and Saenger, W. (1996) Photosystem I at 4 Å resolution reveals the first structural model of a joint photosynthetic reaction centre and core antenna system, *Nature Struct. Biol.* **3**, 965-973.
11. Schubert, W.-D., Klukas, O., Krauß, N., Saenger, W., Fromme, P. and Witt, H.T. (1997) Photosystem I at 4 Å resolution: comprehensive structure analysis, *J. Mol. Biol.* **272**, 741-769.
12. Golbeck, J.H. (1994) Photosystem I in cyanobacteria, in D.A. Bryant (ed.), *Molecular Biology of Cyanobacteria*, Kluwer Academic Publishers, Dordrecht, pp. 179-220.
13. Chitnis, P.R. (1996) Photosystem I, *Plant. Physiol.* **111**, 661-669.
14. Witt, I., Witt, H.T., Di Fiore, D., Rögner, M., Hinrichs, W., Saenger, W., Granzin, J. Betzel, C. and Dauter, Z. (1988) X-ray characterization of single crystals of the rection center I of water splitting photosynthesis, *Ber. Bunsenges. Phys. Chem.* **92**, 1503-1506.
15. Krauß, N., Hinrichs, W., Witt, I., Fromme, P., Pritzkow, W., Dauter, Z., Betzel, C., Wilson, K.S., Witt, H.T. and Saenger, W. (1993) Three-dimensional structure of system I of photosynthesis at 6 Å resolution, *Nature* **361**, 326-331.
16. Fromme, P. and Witt, H.T. (1998) Improved isolation and crystallization of Photosystem I fro structural analysis, *Biochim. Biophys. Acta* **1365**, 175-184.
17. Schubert, W.-D., Klukas, O., Krauß, N., Saenger, W., Fromme, P. and Witt, H.T. (1995) Present state of the crystal structure analysis of photosystem I at 4.5 Å resolution, in P. Mathis (ed.), *Photosynthesis: From Light to Biosphere, Proc. 10th Int. Conf. Photosynth.*, Kluwer Academic Publishers, Dordrecht, vol. 2, pp. 3-11.
18. Mühlenhoff, U., Haehnel, W., Witt, H.T. and Herrmann, R.G. (1993) Genes encoding eleven subunits of photosystem I from the thermophilic cyanobacterium *Synechococcus* sp., *Gene* **127**, 71-78.
19. Adman, E.T., Sieker, L.C. and Jensen, L.H. (1976) The structure of *Peptococcus aerogenes* ferredoxin. Refinement at 2 Å resolution, *J. Biol. Chem.* **251**, 3801-3806.

20. Dunn, P.P., Packman, L.C., Pappin, D. and Gray, J.C. (1988) N-terminal amino acid sequence analysis of the subunitsof pea photosystem I, *FEBS Lett.* **228**, 157-161.
21. Golbeck, J.H. (1993) Shared thematic elements in photochemical reaction centers, *Proc. Natl. Acad. Sci. USA* **90**, 1642-1646.
22. Kamlowski, A., van der Est, A., Fromme, P. and Stehlick, D. (1997) Low temperature EPR on Photosystem single crystals: orientation of the iron-sulphur centers F_A and F_B, *Biochim Biophys. Acta* **1319**, 188-198.
23. Kamlowski, A., van der Est, A., Fromme, P., Krauß, N., Schubert, W.-D., Klukas, O. and Stehlick, D. (1997) The structural organisation of the PsaC protein in Photosystem I from single crystal EPR and X-ray crystallographic methods, *Biochim Biophys. Acta* **1319**, 199-213.
24. Zhao, J., Li, N. Warren, P.V., Golbeck, J.H. and Bryant, D.A. (1992) Site directed conversion of a cystein to aspartate leads to the assembly of a [3Fe-4S]-cluster in PsaC of Photosystem I. The photoreduction of F_A is independent of F_B, *Biochemistry* **31**, 5093-5099.
25. Mannan, R.M., He, W.Z., Metzger, S.U., Whitmarsh, J., Malkin, R. and Pakrasi, H.B. (1996) Active photosynthesis in cyanobacterial mutants with directed modifications in the ligands for the iron-sulphur clusters on the PsaC protein of Photosystem I, *EMBO J.* **15**, 1826-1833.
26. Falzone, C.J., Kao, Y.-H., Zhao, J., Bryant, D.A. and Lecomte, J.T.J. (1994) Three-dimensional structure of PsaE from the cyanobacterium *Synechococcus* sp. strain PCC 7002, a photosystem I protein that shows structural homology with SH3 domains, *Biochemistry* **33**, 6052-6062.
27. Fish, L., Kück, U. and Bogorad, L. (1985) Two partially homolgous adjacent light-inducible maize chloroplast genes encoding polypeptides of the P700 chlorophyll a-protein complex of Photosystem I, *J. Biol. Chem.* **260**, 1413-1421.
28. Brettel, K. (1997) Electron transfer and arrangement of the redox cofactors in Photosystem I, *Biochim. Biophys. Acta* **1318**, 322-373.
29. Valkunas, L., Liuolia, V., Dekker, J.P. and van Grondelle, R. (1995) Description of energy migration and trapping in Photosystem I by a model with two distance scaling parameters, *Photosynth. Res.* **42**, 149-154.
30. Vermaas, W.F.J. (1994) Evolution of heliobacteria: implications for photosynthetic reaction center complexes. *Photosynth. Res.* **41**, 285-294.
31. Fromme, P., Witt, H.T., Schubert, W.-D., Klukas, O., Saenger, W. and Krauß, N. (1996) Structure of photosystem I at 4.5 Å resolution: a short review including evolutionary aspects, *Biochim. Biophys. Acta* **1275**, 76-83.
32. Schubert, W.-D., Klukas, O., Saenger, W., Witt, H.T., Fromme, P. and Krauß, N. (1998) A common ancestor for oxygenic and anoxygenic photosynthetic systems: a comparison based on the structural model of photosystem I, *J. Mol. Biol.* **280**, 297-.314.
33. Rhee, K.-H., Morris, E.P., Zheleva, D., Hankamer, B., Kühlbrandt, W. and Barber, J. (1997) Two-dimensional structure of plant photosystem II at 8-Å resolution
34. de La Fortelle, E. and Bricogne, G. (1997) Maximum-likelihood heavy-atom parameter refinement for multiple isomorphous replacement and multiwavelength anomalous diffraction methods, *Methods Enzymol.* **276**, 472-494.
35. Abrahams, J.P. and Leslie, A.G.W. (1996) Methods used in the structure determination of bovine mitochondrial F_1 ATPase, *Acta Crystallogr.* **D52**, 30-42.
36. Klukas, O., Schubert, W.-D., Jordan, P., Krauß, N., Fromme, P., Witt, H.T. and Saenger, W. (1998) Photosystem I: improved model of the stromal subunits PsaC, PsaD and PsaE, submitted.
37. Klukas, O., Schubert, W.-D., Jordan, P., Krauß, N., Fromme, P., Witt, H.T. and Saenger, W. (1998) Localisation of two phylloquinones Q_K and Q_K' in an improved electron density map of photosystem I at 4 Å resolution, submitted.

ORGANIZATION AND FUNCTION OF PHOTOSYSTEM I TRIMERS AND MONOMERS OF THE CYANOBACTERIUM *SPIRULINA PLATENSIS*

N.V. KARAPETYAN[1], V.V. SHUBIN[1], I.N. BEZSMERTNAYA[1],
M.G. RAKHIMBERDIEVA[1], L.E. MAZHOROVA[1],
I.V. TEREKHOVA[1], D. DORRA[2], A.R. HOLZWARTH[2],
J. KRUIP[3], M. RÖGNER[3], and R.J. STRASSER[4]

[1]*A.N. Bakh Institute of Biochemistry, Leninsky pr. 33, 117071 Moscow, Russia;* [2]*Max-Planck-Institut für Strahlenchemie, 45470 Mülheim and* [3]*Ruhr-Universität-Bochum, 44780 Bochum, Germany;* [4]*Bioenergetic Laboratory, Geneva University, CH-1254 Jussy-Lullier, Switzerland*

1. Abstract

Trimeric complexes of photosystem I (PSI) of *Spirulina* contain the extreme longwave chlorophyll (Chl) form absorbing at 735 nm (Chl735) and emitting at 760 nm (F760). The intensity of F760 is maximal when P700 is reduced and decreases when P700 is oxidized ($P700^+$). Chl735 originates from interaction of Chl molecules, located on the surface of the monomeric subunits forming trimer: (i) deconvolution of 77 K absorption spectrum of PSI trimers shows that each monomer contains two molecules of Chl735; (ii) transfer of PSI trimers into monomers in cyanobacterial membranes at „high salt" is accompanied by decrease of F760 (Chl735 disaggregation), while trimerization of monomers at „low salt" results the recovery of F760 (Chl735 formation); (iii) incubation of PSI monomers in liposomes causes reconstitution of trimers containing Chl735. Illumination of PSI complexes with intense light bleaches bulk Chls but does not effect P700. The non-linear relationship between the kinetics of P700 photooxidation and F760 quenching in PSI trimers at 77 K is interpreted as energy exchange (via Chl735) between antennae of monomers and quenching of the excess energy by $P700^+$. The energy exchange between monomers promotes dissipation of the overexcitation into heat. Thus both PSI trimers and monomers contribute in electron transport whereas PSI trimers in addition protect complex against photodestruction.

J. H. Argyroudi-Akoyunoglou and H. Senger (eds.),
The Chloroplast: From Molecular Biology to Biotechnology, 27–34.

2. Introduction

Photosystem I (PSI) complexes of higher plants and cyanobacteria carry out the same function and have similar organization, but differ in pigment-protein complexes and some polypeptides [1,2]. Cyanobacteria have no chlorophyll (Chl) containing light-harvesting complex (LHC), and whole Chl in cyanobacterial membranes is localized in core complexes of PSI and PSII. As compared with higher plants, cyanobacteria are highly enriched with PSI: the ratio PSI/PSII is about 5-10.

PSI complex in higher plants is organized as monomer, whereas in cyanobacteria it exists as monomer and trimer although both types of complexes contain the same polypeptides [3,4]. The antenna size (Chl/P700) of both complexes was determined to be about 100 indicating that each monomeric subunit within the trimer contains P700. Since the relative amount of PSI reaction centers in cyanobacterial membranes is higher than that of PSII, it was proposed that some PSI reaction centers are involved in cyclic electron transport [5]. Cyanobacterial PSI trimeric and monomeric complexes are very stable and independent of the detergents used for isolation. The equilibrium between PSI trimeric and monomeric complexes in isolated detergent-free membranes could be regulated by changing their electrostatic interaction at various salt conditions [6]. The reason of the existence of PSI trimers and their role in cyanobacterial photosynthesis is not clear.

Core antenna of PSI trimers of various cyanobacteria contains the longwave Chls that differ in the content and spectral characteristics. Two molecules of Chl708 are found in the antenna of PSI trimers of *Synechocystis* sp. [7], whereas trimers of *Synechococcus elongatus* in addition to Chl708 contain six molecules of Chl716 [8]. The most red-shifted Chl absorbing at 735 nm (Chl735) and emitting at 760 nm (F760) presents in the antenna of PSI trimers of *Spirulina* [9]. In contrast to all other longwave Chls, the fluorescence intensity of Chl735 strongly depends on the redox state of P700: it is maximum when P700 is reduced and highly decreased when P700 is oxidized [4,10]. It was calculated that the cation radical of P700 ($P700^+$) is the strongest quencher of F760 as compared with neutral P700 [11]. No energy exchange between the antennae of the monomeric subunits within PSI trimer was found at room temperature that gives the gain in P700 photooxidation. Since Chl735 reflects the redox state of P700 it may be used for study of the dynamics of energy migration from antenna to P700 in dependence on its redox state. In this paper we discuss the organization of PSI trimers in cyanobacteria, the origin of Chl735 and its role in light-induced reactions in *Spirulina*.

3. Materials and Methods

PSI trimeric and monomeric complexes were isolated by means of column chromatography on DEAE-Toyopearl from the membranes of cyanobacterium *Spirulina platensis* treated with Triton X-100 [9], or by Waters system with Poros Butyl column from *Spirilina* membranes treated with n-dodecyl-β,D-maltoside [12]. Low temperature absorption spectra of PSI complexes, membranes and cells were measured with fast-scanning spectrofotometer Omega (Bruins Instruments, Muenchen). 77 K fluorescence emission spectra of PSI complexes were registered with a spectrofluorometer Spex Fluorolog (USA) or luminescence spectrometer SLM-Aminco Spectronic Instruments (USA); spectra were corrected on spectral sensitivity of instruments. To measure 77 K fluorescence spectra of PSI trimers, those were transferred in 200 mM Tris buffer (pH 10.6), then dithionite (20 mg/ml) was added; after short incubation and addition of glycerol (60% v/v) sample was placed in quartz tube (inner diameter 4 mm) and frozen under illumination to accumulate the acceptor part of PSI in reduced state [10]. PSI monomers were frozen without dithionite since fluorescence yield of PSI monomers does not depend on P700 redox state. Concentration of Chl in PSI complexes was 6 μg/ml.

The kinetics of P700 photooxidation and the quenching of F760 at 77 K was simultaneous measured using the combination of Plant Efficiency Analyzer (PEA, Hansatech, King Lynn, UK), Dual Channel Fluorimeter (Hansatech, UK) and home developed multibranched fiberoptic system [13,14]. P700 photooxidation was measured as a light-induced change of reflectance at 820 nm. In these measurements PSI trimers after addition of ascorbate to reduce P700 were settled on a Millipore filter paper as film and cooled in the dark to 77 K [15]. The PSI activity of isolated complexes was measured as a rate of oxygen consumption with Clark type electrode in the presence of reduced dichlorophenol-indophenol (DCPIP) or diaminoduren (DAD), methylviologen and 0.03% dodecyl-maltoside, the intensity of light was 6000 $\mu E\ m^{-2}$. The liposomes were prepared from phosphatidylcholine according to the methods described in [16]; PSI complexes were added to liposomes at lipid:Chl ratio 25 or 125 (w/w).

4. Results and Discussion

Table 1 presents the characteristics of the PSI trimers and monomers isolated from *Spirulina*. Both complexes have similar antenna size, dimensions typical for trimers and monomers from other cyanobacteria, similar polypeptide composition and β-carotene content. These complexes differ mainly in longwave Chls and their content: only PSI trimers contain Chl735 emitting at 760 nm. The PSI complexes isolated using chromatography on DEAE-Toyopearl or Water system contain similar longwave Chls. Fluorescence spectra of purified PSI trimers and monomers show that these complexes contain no uncoupled Chls: no emission at 685 was observed (Fig. 1). The ratio F760/F730 of trimers is an indicator of their purity as well as for the evaluation of the amount of PSI trimers in cells. The main part of PSI complexes in *Spirulina* cells is organized as trimers since this ratio calculated from spectra measured for cells frozen with dithionite is about 3.0 (Fig. 2) as compared with that of 3.2 for isolated trimers (Fig. 1). According to the fluorescence spectrum of *Spirulina* cells that before cooling were adapted to light and measured without dithionite, P700 in that cells is oxidized, whereas P700 in dark adapted cells is partly reduced (Fig. 2). Thus main part of P700 in illuminated cells is accumulated in oxidized state since cyanobacteria contain more PSI than PSII.

It is suggested that the longwave Chls may focus excitation to reaction center, increase the cross section of absorption or protect thylakoids against photodestruction [11,17-19]. Selective excitation of PSI longwave Chls causes photooxidation of P700 even at 77 K [11,20]. Thus energy absorbed by the longwave Chls reaches P700 by uphill energy transfer and is used for photooxidation of P700. About 30% of energy absorbed by bulk Chls migrates to Chl735 in PSI trimers of *Spirulina* [10]. But since only the F760 reflects the redox state of P700 it indicates a special organization of Chl735 as compared with other longwave Chls.

TABLE 1. Characteristics of PSI trimeric and monomeric complexes of the cyanobacterium *Spirulina platensis*.

	trimer	monomer
Chl/P700	90± 5	90± 5
Dimensions (nm)	19x7	15x7.5
Mol. mass (kDa)	900	300
β-carotene/100 Chls	11± 1	10± 1
Polypeptides	PsaA/PsaB, PsaC, PsaD, PsaF, PsaL etc.	
Amount of longwave Chl	6.4%	5.2%
Longwave absorption (nm)	710, 735	710
Longwave fluorescence (nm)	730, 760	730

Several data indicate that the extreme longwave Chl735 emitting at 760 nm originates from pigment-pigment interaction of Chl molecules of core antenna located on different monomeric subunits within the trimer. Deconvolution of 77 K absorption spectra of trimers and monomers allowed to show that both complexes contain similar set of forms of bulk Chl, but only antenna of trimers includes the extreme longwave Chl735 [10. According to the area under the spectral curves of Chl forms, each monomeric subunit forming trimer contains about 2 molecules of Chl735.

The experiments with salt-induced regulation of the ratio of PSI trimers and monomers within the cyanobacterial membranes is another prove of our idea on the intrasubunit pigment-pigment interaction of Chls forming Chl735 [10]. PSI complexes in the membranes at low salt conditions (5 mM) present mainly as trimers since these membranes show intense F760. Transfer of membranes to high salt (more than 150 mM) conditions causes the conversion of trimers into monomers that is accompanied by a decrease of F760 as a result of decomposition of Chl735. Washing of membranes from salts causes trimerization of PSI trimers and recovery of F760 because of formation of Chl735.

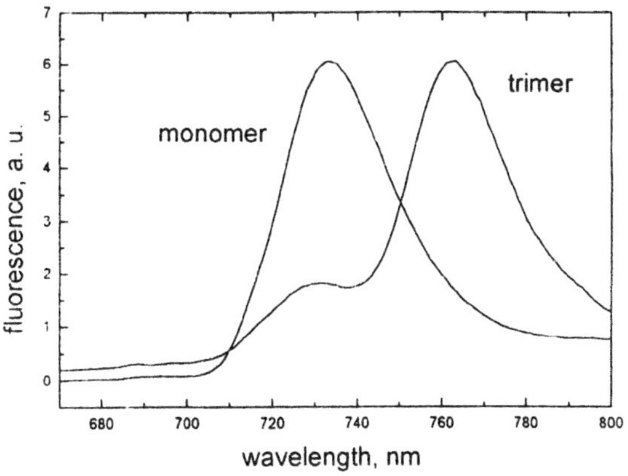

Figure 1. 77 K fluorescence emission spectra of the purified PSI trimeric and monomeric complexes isolated from *Spirulina*. Complexes were frozen with dithionite under illumination, excitation wavelength 440 nm, spectra are normalized to maximum signal.

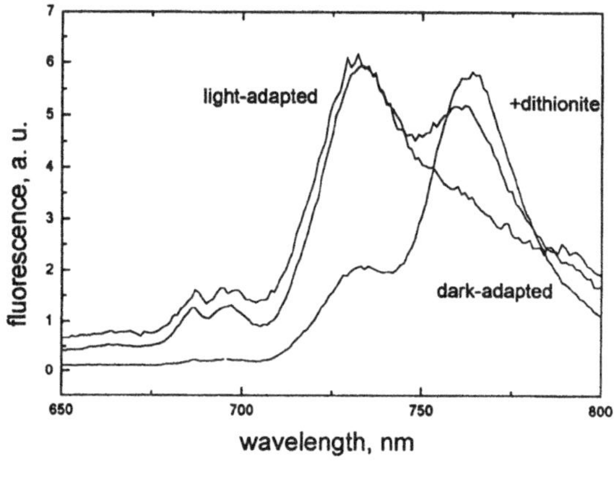

Figure 2. 77 K fluorescence emission spectra of *Spirulina* cells adapted to dark or to light and frozen without dithionite, and cells frozen with dithionite under illumination; excitation wavelength 440 nm, spectra are normalized to maximum signal.

Finally, data obtained in the course of reconstitution of PSI trimers from monomers within the liposomes also indicate that Chl735 appears only in trimers. Incubation of liposomes prepared from phosphatidylcholine with PSI monomers causes partial reconstitution of PSI trimers which are similar to the original trimers since they show F760 (Fig. 3). It is necessary to note that incubation of liposomes with trimers causes their partial transformation into monomers (Fig. 3). According to the ratio F760/F730, about 25-30% of monomers in liposomes is transferred into trimers, this was proved by data from the chromatography of sample from liposomes [21]. Thus reconstitution of trimers is accompanied by formation of Chl735 emitting at 760 nm.

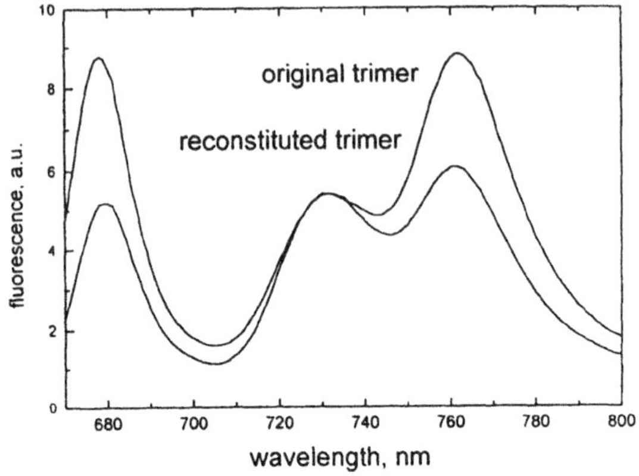

Figure 3. 77 K fluorescence emission spectra of proteoliposomes incubated with original PSI trimers and with monomers (reconstituted trimers). Complexes were frozen with dithoinite under illumination; excitation wavelength 440 nm, spectra are normalized at 733 nm.

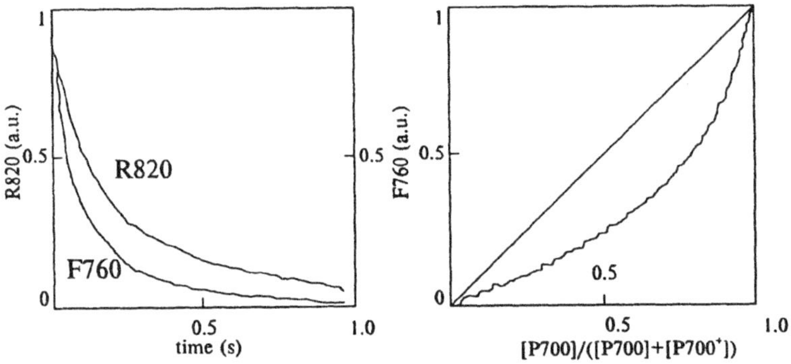

Figure 4. 77 K kinetics of P700 photooxidation (R820) and light-induced decrease of F760 of PSI trimers (left), and the dependence of variable part of F760 on the fraction of reduced P700 (right). P700 was reduced by ascorbate in dark before cooling of trimers.

To learn the role of PSI trimers and monomers in photosynthetic electron transport, their activity was measured as light-induced consumption of oxygen in the presence of methylviologen and electron donors (DAD+ascorbate). Both complexes have shown the same activity-about 200 μM/mg Chl per hour. The activity of cyanobacterial membranes containing variable amount of PSI monomers and trimers measured at different salt conditions was also similar. Thus both PSI compleses are involved in photosynthetic electron transport from PSII to $NADP^+$. The contribution of PSI trimeric and monomeric complexes to electron transport was shown by their ability to cross-linking with soluble ferredoxin [22]. Preillumination of PSI complexes with moderate light intensity (6000 μE m^{-2}) causes the bleaching of bulk Chls determined as decrease of intensity of F730 and F760 in fluorescence spectra and the amount of Chl in treated samples, but had no effect on P700 content in trimers and monomers. It has to be noted that Chl710 is more sensitive to photodestruction than Chl735 (data not shown). The activity of PSI monomers after light treatment is higher because of decomposition of some polypeptides of PSI complex.

To check the possible protective role of PSI trimers, the kinetics of P700 photooxidation and quenching of F760 of PSI trimers was measured simultaneously at 77 K. It was found that fluorescence quenching is faster than accumulation of oxidized P700 (Fig. 4, left), though cation radical of P700 is responsible for F760 quenching [4,11]. Non-linear relationship between F760 quenching and accumulation of oxidized P700 may be interpreted only as energy exchange between the antenna of monomeric subunits (Fig. 4, right). When P700 in one of monomeric subunits is oxidized, energy absorbed by the antenna of other subunits migrates via Chl735 to subunit with oxidized P700 and efficiently quenched since oxidized P700 is the strongest quencher of F760.

In PSII or bacterial photosystems the energy absorbed by antenna with closed reaction center is transferred to the reaction center with open reaction center and is used in photochemistry. $P700^+$ quenches the energy even in PSI monomers of cyanobacteriaand higher plants, but in the case of PSI trimers of *Spirulina* $P700^+$ is able to quench also the energy absorbed by other monomeric subunits because of energy exchange between these subunits. Since the relative content of PSI in cyanobacteria is higher than that of PSII, the electron donation from PSII may be limited, and P700 will accumulate in oxidative state even at moderate light. $P700^+$ in PSI trimers of *Spirulina* decrease overexcitation of the core antenna. It was found that LHCII trimers in higher plants also are contributed in dissipation of the excess energy [23]. Thus energy exchange between antenna of the monomeric subunits within PSI trimer gives gain not in photochemistry but in dissipation of the excess energy into heat. This phenomenon can be considered as a new mechanism of protection of cyanobacterial pigment-protein complexes against photodestruction.

5. Acknowledgement

This work was supported by grants from DFG, Russian Foundation for Basic Research (96-04-48809), and NATO Scientific Affairs Division (HTECH.LG 931646).

6. References

1. Golbeck J.H. (1994) Photosystem I in cyanobacteria, in D.A. Bryant (ed.), *The Molecular Biology of Cyanobactereia*, Kluwer Acadic Publishers, Dordrecht, pp. 319-360.
2. Golbeck J.H. (1992) Structure and function of photosystem I. *Ann. Rev. Plant Physiol. Plant Mol. Biol.* 43, 293-324.
3. Rögner M., Mühlendorf U., Boekema E.J., and Witt H.T. (1990) Mono-, di- and trimeric PSI reaction center complexes from the thermophilic cyanobacterium *Synechococcus* sp. Size, shape and activity. *Biochim. Biophys. Acta* 1015, 415-424.

4. Shubin V.V., Tsuprun V.L., Bezsmertnaya I.N., and Karapetyan N.V. (1993) Trimeric forms of the photosystem I reaction center complex pre-exis in the membranes of the cyanobacterium Spirulina platensis. *FEBS Lett.* 309, 340-342) *FEBS Lettr.* 334, 79-82.
5. Rögner M., Boekema E.J., and Barber J. (1996) How does photosystem 2 split water? The structural basis of efficient energy conversion. *TIBS* 21, 44-49.
6. Kruip J, Bald D, Boekema EJ, and Rögner M (1994) Evidence for the existence of trimeric and monomeric photosystem I complexes in the thylakoid membranes from cyanobacteria. *Photosynth. Res.* 40, 279-286.
7. Gobets B, van Amerongen H., Monshower R., Kruip J., Rögner M., van Grondelle R., and Dekker J.P. (1994) Polarized site-selective fluorescence spectroscopy of isolated photosystem I particles. *Biochim. Biophys. Acta* 1188, 75-85.
8. Palsson L.-O., Dekker J.P., Schlodder E, Monshouwer R, and van Grondelle R. (1996) Polarized site-selective fluorescence spectroscopy of the long-wavelength emitting chlorophylls in isolated photosystem I particles of *Synechococcus elongatus*. *Photosynth. Res.* 48, 239-246.
9. Shubin V.V., Bezsmertnaya I.N., and Karapetyan N.V. (1992) Isolation on from *Spirulina* membranes of two photosystem I-type complexes, one of which contains chlorophyll responsible for the 77 K fluorescence band at 760 nm. *FEBS Lett.* 309, 340-342.
10. Karapetyan N.V., Dorra D., Schweitzer G., Bezsmertnaya I.N., and Holzwarth A.R. (1997) Fluorescence spectroscopy of the longwave chlorophylls in trimeric and monomeric photosystem I core complexes from the cyanobacterium *Spirulina platensis*. *Biochemistry* 36, 13830-13837.
11. Shubin V.V., Bezsmertnaya I.N., and Karapetyan N.V. (1995) Efficient energy transfer from the long-wavelength chlorophylls to P700 in photosystem I complexes from *Spirulina platensis*. *J. Photochem. Photobiol.* 30B, 153-160.
12. Kruip J., Boekema E.J., Bald D., Boonstra A.F., and Rögner M. (1993) Isolation and structural characterization of monomeric and trimeric photosystem I complexes (P700 F_A/F_B and P700 F_X) from the cyanobacterium *Synechocystis* PCC 6803. *J. Biol. Chem.* 268, 23353-23360.
13. Havaux M., Greppin H., and Strasser R.J. (1993) Functioning of photosystems ! and II in pea leaves exposed to heat in the presence or absence of light. *Planta* 186, 88-98.
14. Strasser R.J. , Srivastava A., and Govindjee (1995) Polyphasic chlorophyll fluorescence transients in plants and cyanobacteria. *Photochem. Photobiol.* 61, 32-42.
15. Karapetyan N.V., Shubin V.V., Bezsmertnaya I.N., and Strasser R.J. (1995) Dynamics of primary photochemistry of PSI in membranes, monomeric and trimeric complexes of Spirulina at 77 K: correlation of P700 photooxidation and fluorescence quenching at 760 nm., in P. Mathis (ed.) *Photosynthesis: from Ligh to Biosphere*, Kluwer Acadeim Publishers, Dordrecht, v. 2, pp. 55-58.
16. Gurovskaya K.N., Mamedov M.D., Vasilliev I.R., Golbeck J.H., and Semenov A.Yu. (1997) Electrogenic reduction of the primary electron donor $P700^+$ in photosystem I by redox dyes. *FEBS Lett.* 414, 193-196.
17. van Grondelle R. , Dekker J. , Gilbro T. , and Sundstroem V. (1994) Energy transfer and trapping in photosynthesis. *Biochim. Biophys. Acta* 1187, 1-65.
18. Trissl H.-W. (1993) Long-wavelength absorbing antenna pigments and heterogeneous absorption bands concentrate excitons and increase absorption cross section. *Photosynth. Res.* 35, 247-263.
19. Mukerji I. and Sauer K (1989) Temperature-dependent steady-state and picosecond kinetic fluorescence measurements of a photosystem I preparation from spinach, in Briggs WR (Ed.) *Photosynthesis. Plant Physiology.* Alan R. Liss, v. 8, pp.105-122.
20. Palsson L.-O., Flemming C., Gobets B., van Grondelle R., Dekker J.P., and Schlodder E. (1998) Energy transfer and charge separation in photosystem I. P700 oxidation upon selective excitation of the long-wavelength antenna chlorophylls of *Synechococcus elongatus*. *Biophys. J.* 74, 2611-2622.
21. Kruip J., Karapetyan N.V., Berry S., Terekhova I.V., and Rögner M. (1998) Reconstitution in liposomes of trimers from the photosystem I monomeric complexes of *Spirulina* (in preparation).
22. Lelong C., Boekema E.J., Kruip J., Bottin H., Rögner M., and Setif P. (1996) Characterization of a redox active cross-linked complex between cyanobacterial photosystem I and soluble ferredoxin. *EMBO J.*, 15, 2160-2168.
23. Horton P., Ruban A.V., and Walters R.G. (1996) Regulation of light harvesting in green plants. *Ann. Rev. Plant Physiol. Plant Mol. Biol.* 47, 655-684.

PROGRESS IN THE CRYSTALLIZATION OF PHOTOSYSTEM II COMPONENTS

Noam Adir, Rina Anati, Valeria Cheredman and Natalia Lerner
Department of Chemistry, Technion, Technion City, Haifa 32000 Israel

1. Introduction

The biochemical processes involved in photosynthetic electron transfer are being studied at increasingly higher detail. Structural information on the atomic and molecular level are needed in order to elucidate biochemical reaction mechanisms. Determination of the three dimensional structures of proteins and protein complexes has become essential for any meaningful discussion on the photosynthetic reactions (1-3). Since most of the protein complexes found in photosynthetic systems are larger than can be dealt with by NMR techniques, crystallographic techniques must be utilized. In most cases obtaining high quality crystals remains the most problematic step.

Photosystem II (PSII) is one of the most complex of the different photosynthetic systems (4-5). Found in all higher plants, green algae and cyanobacteria, it is the site of the initiation of linear electron flow, proton gradient formation and oxygen evolution. It's reaction center (RCII), has a molecular weight of about 250,000, and is surrounded by a highly heterogeneous mixture of various antenna pigment protein complexes (LHCII and CPs in eukaryots, phycobilisomes in cyanobacteria) and other proteins with regulatory roles. We are attempting to obtain structural information on PSII by X-ray crystallographic methods. Because of its complexity, we are trying to obtain diffraction quality crystals from the entire RCII complex, and also of several isolated components of PSII. We describe here our progress in these areas.

2. Results and Discussion

2.1. CRYSTALLIZATION OF RCII

We have crystallized isolated oxygen evolving RCII from spinach and pea. Since our original communication of these results (6) we have been working steadily to try and improve the quality of the crystals. by

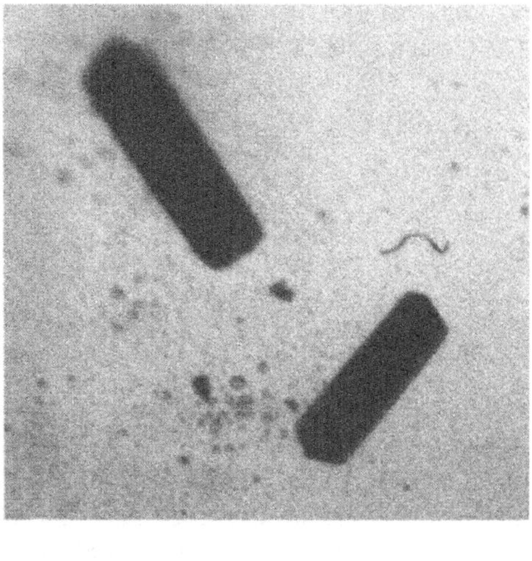

Figure 1: Spinach RCII crystal

J. H. Argyroudi-Akoyunoglou and H. Senger (eds.),
The Chloroplast: From Molecular Biology to Biotechnology, 35–40.

refining our isolation and crystallization methods while at the same time understand the biochemical reasons for the difficulty behind its crystallization.

RCII contains 9 protein subunits that can be subdivided into 4 subcomplexes: the reaction center core (D1, D2, cyt. b_{559} and two small polypeptides), the two internal antennas (CP47 and CP43) and the 33 kDa protein of the oxygen evolving complex (OEC). A large variety of various covalent modifications occur to RCII protein subunits, increasing further the heterogeneity of the system. One of the most difficult problems has been the heterogeneity of the system due to the turnover of the D1 protein. It has been well established that this phenomenon, includes covalent modifications to the D1 protein at both high and low light levels (7,8). Modification and degradation processes continue *in vitro* in different types of isolated RCII (9). Modifications and heterogeneities induced by the isolation process were also identified for RCII prior to (6) and after crystallization (10). It is our belief that the heterogeneous nature and the intrinsic instability of RCII are probably the primary reason behind the difficulty in obtaining high quality crystals.

We have crystallized RCII from crystallization liquors that contain detergent mixtures. The crystals that have been obtained have a hexagonal morphology and reach dimensions of 1mm x 0.3mm x 0.3mm (Fig. 1). The crystals diffract to ~8Å on our in house diffractometer. Recently, crystals analyzed on beamline ID2 at the ESRF facility in Grenoble showed weak diffraction to below 7Å. Analysis using the DENZO (11) autoindexing software package show that the crystals belong to an as yet undetermined hexagonal space group with unit cell dimensions of $a = 495Å$, $b = 495Å$, $c = 115Å$, $\alpha = \beta = 90°$, $\gamma = 120°$.

We have now found nine different detergent mixtures, which all promote the growth of the same crystal form (Table I).

Table 1: Summary of detergent mixtures that promoted RCII crystal growth.

Detergent mixes	Detergent 1[a]	cmc (mM)[b]	Detergent 2[a]	cmc (mM)
1	DM	0.17	HTG	29
2	DM	0.17	MEGA-8	79
3	DM	0.17	MEGA-9	25
4	Cymal-6	0.56	Cymal-3	34.5
5	Cymal-6	0.56	HTG	29
6	Cymal-6	0.56	MEGA-9	25
7	DG	0.16	OGal	29.5
8	Triton x-100	0.23	MEGA-9	25
9	DecM	1.8	HTG	29

[a]Cymal-3, Cyclohexyl-propyl-β-D-maltopyranoside; Cymal-6,Cyclohexyl-hexyl-β-D- maltopyranoside; DecM, n-decyl-β-D-maltopyranoside; DG, Dodecyl-β-D-glucoside; DM, n-Dodecyl-β-D maltopyranoside; HTG, 1-S-Heptyl-β-D-thioglucopyranoside; OGal, Octyl- β-D-galactoside. [b]cmc, critical micelle concentration

The detergents have been shown to interact with RCII by HPLC analysis. RCII crystals were solubilized, and analyzed by size-exclusion HPLC (Fig. 2), which showed the same elution profile as detergent treated isolated RCII, prior to crystallization.

2.2. CRYSTALLIZATION OF THE OEC 33 KDA PROTEIN

Oxygen evolving RCII contains a single non-transmembrane subunit, the 33 kDa protein. This subunit is the major protein of the oxygen-evolving complex (OEC), and is found in all RCII containing organisms. It has been proposed that the 33kDa protein is involved in the creation of the environment needed for water oxidation by the tetramaganese center (12). It appears however not to be the manganese binding protein. Removal of the protein from RCII is usually accompanied by the loss of manganese and water oxidation.

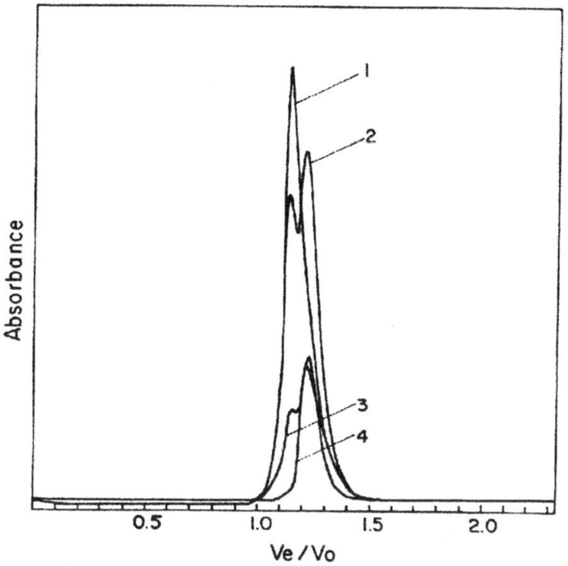

Figure 2. Size-exclusion HPLC analysis of RCII before and after crystallization. Trace 1, Isolated spinach RCII in 0.025% dodecyl maltoside. Trace 2, Isolated spinach RCII treated with detergent mix #1 (Table I) for 3 days prior to HPLC analyses. Trace 3, the same sample as in trace 2, after an additional 7 days incubation. Trace 4, spinach RCII from solubilized crystals grown using detergent mix #1.The column was equilibrated with 0.025% dodecyl maltoside, and run at 0.25 ml/min. Detection of elution peaks was performed at 280nm. The elution pattern of the column was calibrated using a molecular weight standard kit (Sigma). Vo (the column void volume) was determined using dextran blue.

Unlike the other RCII subunits, there exists a high degree of variability in both the sequence and size of the 33 kDa protein in different organisms, even those with close evolutionary ties. The 33 kDa protein has been shown to associate with RCII on it's luminal side. Interactions have been proposed between the 33 kDa protein and almost all of the other subunits. It's position is unknown, but it has been proposed to bind along a pseudo-C_2 axis of symmetry which may pass through RCII in a manner similar to that found in bacterial reaction centers. Other experimental results allude to a binding site off center, and closer to the D1 protein side of RCII, containing the redox active tyrosine Y_Z. of RCII.

We are attempting to solve the three-dimensional structure of the 33 kDa protein. While we are able to isolate relatively large amounts of very pure protein we have only succeeded in obtaining different types of microcrystals (fig. 3). The crystals in fig. 3 were treated with the dye Izit ™ (Hampton Research) that penetrates and stains protein crystals, but not crystals that contain small molecules. Another type of microcrystal was used for

SDS-PAGE analysis (by harvesting, washing and solubilization), and showed the presence of 33 kDa protein.

The growth of microcrystals could be due to heterogeneities. To assess this possibility, reverse phase HPLC coupled to electronspray ionization-mass spectroscopy (ES-MS) was performed (fig. 4) The result of the HPLC analysis was that the protein sample was essentially pure (fig. 4a). Three small secondary peaks were also resolved by HPLC. All three satellite peaks were identical in composition to the main 33 kDa peak (by ES-MS analysis, data not shown) and thus could be the result of either oligomerization or a conformational difference. The former possibility was discounted by size-exclusion HPLC analysis (data not shown), and thus we believe that the presence of more than one conformation of the 33 kDa protein may be one of the reasons for the small crystal size.

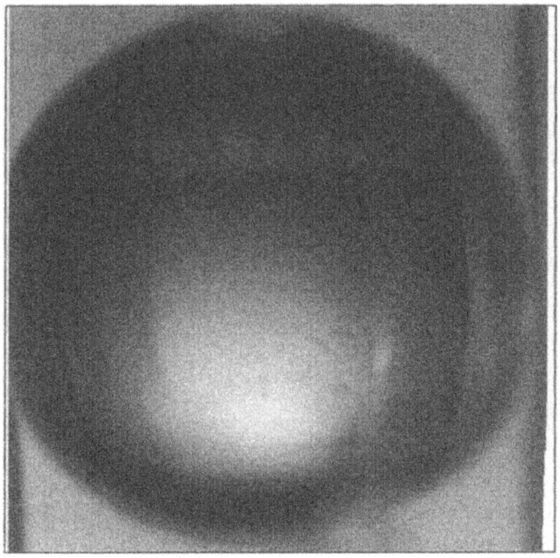

Figure 3: Spinach 33 kDa protein microcrystals, following treatment with protein specific dye.

The ES-MS analysis resulted in a molecular mass of 26,535 (fig. 4b). The sequence of spinach 33kDa protein (13) indicates that following processing of the precursor by removal of the transit peptide, the molecular mass should be one amino acid longer (26,635). Automated Edman degradation sequencing of the protein showed that as proposed, the isolated mature protein begins, as expected, with the sequence Glu – Gly – Gly – Lys – Arg. We thus conclude that the 33 kDa protein is processed on it's C-terminal end, resulting in the loss of a single glutamine residue. Oh-oka *et al.* 1986 (14), performed a complete Edman degradation sequencing of this protein, and suggested that the 33kDa is heterogeneous in its C-terminus, ending in both –Gln and -Gln-Gln forms. By our analysis, this is not the case, only the processed –Gln form is present

2.3 THE INTERMEDIATE ANTENNA COMPLEX CP29

RCII is connected to the bulk antenna complex, LHCII, by a number of intermediate antenna complexes. All of this chlorophyll/protein complexes are members of the cab gene family (15). One of the most important is CP29. This protein has been shown to bind rather strongly to RCII, and it thus has been proposed to reside in close proximity to the reaction center. As with other RCII associated proteins, this complex appears to be heterogeneous, including a partially phosphorylated population.

a b

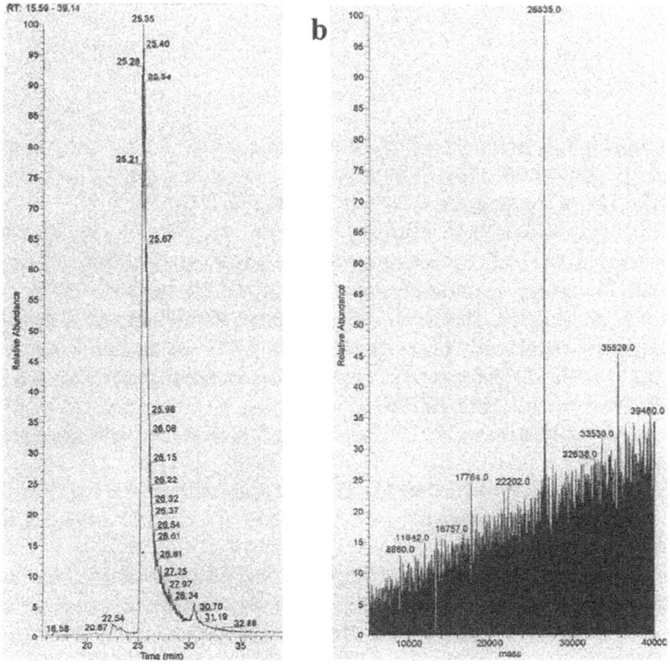

Figure 4: Electronspray-mass spectroscopy (ES-MS) analysis of the isolated spinach 33 kDa protein. Panal a, reverse-phase HPLC chromatographic separation of the isolated 33 kDa protein. Panel b, transformed ES-MS mass spectrum of the major peak from panel a, showing molecular mass of the 33 kDa protein of 26,535.

We have isolated CP29 from crude RCII complex. The isolated CP29 was crystallized in the presence of dodecyl maltoside, resulting in long thin needles. These crystals diffract X-rays to ~20Å (Fig 5).

Figure 5: Spinach CP29 crystals.

3. Acknowledgments

We thank Dr. Gadi Schuster for providing us with hydroponically grown spinach. We would like to that the Women's Division of the American Technion Society for the endowment of MEPXVIII, The Macromolecular Structure Research Laboratory. This work is supported by the Israel Science Foundation founded by the Israel

Academy of Sciences and Humanities (374-96-1).

4. References

1. Stowell, M.H.B., McPhillips, T.M., Rees, D.C., Soltis, S.M., Abresch, E. and Feher, G. (1997) Light induced structural changes in photosynthetic reaction center: implications for mechanism of electron-proton transfer. *Nature* **276**, 812-816.
2. Krauß, N., Schubert, W.-D., Klukas, O., Fromme, P., Witt, H.T. and Saenger, W. (1996) Photosystem I at 4Å resolution represents the first structural model of a joint photosynthetic reaction centre and core antenna system. *Nat. Struc. Biol.* **3**, 965-973.
3. Rhee, K.H., Morris, E. P., Zheleva, D., Hankamer, B., Kuhlbrandt, W. and Barber, J.(1997) Two-dimensional structure of plant photosystem II at 8-Å resolution *Nature* **389**, 522-526.
4. Debus, R.J., (1992) The manganese and calcium ions of photosynthetic oxygen evolution. *Biochim. Biophys. Acta* **1102**, 269-352.
5. Barber, J. and Andersson, B. (1994) Revealing the blueprint of photosynthesis. *Nature* **370**, 31-34.
6. Adir, N., Okamura M.Y. and Feher; G. (1992) Crystallization of the PSII-Reaction Center, in: *Research in Photosynthesis*, (N. Murata, ed.) Vol. II, pp. 5.195-5.198, Kluwer Academic Publishers, The Netherlands.
7. Prasil, O., Adir N. and Ohad, I. (1992) Dynamics of Photosystem II: Mechanism of photoinhibition and recovery processes, in *The Photosystems: Structure, Function and Molecular Biology*, (J. Barber ed.), Elsievier Science Publishers, B.V., Amsterdam, pp.295-348.
8. Keren, N.,Berg, A., van Kan, P. J. M., Levanon, H. and Ohad, I. (1997) Mechanism of photosystem II photoinactivation and D1 protein degradation at low light: The role of back electron flow*Proc. Natl. Acad. Sci.* U.S.A. **94**, 1579.
9. Shipton, C.A. and Barber, J. (1991) Photoinduced degradation of the D1 polypeptide in isolated reaction centers of photosystem II: Evidence for an autoproteolytic process triggered by the oxidizing side of the photosystem. *Proc. Natl. Acad. Sci.* U.S.A.**88**, 6691-6695.
10. Adir, N. in preparation.
11. Otwinowski, Z., (1993) in *Data Collection and Processing*, (Sawyer, L., Isaacs, L., Baily, S., eds.), pp. 56-62, SERC Daresbury Laboratory, Daresbury, UK.
12. Ghanotakis, D.F. and Yocum, C.F. (1990) Photosystem II and the oxygen-evolving complex *Ann. Rev. Plant Physiol. Plant Mol.Biol.* **41**, 255-276.
13. Tyagi A., Hermans J., Steppuhn, R.C.J., Jansson C., Vater F., and Herrmann R., (1987) G.Mol. Gen. Genet. **207**, 288-293.
14. Oh-Oka H., Tanaka S., Wada K., Kuwabara T., and Murata N. (1986) Complete amino acid sequence of 33 kDa protein isolated from spinach photosystem II particles. FEBS LETT. **197**, 63-66.
15. Green, B.R., Pichersky, E. and Kloppstech, K. (1991) Chlorophyll a/b-binding proteins: an extended family. *Trends Biochem. Sci.* **16**, 181-186.

THE PHOTOSYNTHETIC APPARATUS OF *PROCHLORON*-LIKE CYANOBACTERIA

JÜRGEN MARQUARDT[1,2] and ERHARD MÖRSCHEL[1]
[1]*Department of Biology, Philipps University, Marburg, Germany*
[2]*ICBM, Carl von Ossietzky University, Oldenburg, Germany*

1. Introduction

Cyanobacteria are photosynthetic prokaryotes with two photosystems (PSI and PSII). As higher plants, they use water as primary electron donor of the photosynthetic electron transport chain. The organization of their thylakoid membranes, however, differs significantly from that in higher plants. Usually, cyanobacterial thylakoids are unstacked and covered with phycobilisomes (PBS) which serve as antenna complexes of PSII. Phycobilisomes are normally constructed of a three-cylindrical core unit containing allophycocyanin (AP) from which several peripheral rods radiate. These rods are composed of phycocyanin, either alone or in combination with phycoerythrin (PE) or phycoerythrocyanin (PEC). For a schematic representation of a phycobilisome, see Fig. 1A.

There are, however, a few prokaryotes capable of oxygenic photosynthesis which contain partially stacked thylakoids and no phycobilisomes, similar to higher plant chloroplasts. Some of them, *Prochloron didemni*, an extracellular symbiont of ascidians, the filamentous freshwater species *Prochlorothrix hollandica*, and the free-living coccoid species *Prochlorococcus marinus* even contain Chl *a* and *b* or divinyl (DV)-Chl *a* and *b*, respectively [1-3]. They were unified in the phylum Prochlorophyta as proposed by Lewin [4]. Because of their ultrastructure and their pigment composition, the prochlorophytes were discussed as direct descendants of the progenitor of higher plant chloroplasts (e.g. [5]). However, 16S rRNA analysis revealed that they form a polyphyletic group within the cyanobacterial radiation and are more closely related to regular cyanobacteria than to chloroplasts [6-8].

Recently, another photosynthetic prokaryote with stacked thylakoid membranes and without phycobilisomes, *Acaryochloris marina*, was isolated by Miyashita and co-workers [9]. Like *Prochloron didemni*, *Acaryochloris marina* is an extracellular symbiont of tropical ascidians. But it differs significantly from *Prochloron* and the other prochlorophytes because of its unique pigment composition. *Acaryochloris* is the only organism yet known that contains Chl *d* as major pigment. Analysis of the 16S rRNA indicates that also this organism has derived from normal-type cyanobacteria [10]. *Acaryochloris* and the prochlorophytes will be summarized as "*Prochloron*-like cyanobacteria" in this communication.

J. H. Argyroudi-Akoyunoglou and H. Senger (eds.),
The Chloroplast: From Molecular Biology to Biotechnology, 41–46.

2. Pigments

As higher plants and green algae, *Prochloron* and *Prochlorothrix* are characterized by the occurence of Chl *a* and Chl *b* [1, 2]. In contrast, *Prochlorococcus* contains the divinyl forms of these chlorophylls [3, 11]. The Chl *a*/*b* ratios can be significantly higher than in green eukaryotes. Values of 4.36 - 6.9 and of 8 were reported for *Prochloron* and *Prochlorothrix*, respectively [12, 2]. In *Prochlorococcus* the DV-Chl *a*/*b* ratio varied between 1 and 15, depending on strain and culture conditions [13]. *Acaryochloris* differs significantly from these organisms since it contains Chl *a* and Chl *d* with Chl *d* as the predominating pigment. The Chl *a*/*d* ratio of the cells varies from 0.027 to 0.092 [9]. Prior to the discovery of *Acaryochloris*, Chl *d* was isolated as a minor component from some red algae only [14], and it remained unclear whether or not it really occured *in vivo*. The occurence of a Chl *c*-like pigment (probably Mg divinyl 2,4-pheoporphyrin a_5 monomethyl ester) was reported for all *Prochloron*-like organisms [3, 11, 15]. Larkum and co-workers showed that it even contributes to light-harvesting in *Prochloron* [15].
The major xanthophyll of all four organisms is zeaxanthin, a pigment common in cyanobacteria [2, 3, 9, 12,]. The carotene species, however, differs. *Prochloron* and *Prochlorothrix* contain β-carotene like usual cyanobacteria [12, 2]. In contrast, *Prochlorococcus* and *Acaryochloris* harbor α-carotene that is not found in other prokaryotes [3, 9].

3. Chlorophyll-binding proteins

Despite their sometimes different pigment complement (DV-Chl in *Prochlorococcus* and Chl *d* in *Acaryochloris*), the PSI and PSII core complexes of the *Prochloron*-like organisms seem to be very similar to those of regular cyanobacteria. As from many other cyanobacteria, PSI has been isolated in the form of trimers also from *Prochlorothrix* [16] and from *Acaryochloris* (Q. Hu, personal communication).
A specific characteristic of the prochlorophytes is the presence of Chl *a*/*b*-binding antenna complexes which consist of polypeptides of approximately 30 to 38 kDa [12, 13, 17, 18]. Their Chl *a*/*b* ratios are usually higher than those of higher plant Cab proteins. Van der Staay et al. [19] found a value of 4 for *Prochlorothrix hollandica*, while the DV-Chl *a*/*b* ratio of isolated *Prochlorococcus* antenna complexes varied between 0.86 and 10, depending on strain and light conditions [13]. Some of the Chl *a*/*b*-binding polypeptides can be phosphorylated at least *in vitro* [20-22]. Post *et al.* [22] even demonstrated a light-dependent phosphorylation *in vivo*, similar to state 1-state 2 transitions in higher plants.
As shown by earlier investigations [13, 17, 23] antenna proteins from different prochlorophytes are immunologically related to each other, but distinct from Chl a/b proteins of higher plants. These findings were confirmed by more recent analyses of the genes encoding for prochlorophytic Chl *a*/*b* proteins (*pcb* genes) [24, 25]. Indeed, they show no homology to *cab* genes, but cluster with the genes encoding for CP43 and IsiA, a Chl a binding protein induced by iron-deficiency in many cyanobacteria. In *Acaryochloris marina* no specific Chl *d*-binding light-harvesting complexes have been found to date.

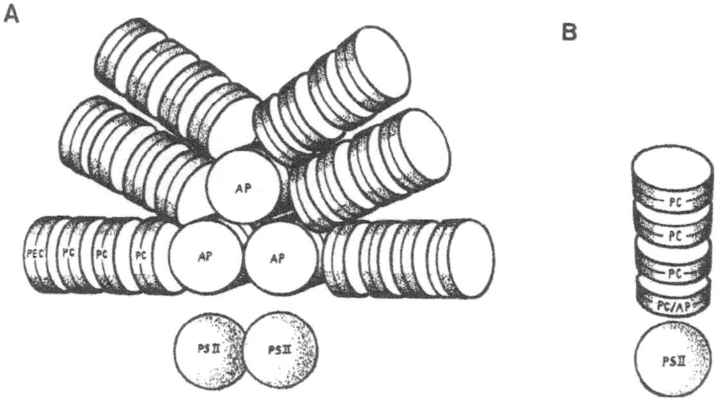

Figure 1. Model of a typical cyanobacterial phycobilisome (A) and of a phycobiliprotein aggregate from *Acaryochloris marina* (B).

4. Phycobiliproteins

Though it was originally assumed that the prochlorophytes contain no phycobiliproteins (PBP), Hess *et al.* [26] could show the presence of PE in a *Prochlorococcus* strain. Also Miyashita and co-workers reported the existence of small amounts of a PBP-like pigment in *Acaryochloris* [9]. The organization of the PBP in these organisms remained unclear. Recently, we could isolate and analyze the biliprotein aggregates from this organism *Acaryochloris* [27]. Here, PBP are not organized as PBS as in regular cyanobacteria but as smaller rod-shaped units of 26.0 x 11.3 nm. They consist of four PBP hexamers and resemble the peripheral rods of PBS. In contrast to these rods, however, they contain PC as well as AP. Also traces of PE were found. Upon partial dissociation, pure PC hexamers as well as mixed hexamers containing AP and PC could be isolated [28]. The aggregates are structurally and functionally coupled to PSII [29]. From these results we propose the following model for the supramolecular organization of PBP in *Acaryochloris* (Fig. 1B): The organism contains aggregates consisting of three PC hexamers and one PC/AP heterohexamer. This heterohexamer connects the aggregate with the thylakoid membrane and funnels light energy to PSII. A certain portion of the aggregates also contains PE, either in the form of additional peripheral PBP units or replacing the peripheral PC hexamer.

Immunogold-labelling of the cells revealed that the PBP are concentrated at certain thylakoid domains. From contrasted ultrathin sections we found evidence that they are located at the cytoplasmic side of unstacked thylakoids: the space between these thylakoids is filled with electron-dense material, and the distance between unstacked thylakoids corresponds well to the size of the biliprotein aggregates [29].

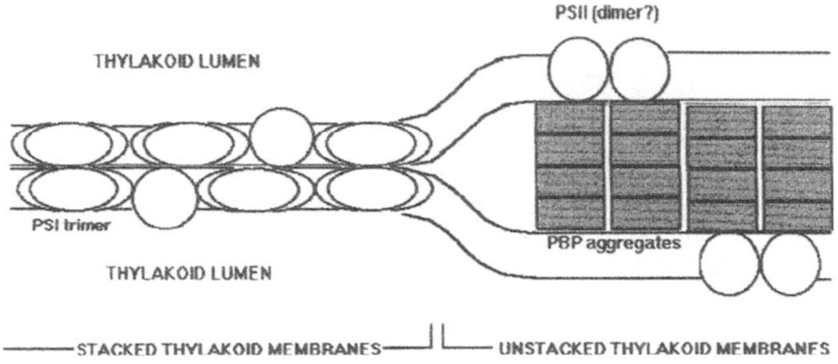

Figure 2. Preliminary model of the organization of thylakoids in *Acaryochloris marina*

5. Organization of thylakoids

It is still a matter of debate to which photosystem the antenna proteins are bound. Some investigators found evidence that both photosystems are connected with the light-harvesting complex. According to a model by Post and Bullerjahn [30] both photosystems share a common light-harvesting antenna under state 1 conditions, while under state 2 conditions the bulk of the antenna is connected to PSI. Others claim that the LHC is exclusively bound to PSII [19]. Van der Staay and Staehelin [31] fractionated stacked and unstacked thylakoids from *Prochlorothrix*. They found PSII and the Chl *a/b* antennae in the stacked thylakoids and PSI core complexes in the unstacked. This distribution was not altered by phosphorylation of the LHC. In contrary to these findings there is some evidence that DV-Chl *b* contributes to light-harvesting of PSI in *Prochlorococcus* [13]. However, the organization of the thylakoids in *Prochlorothrix* and *Prochlorococcus* might be quite different.

According to our preliminary results the organization of the thylakoids seems to be rather different in *Acaryochloris marina*, as shown in Fig. 2. The localization of biliproteins in the unstacked thylakoid regions indicates that also PSII centers are concentrated here. Obviously there is a spatial segregation of PSI and PSII, as it is in *Prochlorothrix* and higher plants. In contrast to *Prochlorothrix* and higher plants, however, PSI seems to be located in the stacked and PSII in the unstacked thylakoid regions.

6. References

1. Lewin, R.A. and Withers, N.W. (1975) Extraordinary pigment composition of a prokaryotic alga, *Nature*, 256, 735-737.
2. Burger-Wiersma, T., Veenhuis, M., Korthals, H.J., van de Wiel, C.C.M., and Mur, L.R. (1986) A new prokaryote containing chlorophylls *a* and *b*, *Nature* 320, 262-264.

3. Chisholm, S.W., Frankel, S.L., Goericke, R., Olson, R.J., Palenik, B., Waterbury, J.B., West-Johnsrud, L., and Zettler, E.R. (1992) *Prochlorococcus marinus* nov. gen. nov. sp.: an oxyphototrophic perokaryote containing divinyl chlorophyll *a* and *b*, *Arch Microbiol.* 157, 297-300.
4. Lewin, R.A. (1976) Prochlorophyta as a proposed new division of algae, *Nature* 261, 679-680
5. Stanier, R.Y. and Cohen-Bazire, G. (1977) Phototrophic Prokaryotes: The Cyanobacteria, *Ann Rev. Microbiol.* 31, 225-274
6. Seewaldt, E. and Stackebrandt, E. (1982) Partial sequence of 16S ribosomal RNA and the phylogeny of *Prochloron*, *Nature* 295, 618-620.
7. Palenik, B., and Haselkorn, R. (1992) Multiple evolutionary origins of prochlorophytes, the chlorophyll *b*-containing prokaryotes, *Nature* 355, 265-267.
8. Urbach, E., Robertson, D.L., and Chisholm, S.W. (1992) Multiple evolutionary origins of prochlorophytes within the cyanobacterial radiation, *Nature* 355, 267-270.
9. Miyashita, H., Adachi, K., Kurano, N., Ikemoto, H., Chihara, M., and Miyachi, S. (1997) Pigment composition of a novel oxygenic photosynthetic prokaryote containing chlorophyll *d* as the major chlorophyll, *Plant Cell Physiol.* 38, 274-281.
10. Miyashita, H., Shimada, A., Kurano, N., Ikemoto, H., Miyachi, S., and Chihara, M. (1997) Phylogenetic analysis of a new photosynthetic prokaryote containing chlorophyll *d* as a major pigment, *Phycologia* 36 Suppl., 73.
11. Goericke, R., and Repeta, D.J. (1992) The pigments of Prochlorococcus marinus: The presence of divinyl chlorophyll *a* and *b* in a marine prokaryote, *Limnol. Oceanogr.* 37, 425-433.
12. Withers, N.W., Alberte, R.S., Lewin, R.A., Thornber, J.P., Britton, G., and Goodwin, T.W. (1978) Photosynthetic unit size, carotenoids, and chlorophyll-protein composition of *Prochloron sp.*, a prokaryotic green alga, *Proc. Natl. Acad. Sci. USA* 75, 2301-2305.
13. Partensky, F., La Roche, J., Wyman, K., and Falkowsky, P.G. (1997) The divinyl-chlorophyll *a/b*-protein complexes of two strains of the oxophototrophic marine prokaryote *Prochlorococcus* - characterization and response to changes in growth irradiance, *Photosynth. Res.* 51, 209-222.
14. Manning, W.M. and Strain, H.H. (1943) Chlorophyll *d*, a green pigment of red algae, *J. Biol. Chem.* 151, 1-19
15. Larkum, A.W.D., Scaramuzzi, C., Cox, G.C., Hiller, R.G., and Turner, A.G. (1994) Light-harvesting chlorophyll c-like pigment in Prochloron, *Proc. Natl. Acad. Sci. USA* 91, 679-683.
16. van der Staay, G.W.M., Boekema, E.J., Dekker, J.P., and Matthijs, H.C.P. 81993) Characterization of trimeric photosystem I particles from the prochlorophyte *Prochlorothrix hollandica* by electron microscopy and image analysis, *Biochim. Biophys. Acta* 1142, 189-193
17. Hiller, R.G. and Larkum, A.W.D. (1985) The chlorophyll-protein complexes of *Prochloron sp.* (Prochlorophyta), *Biochim. Biophys. Acta* 806, 107-115.
18. Bullerjahn, G.S., Matthijs, H.C.P., Mur, L.R., and Sherman, L.A. (1987) Chlorophyll-protein composition of the thylakoid membrane from *Prochlorothrix hollandica*, a prokaryote containing chlorophyll *b*, *Eur. J. Biochem.* 168, 295-300.
19. van der Staay, G.W.M., Brouwer, A., Baard, R.L., van Mourik, F., and Matthijs, H.C.P. (1992) Separation of photosystems I and II from the oxychlorobacterium

(prochlorophyte) *Prochlorothrix hollandica* and association of chlorophyll *b* binding antennae with photosystem II, *Biochim. Biophys. Acta* 1102, 220-228

20. van der Staay, G.W.M., Matthijs, H.C.P., and Mur, L.R. (1989) Phosphorylation and dephosphorylation of membrane proteins from the prochlorophyte *Prochlorothrix hollandica* in fixed redox states, *Biochim. Biophys. Acta* 9975, 317-324
21. Schuster, G., Owens, G.C., Cohen, Y., and Ohad, I. (1984) Thylakoid polypeptide composition and light-dependent phosphorylation of the chlorophyll *a,b*-protein in *Prochloron*, a prokaryote exhibiting oxygenic photosynthesis, *Biochim. Biophys. Acta* 767, 596-605
22. Post, A.F., Gal, A., Ohad, I., Milbauer, K.M., and Bullerjahn, G.S. (1992) Characterization of light-activated reversible phosphorylation of a chlorophyll *a/b* antenna apoprotein in the photosynthetic prokaryote *Prochlorothrix hollandica*, *Biochim. Biophys. Acta* 1100, 75-82
23. Bullerjahn, G.S., Jensen, T.C., Sherman, D.M., and Sherman, L.A. (1990) Immunological characterization of the *Prochlorothrix hollandica* and *Prochloron sp.* chlorophyll *a/b* antenna proteins, *FEMS Microbiol. Lett.* 67, 99-106.
24. La Roche, J., van der Staay, G.W.M., Partensky, F., Ducret, A., Aebersold, R., Li, R., Golden, S.S., Hiller, R.G., Wrench, P.M., Larkum, A.W.D., and Green, B.R. (1996) Independent evolution of the prochlorophyte and green plant chlorophyll *a/b* light-harvesting proteins, *Proc. Natl. Acad. Sci. USA* 93, 15244-15248.
25. van der Staay, G.W.M., Yurkova, N., and Green, B.R. (1998) The 38 kDa chlorophyll *a/b* protein of the prokaryote *Prochlorothrix hollandica* is encoded by a divergent *pcb* gene, *Plant Mol. Biol.* 36, 709-716.
26. Hess, W.R., Partensky, F., van der Staay, G.W.M., Garcia-Fernandez, J.M., Börner, T., and Vaulot, D. (1996) Coexistence of phycoerythrin and a chlorophyll *a/b* antenna in a marine prokaryote, *Proc.Natl. Acad. Sci. USA* 93, 11126-11130.
27. Marquardt, J., Senger, H., Miyashita, H., Miyachi, S., and Mörschel, E. (1997) Isolation and characterization of biliprotein aggregates from *Acaryochloris marina*, a *Prochloron*-like prokaryote containing mainly chlorophyll *d*, *FEBS Lett.* 410, 428-432
28. Marquardt, J., Hu, Q., Iwasaki, I., Miyashita, H., Kurano, N., Miyachi, S., and Mörschel, E. (1998) The structure of the biliprotein aggregates of the Chl *d*-containing prokaryote *Acaryochloris marina*, to be published in G. Garab and J. Pusztai (eds.), *Proceedings of the XIth International Congress on Photosynthesis*, Kluwer Academic Publishers, Dordrecht
29. Hu, Q., Marquardt, J., Iwasaki, I., Miyashita, H., Kurano, N., Mörschel, E., and Miyachi, S. (1998) Localization and function of phycobiliproteins in a chlorophyll *a/d* containing oxygenic photosynthetic prokaryote, *Acaryochloris marina*, to be published in G. Garab and J. Pusztai (eds.), *Proceedings of the XIth International Congress on Photosynthesis*, Kluwer Academic Publishers, Dordrecht
30. Post, A.F. and Bullerjahn, G.S. (1994) The photosynthetic machinery in Prochlorophytes: Structural properties and ecological significance, *FEMS Microbiol. Rev.* 13, 393-414.
31. van der Staay, G.W.M. and Staehelin, L.A. (1994) Biochemical characterization of protein composition and protein phosphorylation patterns in stacked and unstacked thylakoid membranes of the Prochlorophyte *Prochlorothrix hollandica*, *J. Biol. Chem.* 269, 24834-24844

PHYCOBILISOME-TO-PHOTOSYSTEM I EXCITATION TRANSFER IS ENHANCED IN WATER-DEPLETED CELLS AND DEPRESSED IN WATER-REPLETE CELLS OF CYANOBACTERIUM *Synechococcus* sp. PCC7942

K. STAMATAKIS, G. C. PAPAGEORGIOU
NSCR Demokritos, Institute of Biology,
Athens, Greece 153 10

1. Introduction

Cyanobacteria emit chlorophyll (Chl) *a* fluorescence whose intensity depends on the light history of cells: it is weaker in dark-acclimated cells and stronger in light-acclimated cells. These light/dark-induced fluorescence intensity shifts are not caused by the oxidation or reduction of Q_A, the primary plastoquinone electron acceptor of photosystem (PS) II, since they occur also in the presence of the electron transport inhibitor 3-(3,4-dichlorophenyl)-1,1'-dimethylurea (DCMU) which blocks the oxidation of Q_A^-. The phenomenon was reported originally for cyanobacteria [1], red algae [2], green algae [3], and higher plants [4]. Murata [2] and Bonaventura and Myers [3] named the state of dark acclimation *light state 2* and the state of light acclimation *light state 1*.

Hyper-osmotic suspension media suppress the Chl *a* fluorescence of light-acclimated cyanobacteria [5, 6]. The extent of *hyper-osmotic quenching* varies with the osmolality of the cell suspension according to the Boyle-van t' Hoff law. Only excluded solutes, or solutes that pass the plasma membrane (PM) slowly compared to the assay time, contribute to the osmotic pressure perceived by cells. At maximal hyper-osmotic quenching, the Chl *a* fluorescence of light-acclimated and of dark-acclimated cells are equal. No light-acclimative fluorescence rise (*i.e.*, no state 2 - state 1 transition) occurs in dark-acclimated cells if the suspension is hyper-osmotic. When dark-acclimated cells are swollen, after taking in NaCl from the medium [7, 8], then Chl *a* fluorescence becomes sensitive to hyper-osmotic quenching to an extent that is again a Boyle-van t' Hoff function of the external osmolality the cells perceive [9].

PM of cyanobacterial cells allow water to pass freely but present an effective barrier to most solutes. Water moves in and out of the cells in response to changes in the concentration of excluded solutes in the suspension. Cell volumes increase or decrease accordingly [8]. Thus, at room temperature Chl *a* fluorescence is quenched when cells contract by losing water, and enhanced when cells swell by taking in water.

In this research we examined if depletion of cytoplasmic water influences radiationless excitation energy transfers from phycobilisomes (PBS), the extra-thylakoid light harvesting organelles of cyanobacteria [10] to the Chl *a* holochromes of photosystems I (PS I) and II (PS II). We compared normal *Synechococcus* sp. PCC7942 cells with cells treated with the selective sulfhydryl alkylating reagent N-ethyl maleimide (NEM). Glazer et al. [11] reported that NEM blocks excitation transfers from PBS to PS I, but not from PBS to PS II, and has no effect on the size of the Chl *a* antenna

J. H. Argyroudi-Akoyunoglou and H. Senger (eds.),
The Chloroplast: From Molecular Biology to Biotechnology, 47–54.

available to PS II. NEM-treated cyanobacterial cells are unable to photo-oxidize P700 with light absorbed by PBS, but able to do so with light absorbed by Chl *a*. We shall show that hyper-osmotic conditions suppress the Chl *a* fluorescence of light-acclimated control cells, but not the Chl *a* fluorescence of light-acclimated NEM–treated cells. We shall argue that the site of osmotic regulation of Chl *a* fluorescence is the PBS-to-PS I excitation transfer route. This route is functional in hyper-osmotically-contracted (water-depleted) cyanobacterial cells, but not in hypo-osmotically-swollen (water-replete) cells.

2. Materials and Methods

Synechococcus sp. PCC7942 was cultured in BG11 medium buffered with 20 mM Hepes-NaOH pH 7.5 as in [2]. Cells were harvested from exponentially grown cultures and were resuspended in buffered BG11 in the presence of 10 μM DCMU to ensure closure of PS II reaction centers on illumination. Cell suspensions were treated with N-ethyl maleimide (NEM) as in [11]. Unreacted NEM was removed with two washes.

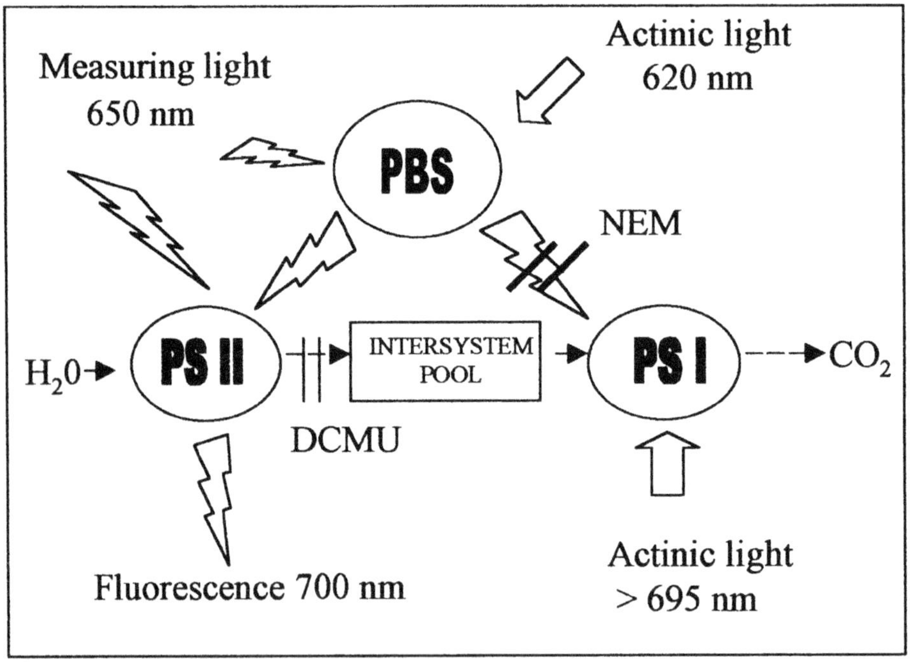

Figure 1. Excitation energy transfer routes in cyanobacteria. Wavy thick arrows depict routes of modulated measuring excitation, radiationless excitation transfer and Chl *a* fluorescence. Straight thick arrows depict routes of actinic excitation. The fluorescence, which the actinic light excites, is not detected by the fluorometer and, therefore, it does not appear on the figure. The diagram depicts also the noncyclic photosynthetic electron transfer route from water to carbon dioxide, through PSII, the intersystem pool and PS I.

Fluorescence kinetics were measured with a PAM-fluorometer which detects signals synchronous with the modulated *measuring excitation* (650 nm; $\Delta\lambda$ = 25 nm; 1.6 kHz; 1 µs flashes). The excitation was absorbed both by Chl *a* and PBS and generated Chl *a* fluorescence partly directly and partly *via* excitation transfers from PBS. Its intensity was too weak (70 nmoles m^{-2} s^{-1}) to light-acclimate the cells.

Cells were acclimated to light with various kinds of continuous actinic light obtained from an incandescent source through colored glass filters: (a) *Far-red actinic light*; filter with 695 nm threshold transmission, half-maximal transmission at 715 nm and steady (72%) transmission above 770 nm. (b) *Narrow-band orange actinic light*; interference filter λ_{max} = 620 nm, $\Delta\lambda$ = 12 nm). This light excites PBS directly, and PS II and PS I indirectly. (c) *Broad-band blue light*; filter with λ_{max} = 410 nm and 380 – 480 nm transmission. This light excites primarily Chl *a* of PS II and PS I. (d) *Broad-band orange-red actinic light*; cut-off filter, transmission threshold 620 nm, steady transmission (87%) beyond 580 nm. This light excites PSII, PSI and PBS holochromes directly.

Media of defined osmolality were prepared by dissolving sorbitol in buffered BG11. Osmolalities were measured cryoscopically as in [6]. Cell suspensions are designated as hypo-osmotic or hyper-osmotic by reference to the *turgor threshold* of the cells (maximal external osmolality at which cells maintain turgor; 0.22 Osm kg^{-1} for *Synechococcus* sp. PCC7942; see ref. 5).

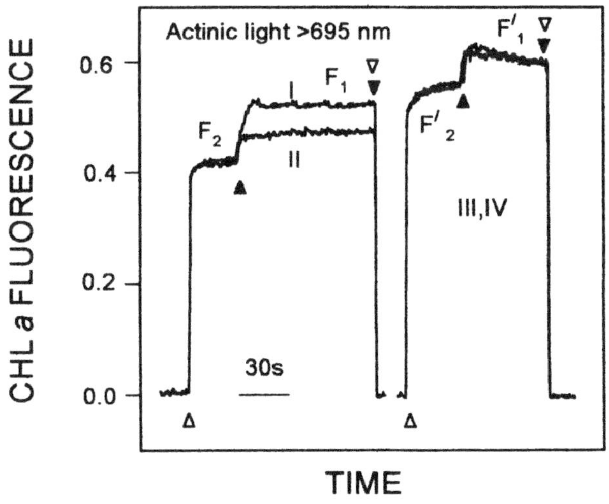

Figure 2. Light-induced kinetics of Chl *a* fluorescence emitted by suspensions of control (traces I and II) and NEM-treated (traces III and IV) *Synechococcus* sp. PCC7942 cells. The cells were suspended either in hypo-osmotic medium (BG11 plus 0.02 Hepes-NaOH, pH 7.5, 0.08 Osm kg^{-1}; traces I and III) or in hyper-osmotic medium (same plus 0.24 M sorbitol, 0.34 Osm kg^{-1}; traces II and IV) and were dark-acclimated for 4 min prior to recording the fluorescence traces. Δ, ∇ denote on/off times of the weak modulated excitation (measuring light); ▲, ▼ on/off times of the continuous far-red actinic light (λ > 695 nm). Fluorescence magnitudes F_1, F_2, F_1', F_2' are defined on the figure.

3. Results

We shall discuss results by reference to Fig. 1, which shows routes of light absorption, emission, excitation energy transfer, and electron transfer in cyanobacteria. The detected (modulated) fluorescence originated exclusively from the Chl *a* of PS II. The modulated excitation was absorbed both by Chl *a* and by PBS, so Chl *a* fluorescence was excited both directly, and indirectly, *via* radiationless excitation transfers. The fluorescence that the continuous actinic light excited was not detected by the fluorometer.

Figure 2 shows typical light-induced kinetics of Chl *a* fluorescence of control cells (traces I and II) and of NEM-treated cells (traces III and IV). The medium was either buffered BG11 (hypo-osmotic suspension; 0.08 Osm kg^{-1}) or buffered BG11 that contained 0.21 M sorbitol (hyper-osmotic suspension; 0.34 Osm kg^{-1}). The cells were acclimated to darkness for 4 min before the measuring light was turned on. In control samples this generated the F_2-level fluorescence. Subsequent illuminations with continuous far-red actinic light ($\lambda > 695$ nm; absorbed by PS I chlorophylls) drove fluorescence to the higher levels denoted as F_1, evidencing a light-acclimative (state 2 to state 1) transition. Characteristically, the fluorescence induction amplitude ($\Delta F = F_1 - F_2$) was larger in the hypo-osmotic suspension than in the hyper-osmotic suspension.

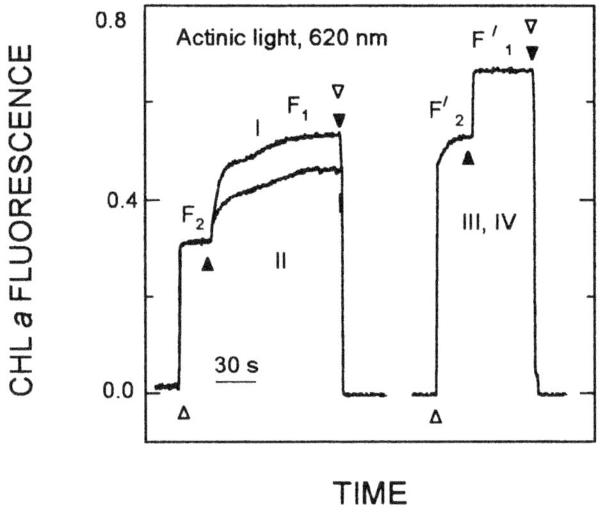

Figure 3. Light-induced kinetics of Chl *a* emitted by suspensions of control (traces I and II) and NEM-treated cells (traces III and IV) *Synechococcus* sp. PCC7942 cells. The cells were suspended either in hypo-osmotic medium (traces I and III) or in hyper-osmotic medium (traces II and IV). Narrow-band orange actinic light (λ_{max} = 620 nm, $\Delta\lambda$ = 12 nm) was used. Other details as in Fig. 2.

NEM-treated cells responded differently. First, the fluorescence excited by measuring light reached a higher steady level $F_2^{/}$ ($F_2^{/} \approx F_1 > F_2$; traces III and IV) but only after some induction. This induction was absent or at least not as pronounced, in control cells (traces I and II). Subsequent illumination with far-red light drove fluorescence to the higher level $F_1^{/}$. Thus, the NEM-treated cells proved also capable of a light-acclimative fluorescence rise (state 2-to-state 1 transition?), although PS I could not photo-oxidize the intersystem intermediates with light absorbed by PBS [11]. However, in contrast to the osmotically sensitive control cells, the fluorescence of light-acclimated NEM-treated cells was insensitive to hyper-osmotic conditions: the $F_1^{/}$ levels in the hypo-osmotic cell suspension (trace III) or in the hyper-osmotic suspension (trace IV) were equal.

Figure 3 shows a similar experiment in which we used narrow-band orange actinic light (λ_{max} = 620 nm; $\Delta\lambda$ = 12 nm; absorbed by C-phycocyanin). In control cells it sensitizes both PS II and PS I chlorophylls by radiationless excitation transfers; in NEM-treated cells it sensitizes PS II chlorophylls only (see Fig. 1). Recorded kinetic traces were similar to those of Fig. 2. Thus, NEM-treated cells were capable of a light acclimative rise of Chl *a* fluorescence although only PS II chlorophylls were activated. As in the case of far-red actinic light (Fig. 1) the 620 nm light-induced fluorescence rise in NEM-treated cells was osmotically-insensitive.

Hypo-osmotic suspensions of DCMU-treated control cells, as well as of DCMU-treated NEM-treated cells, were capable of light acclimative rises of Chl *a* fluorescence with various kinds of actinic illumination (Table 1). All these illuminations activated PS I and oxidized the pool of intersystem intermediates in control cells, but not in NEM-treated cells.

Table 1. Effects of various types of continuous actinic illuminations on the Chl *a* fluorescence emitted by DCMU-treated *Synechococcus* sp. PCC7942 cells. The detected fluorescence was excited by 650 nm modulated light (1.6 kHz; 70 nmol m^{-2} s^{-1}).

Actinic light	Direct activation (Indirect activation)	Occurrence of light-acclimative fluorescence rise: Control cells (Osmo-sensitive)	NEM-treated cells (Osmo-insensitive)
λ > 695 nm	PS I	+	+
λ = 620 nm ($\Delta\lambda$ = 12 nm)	PBS (PS II, PS I)	+	+
λ = 350 – 480 nm	PS II, PS I	+	+
λ > 540 nm	PBS, PS II, PS I (PS II, PS I)	+	+

We have shown $\Delta F\ F_2^{-1}$ to diminish asymptotically with increasing suspension osmolality. This property can be used to estimate turgor thresholds of cyanobacterial cells (5, 6). Fig. 3 shows that $\Delta F\ F_2^{-1}$, measured after acclimation of control cells to 695 nm actinic light (as in Fig. 1), to diminish with increasing suspension osmolality In contrast, $\Delta F^{/}\ F_2^{/-1}$ measured with NEM-treated cells was osmotically indifferent. Analogous results were obtained with 620 nm actinic light (data not shown).

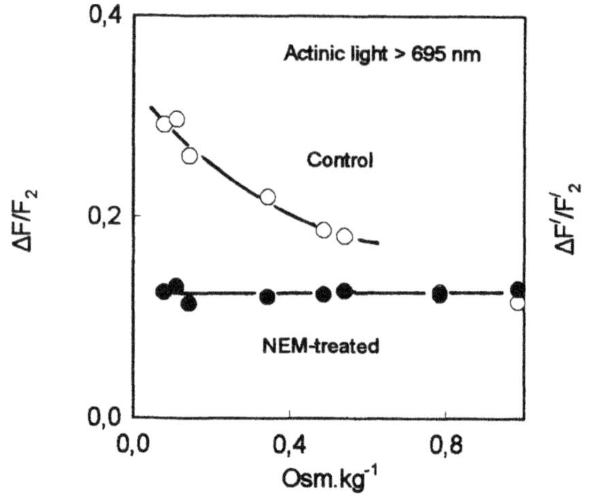

Figure 4. Normalized light-acclimative rises of Chl *a* fluorescence obtained by far-red actinic illumination (λ > 695 nm) of control *Synechococcus* sp. PCC7942 cells [$\Delta F\ F_2^{-1} = (F_1 - F_2)\ F_2^{-1}$] and of NEM-treated cells [$\Delta F'\ F_2^{\prime -1} = (F_1' - F_2')\ F_2^{\prime -1}$] as a function of the osmolality of the cell suspension.

Finally, Fig. 5 shows the high fluorescence level (F_1') of light-acclimated NEM-treated cells to revert to the low fluorescence level (F_2') upon dark incubation. This proves the reversibility of the light-induced transition of NEM-treated cells.

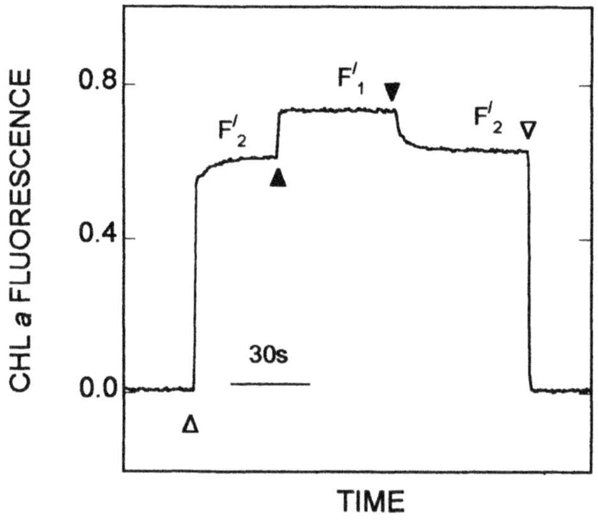

Figure 5. Light/dark-induced Chl *a* fluorescence kinetics in NEM-treated cells. Actinic light, 620 nm

4. Discussion

According to Fig. 1, the detected fluorescence originated from the Chl *a* of PS II and was excited only by the modulated measuring light. This fluorescence comprised two components: (a) emissions by directly excited chlorophylls; and (b) emissions by chlorophylls that were sensitized by radiationless transfers from PBS. Both DCMU-treated control cells and DCMU-treated NEM-treated cells were capable of light-induced rises of Chl *a* fluorescence, and in both cases these fluorescence rises were reversed by darkness. Under actinic illumination, DCMU-treated control cells could photooxidize the pool of intersystem intermediates, but DCMU-treated NEM-treated cells could not. We are thus led to conclude that the redox poise of the intersystem pool plays is irrelevant to the light-acclimative transition of the NEM-treated *Synechococcus* cells.

The most striking difference between the light-acclimated NEM-treated cells and the light-acclimated control cells was that the fluorescence of NEM-treated cells could not be quenched by hyper-osmotic conditions. In contrast, the Chl *a* fluorescence of light-acclimated control cells was osmotically sensitive, as usual. Since their difference was only in the PBS-to-PS I excitation transfer route, which was functioning in control cells, but not in NEM-treated cells [11] we are led to conclude that this excitation transfer route is sensitive to hyper-osmotic conditions.

In a hyper-osmotic suspension of control cells (water-depleted cells) the PBS-to-PS I excitation transfer route is functional. PBS excitation, generated by the absorption of measuring light, is channeled both to the fluorescing Chl *a* of PS II and to the nonfluorescing Chl *a* of PS I, thereby producing weaker fluorescence. In a hypo-osmotic suspension of control cells (water-replete cells) the PBS-to-PS I route is blocked. PBS excitation, generated by the absorption of measuring light, is channeled only to the fluorescing Chl *a* of PS II, thereby producing stronger fluorescence. After treatment with NEM, the PBS-to-PS I excitation transfers was blocked permanently [11], so the stronger fluorescence of the light-acclimated could not be suppressed by external hyper-osmotic conditions.

The idea that the PBS-to-PS I excitation transfer route is functional in hyper-osmotic suspensions but not in hypo-osmotic suspensions is supported also by 77-K fluorescence spectra of light-acclimated *Synechocystis* sp. PCC6803 cells [14]. The difference fluorescence spectrum between a hypo-osmotic suspension and a hyper-osmotic suspension contained excess fluorescence emissions at 666 nm (allophycocyanin), 685 nm (CP43; D1D1Cyt b559) and 696 nm (CP47; see ref. 15 fluorescence band assignments). Our results are in favor of models that visualize simultaneous electronic excitation transfers from PBS to both PS I and PS II (PS II $\leftarrow$ PBS $\rightarrow$ PS I; see ref. 16).

5. References

1. Papageorgiou, G. C. and Govindjee (1967) Changes in the intensity and spectral distribution of fluorescence. *Biochim. Biophys. Acta* **7**, 375-389.
2. Murata, N. (1969) Control of excitation transfer in photosynthesis. I. Light-induced changes of chlorophyll *a* fluorescence in *Porphyridium cruentum*. *Biochim. Biophys. Acta* **172**, 242-251.
3. Bonaventura, C. and Myers, J. (1969) Fluorescence and oxygen evolution from *Chlorella pyrenoidosa*. *Biochim. Biophys. Acta* **189**, 366-383.

4. Allen, J. F., Bennett, J., Steinback, K. E. and Arntzen, C. J. (1981) Chloroplast protein phosphorylation couples plasto-quinone redox state to distribution of excitation energy between photosystems. *Nature* **291**, 25-29.
5. Papageorgiou, G. C. and Alygizaki-Zorba, A. (1997) A sensitive method for the estimation of the cytoplasmic osmolality of cyanobacterial cells using chlorophyll *a* fluorescence. *Biochim. Biophys. Acta* **1335**, 1-4.
6. Papageorgiou, G. C., Alygizaki-Zorba, A., Ladas, N. and Murata, N. (1998) A method to probe the cytoplasmic osmolality and osmotic water and solute fluxes across the cell membrane of cyanobacteria with chlorophyll *a* fluorescence: Experiments with *Synechococcus* sp. PCC7942. *Physiol. Plant.***103,** 215-224.
7. Blumwald, E. R., Melhorn, R. J. and Packer, L. (1983) Ionic osmoregulation during salt adaptation of the cyanobacterium *Synechococcus 6311*. *Plant Physiol.* **73,** 377-380.
8. Blumwald, E. R., Melhorn, R. J. and Packer, L. (1983) Studies on osmoregulation in salt adaptation of cyanobacteria with ESR spin-probe techniques. *Proc. Natl. Acad. Sci. USA* **80,** 2599-2602.
9. Stamatakis, K., Ladas, N., Alygizaki-Zorba, A. and Papageorgiou, G. C. (1998) Chlorophyll a fluorescence senses water and NaCl fluxes across the plasma membranes of cyanobacterium *Synechococcus* sp. PCC7942. Proceedings *of XIth International Photosynthesis Congress,* (G. Garab, edit.) in press.
10. Sidler, W. (1995) Phycobilisome and phycobiliprotein structures, D. Bryant (edit) *Molecular Biology of Cyanobacteria,* Kluwer Academic Publishers, Dordrecht, pp. 139-216.
11. Glazer, A. N., Gindt, Y. M, Chan, C, F. and Sauer, K. (1994) Selective disruption of energy flow from phycobilisomes to photosystem I. *Photosynth. Research* 40, 167-173.
12. Fork, D. C. and Satoh, K. (1986) The control by state transitions of the distribution of excitation energy in photosynthesis. *Ann. Rev. Plant Physiol.* **37,** 355-361.
13. Williams, W. P. and Allen, J. F. (1987) State 1/state 2 changes in higher plants and algae. *Photosynth. Research* **13,** 19-45.
14. Papageorgiou, G. C., Govindjee, Govindjee, R., Mimuro, M., Stamatakis, K. and Murata, N. (in press) Temperature and lipid unsaturation effects on plasma and thylakoid membranes of *Synechocystis* sp. PCC6803, in G. Garab (ed.) *Proceedings of 11th International Congress on Photosynthesis,* in press.
15. Papageorgiou, G. C. (1996) The photosynthesis of cyanobacteria (blue bacteria) from the perspective of signal analysis of Chl *a* fluorescence.
16. Zhao, J. Zhou, J. and Bryant, D. (1992) Energy transfer processes in phycobilisomes as deduced from analyses of mutants of *Synechococcus* sp. PCC7002, in N. *Murata Research in Photosynthesis,* Kluwer Academic Publishers, Dordrecht, **1,** 25-32.

STRUCTURAL-FUNCTIONAL ORGANIZATION OF THYLAKOIDS IN DEVELOPING CHLOROPLASTS

M. Droppa[1], Z. Kóta[2], T. Páli[2], B. Szalontai[2], L.I. Horváth[2], G. Horváth[1]

[1]Department of Plant Physiology, University of Horticulture and Food Industry, P.O.B. 53, Budapest, Hungary, H-1518 and [2]Institute of Biophysics, Biological Research Center, P.O.B. 521, Szeged, Hungary, H-6701

Abstract

Functional and structural changes of greening thylakoid membranes were investigated by comparing the fluorescence induction characteristics, and the structural organization of thylakoids at the level of lipid structure and protein conformations studied by electron spin resonance (ESR), and Fourier transform infrared (FTIR) spectroscopies.

After illumination of etiolated leaves, the photosynthetic electron transport rapidly intensified as chlorophylls were accumulated. The protein/lipid ratio, however, did not change during the first 24 hours and showed only slight increase at the later steps of greening. The integration of TEMPO, a small lipophil spin labelled compound, into the membrane was stepwise increased indicating progressive structural rearrangement of lipid bilayers during thylakoid membrane development. In the FTIR spectra of greening etio-chloroplasts, an Amide I component band exhibited characteristic bandshift from 1652 cm^{-1} to 1656 cm^{-1} corresponding to the enhanced α-helix content of the newly synthetised proteins. An other Amide I component band was shifted from 1624 to 1632 cm^{-1} which indicates a definite increase of the β-structure content of new proteins inserted into the developing chloroplasts.

1. Introduction

The development of proplastids into photochemically active chloroplasts is arrested at an intermediate stage, when plants are kept in the dark. These so called etioplasts contain no chlorophylls and differ in structure and composition from mature chloroplasts [1]. Upon illumination, these etioplasts differentiate

J. H. Argyroudi-Akoyunoglou and H. Senger (eds.),
The Chloroplast: From Molecular Biology to Biotechnology, 55–60.

into chloroplasts which involves the synthesis of chlorophylls and many polypeptides and the reorganization of plastid structure [2-5] which are prerequisites for the optimal function of the photosynthetic machinery. Although during the past decade interest has been increasing in studying the differentiation of thylakoids, our knowledge about the physical background of structural changes of main membrane constituents, such as lipids and proteins is still not complete.

The main goal of the present study was, therefore, to obtain further detailed information about the functional-structural rearrangements of the lipid and protein components of developing thylakoids. For this purpose FTIR and ESR spectroscopies were used to analyze structural changes with the simultaneous measurements of fluorescence induction kinetics, as a functional parameter of greening barley chloroplasts. Our data indicate that a complete rearrangement of both the lipid and protein components is required to optimize photosynthetic functions of fully developed chloroplast thylakoids.

2. Material and Methods

Plant material. Barley (*Hordeum vulgare* L. cv. Triangel) was grown from seed in vermiculate in complete darkness at 25 °C. After 6 days of growth, seedlings were illuminated continuously with 100 $\mu mol \cdot m^{-2} \cdot s^{-1}$ white light. Samples were taken at 0, 3, 6, 9, 12, 24 and 48^{h} of illumination and only the middle part of the leaves, obtained by cutting off 0.5 cm segments from both ends of leaves were used for analyses.

Chlorophyll determination. Chlorophyll content was determined in 80 % acetone extract of 0.2-0.4 g of leaf segments according to the method of Arnon [6].

Thylakoid isolation. Plastids were isolated at 0 °C in a medium containing 0.4 M sorbitol, 10 mM NaCl, 5 mM $MgCl_2$, 1 mM $MnCl_2$, 1 mM PMSF and 50 mM MES (pH 6.5). The isolated plastids were disrupted by osmotic shock in a medium containing 40 mM sucrose, 10 mM NaCl, 5 mM $MgCl_2$, 1 mM PMSF and 20 mM Tricine-KOH buffer (pH 7.8).

Determination of the lipid and protein content. Extraction and analysis of lipids were carried out as described in [7]. The total protein content of thylakoids was determined according to Markwell et al. [8].

Fluorescence induction measurements. Chlorophyll fluorescence measurements were carried out by using a portable Hansatech Plant Efficiency Analyser in 6 repetions on 5 leaves. After 20 min dark incubation, samples were excited with the light intensity of 1600 $\mu mol \cdot m^{-2} \cdot s^{-1}$ at 650 nm and fluorescence was detected for 60 s.

FTIR spectroscopy. 50 µl of thylakoid suspension (corresponding to 600-750 µg protein) were spinned down and resuspended in the same but D_2O based buffer and the suspension was centrifuged again. The pellet was placed between CaF_2 windows using a spacer of 20 µm. FTIR spectra were recorded in a Philips PU 9800 Fourier transform infrared spectrometer purged with dry air to eliminate water wapor. For one spectrum 128 interferograms were accumulated and averaged. Data processing was performed using SPSERV© spectrum handling software (Bagyinka Cs., Inst. Biophys., BRC, Szeged, Hungary). Bands of interest were analysed by fitting Lorentzian components to the experimental spectrum after substracting the baseline.
ESR measurements. For spin labeling, 10 µL aqueous TEMPO (2,2,6,6-tetramethyl-piperidine-1-oxyl) solution (10 mM) was added to 100 µL thylakoid suspension (corresponding to 1.2-1.5 mg protein). Samples were incubated for 10 min in the dark and repeatedly vortexed. The labeled suspension was then filled into 1 mm I.D. ESR sample capillaries and concentrated in a benchtop centrifuge. Care was taken to keep the membrane/water ratio constant. The EPR spectra were recorded using an X-band Bruker (Rheinstetten, Germany) ECS 106 spectrometer equipped with a computer controlled thermostat and TE102 cavity using 100kHz field modulation with the amplitude of 0.3 Gauss, and first harmonic in-phase detection.

3. Results and Discussion

The time course of greening strongly depends on the environmental conditions such as the intensity of illumination, light quality, water supply and humidity etc. [9]. These factors markedly influence the onset and velocity of pigment and protein synthesis and the membrane content of chloroplasts.
Figure 1. shows that in our experiments, the chlorophyll accumulation had a short lag period followed by a rapid accumulation of pigments which slowed down after the 24th hour of greening. In agreement with earlier findings [3] the protein/chlorophyll and lipid/chlorophyll ratios drastically decreased during the first 24 hours of greening (data not shown) which indicate enhanced synthesis of these membrane constituents. Although in earlier works, both protein and lipid compositions of the membrane showed significant alterations during greening [3, 10, 11], in our experiments the protein/lipid ratio, an important parameter of the membrane structure, remained unchanged during the first 24 hours and only a slight increase could be observed for the 48th hour of greening (Figure 1.)

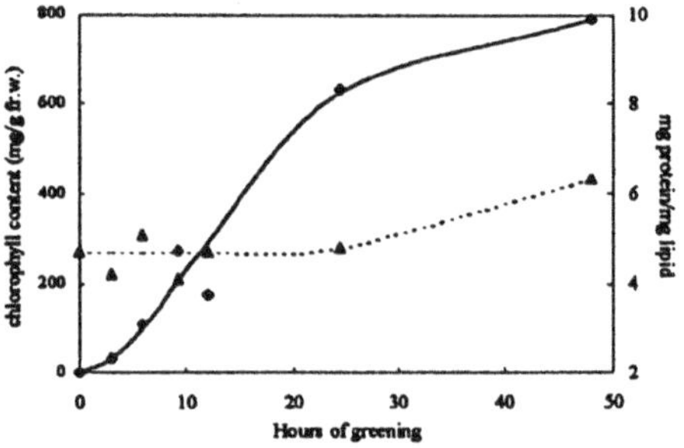

Figure 1. Change of the chlorophyll content and the protein/lipid ratio during greening of barley.

Most membrane proteins synthetised during greening have double function, which requires a specific structure to ensure both the optimal operation of phototosynthesis and the structural stability of the membrane. The Amide I region (between 1700-1600 cm^{-1}) of the FTIR spectra is very informative about the secondary structures of proteins [12]. We have decomposed the Amide I band into two major components which gave sufficient fit for the above region. (Figure 2.) The insert shows the component bands of the Amide I region. The component band centered around 1651 cm^{-1} in etioplasts, showed a freqency shift to 1656 cm^{-1} upon greening. This shift was interpreted as the result of increasing amount of α-helical structures in the newly synthetised proteins. The other component band was around 1624 cm^{-1} in etioplasts and shifted to 1632 cm^{-1} in fully developed thylakoids. This shift was attributed to the increasing amount of intramolecular β-structures in the new membrane proteins (Figure 2).

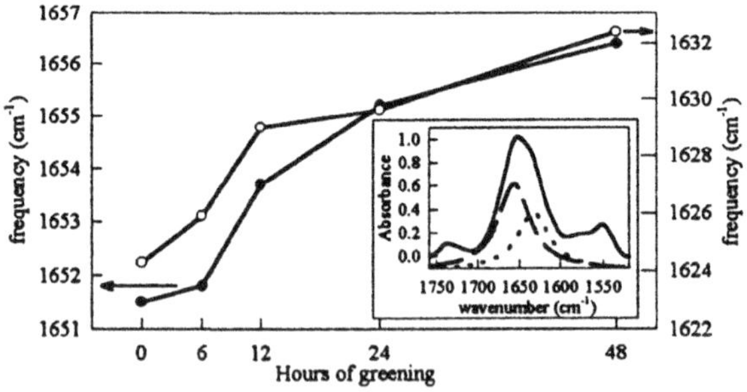

Figure 2. The frequency shift of the 1651 cm^{-1} and 1624 cm^{-1} components of the Amide I band of FTIR spectrum measured in greening barley thylakoids. The insert shows the position of these components obtained by mathematical decomposition of the spectrum.

The rise of the photosynthetic function represented by the Fv/Fm ratio of fluorescence induction kinetics (Figure 3) showed good correlation with the appearence of the above discussed highly organised structural elements of membrane proteins. The fact that both the Fv/Fm ratio and the frequency shift of the components of the Amide I band exhibited the majority changes within the first 24 hours of greening and later the rate of the changes has been slowed down considerably indicates that the structural elements required for optimal functions are readily formed for this time.

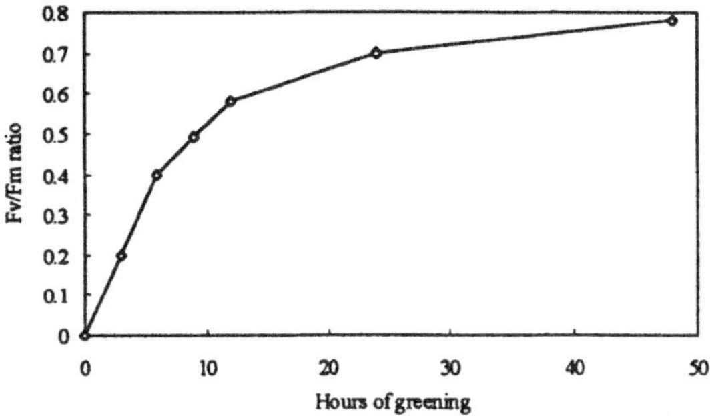

Figure 3. Change of the Fv/Fm ratio in greening barley leaves.

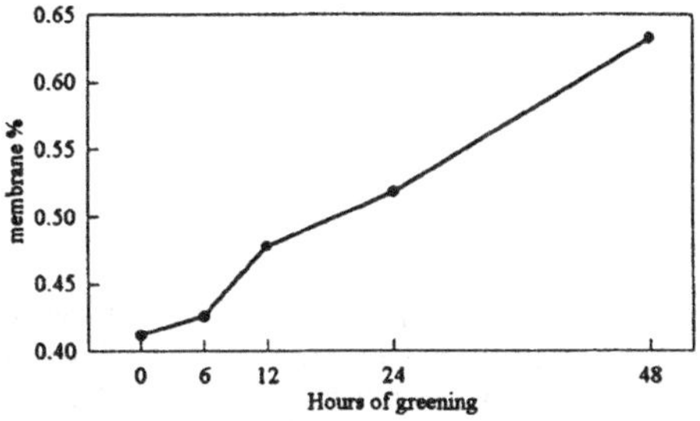

Figure 4. Change of the membranous fraction of TEMPO (2,2,6,6-tetramethyl-piperidine-1-oxyl) in greening barley thylakoids.

The rearrangement of lipid bilayers, was well reflected by the enhanced integration of TEMPO into the membrane fraction during greening (Figure 4.)[13]. We assume that this rearrangement of the lipid bilayer through changes in lipid composition [11] is a basic requirement to establish the appropriate molecular microenvironment for the optimal function of the newly synthesized polypeptides in developing thylakoids. The time course of strucutral changes of lipid bilayers was, however, different from that of the protein structural and photosynthetic activity changes, which were almost completed after 24 hours. The massive increase of membranous TEMPO fraction between 24 and 48 hours of greening, indicated that the formation of the functionally and structurally optimised thylakoid membrane needs longer time. Since TEMPO integrates primarily into the hydrophobic interior of the lipid parts of the membrane, the ongoing increase of its binding fraction also means that the synthesis of the appropriate lipids and their integration into the membrane architecture is the final requirement for the functional-structural optimalization of the thylakoid .

Acknowledgments

This work was supported by the Ministry of Culture and Education (FKFP 0096/1997) and by the Hungarian National Science Found (OTKA T019130)

References

[1] Kirk, J. T. O. and Tilney-Basset, R. A. E. (1978) *The Plastids. Their Chemistry, Sturcture, Growth and Inheritance.* Elsevier North-Holland Biomedical Press, Amsterdam.

[2] Virgin, H. I. (1993) Effectiveness of light of different wavelengths of induced chlorophyll biosynthesis in rapidly and slowly greening tissues. *Physiol. Plant.* **89**, 761-766.

[3] Droppa, M., Masojidek, J. and Horváth, G. (1990) Changes of the polypeptide composition in thylakoid membranes during differentiation. *Z. Naturforsch.* **45c**, 253-257.

[4] Pauncz, Y., Gepstein, S. and Horowitz, B. A. (1992) Photocontrol of the accumulation of plastid polypeptides during greening of tomato cotyledon. *Plant Physiol.* **100**, 1934-1939.

[5] Horváth, G., Garab, Gy., Mustárdy, L. A., Halász, N. and Faludi-Dániel, Á. (1975) The development of thylakoids and photochemical properties of greening maize leaves. *Plant Sci. Lett.* **5**, 239-244.

[6] Arnon, D. I. (1949) Copper enzymes in isolated chloroplasts. Polyphenol oxidase in *Beta vulgaris. Plant Physiol.* **24**, 1-15.

[7] Horváth, G., Droppa, M., Hideg, É. and Rózsa, Zs. (1989) The role of phospholipids in regulating photosynthetic electron transport activities: treatment of chloroplasts with phospholipase A_2. *J. Photochem. Photobiol. B. Biology* **3**, 515-527.

[8] Markwell, M. A. K., Haas, S. M., Tolbert, N. E., Bieber, L. L. (1981) Protein determination in membrane and lipoprotein samples: manual and automated procedures. *Methods Enzymol.* **72**, 296-303.

[9] Horváth, G., Mustárdy, L. A. and Faludi-Dániel, Á. (1978) Cellular environment affecting granum formation. A working hypothesis, in G. Akoyunoglou and J. Argyroudi-Akoyunoglou (eds), *Chloroplast Development*, Elsevier North-Holland Biomedical Press, Amsterdam, pp. 863-869.

[10] Pötter, E. and Kloppstech, K. (1993) Effect of light stress on the expression of early light-inducible proteins in barley. *Eur. J. Biochem.* **214**, 779-786.

[11] Bhardwaj, L., Merillon, J. M. and Ramawat, K. G. (1995) Changes in the composition of membrane lipids in relation to differentiation in *Aegbe marmelos* callus cultures. *Plant Cell Tissue Organ Culture* **42**, 33-37.

[12] Arrondo, J. L. R., Muga, A., Castresana, J. and Goni, F. M. (1993) Quantitative studies of the structure of proteins in solution by Fourier-transform infrared spectroscopy. *Prog. Biophys. Molec. Biol.* **59**, 23-56.

[13] McConnel, H. M. (1976) Spin labelling: Theory and applications. in L.J. Berliner (ed), *Molecular Motion in Biological Membranes*, Acad. Press New York, pp. 525-560.

CHLOROPLAST GENE EXPRESSION / REGULATION

THE CHLOROPLAST: PART OF THE INTEGRATED GENETIC SYSTEM OF THE PLANT CELL

R.G. HERRMANN
Botanisches Institut der Ludwig-Maximilians-Universität, Menzinger Str. 67, D-80638 Munich, Germany

1. Introduction

Life on earth ultimately depends on energy derived from the sun. Biological photosynthesis is globally the only process of biological importance that can harvest this energy. It provides energy, organic matter, in its oxygenic version also oxygen, for nearly all biotic processes, and is the only renewable energy source of our planet. In plants, photosynthesis takes place in a distinct cellular organelle, the chloroplast (rhodoplast, cyanoplast, phaeoplast etc.). The fundamental role of this organelle, the enormous versatility of the photosynthetic process and of its underlying structures, and intriguing phylogenetic aspects that have emerged recently, have been primarily responsible for the extraordinary interest in the analysis of this organelle.

The typical chloroplast is a double membrane-bound organelle that houses the functional components required both for the light-driven and enzymatic reactions of photosynthesis, namely thylakoid membranes and stroma. Thylakoid membranes are unique biomembranes, capable of converting inorganic substrates with solar energy to utilizable high-energy chemical metabolites, initially in the forms of ATP and NADPH, which in turn provide the necessary energy for the „dark" reactions involved in the fixation of atmospheric CO_2 into sugar intermediates. In chloroplasts, they are lipid bilayers which contain several different multisubunit protein complexes (reviewed in [1,2]). These protein assemblies together with mobile electron carriers cooperate in the conversion of radiant energy. A second fundamental feature of the chloroplast is that it houses a genetic system. Plastids, as mitochondria, contain DNA, the mechanisms to maintain this information and to convert it into function. Consequently, the plant cell is characterized by a *compartmentalized genome* which adds a substantial degree of complexity to the system.

The biogenesis, function and modification of the chloroplast are embedded into the outlined genetic context of the plant (autotrophic eukaryotic) cell. Plant cells and their genomes of are beyond doubt the result of endocytobioses. They trace back to associations of three, four or even five cells, arisen through the incorporation of a bacterial or a unicellular eukaryotic cell (microalga) into a host organism, which followed the α-proteobacterial endocytobiosis that led to the generation of mitochondria. Both,

J. H. Argyroudi-Akoyunoglou and H. Senger (eds.),
The Chloroplast: From Molecular Biology to Biotechnology, 63–72.

chloroplasts and mitochondria have preserved remnants of the ancestral (eubacterial) genomes. The initial cell associations developed a compartmentalized genetic system with a *common* metabolism and a *common* inheritance. The approach of molecular phylogeny has shown that the phylogenetically important changes occurred gradually and include a massive intracellular restructuration of genetic material. This work has led to a new understanding of the chloroplast, the plant genome and plant cell, which appear in a new light, phylogenetically and functionally. It suggests that the genome restructuration generated an integrated genetic system, rather than caused semiautonomous organelles and nucleus ([3], see below). It becomes more and more evident that major aspects in design, biogenesis and regulation of chloroplast structure cannot be understood without the history of the plant cell.

2. Plastid Chromosomes; Origin of Photosynthetic Organelles

Sequence comparisons and ultrastructural information have been instrumental in restructuring all major groups of plants and provided an outline of a global, coherent genealogical system of photoautotrophs. The chronometer molecules used to deduce a reasonably accurate plant genealogy were initially rRNA sequences, then organelle DNAs (e.g. [4,5]), especially operon structures and genes involved in photosynthesis or in the flux of genetic information in organelles, more recently nuclear genes involved in plastid biology, and nowadays entire organismic coding potentials [2,6,7]. Plastid chromosomes are circular duplex DNA molecules in the range of 70 to 250 kb depending on the organism. The complete nucleotide sequences, and implicitly coding potentials, of entire plastid chromosomes from nearly 20 species including representatives of widely distant chlorophytic, glaucocystophytic, rhodophytic and chromophytic proto- and thallophytes (algae), bryophytes and vascular plants, and information for >300 nuclear genes encoding plastid proteins are now available.

Available evidence suggests that all plastids trace back to a single, primary (eukaryotic/prokaryotic) endocytobiosis [4]. The striking similarity between corresponding cyanobacterial and plastid gene and operon structures, nuclear genes encoding chloroplast proteins that plastid operons are missing, substantial sequence conservation, and the almost identical design of the basic photosynthetic machinery (but not of antenna systems) in the various plant lineages, are all consistent with the idea that the basic gene arrangement of plastid chromosomes was established prior to the divergence of the principal plant lineages. This supports a common, *monophyletic* ancestry of the organelle and host cell [3-5]. Plastid genes fall basically into three principal classes: (a) those that contribute to the flow of genetic information within the organelle, including four genes for subunits of one of the organelle RNA polymerases (see below), (b) genes involved in establishing the photosynthetic energy transduction machinery, and (c) open reading frames of unknown function. Approximately two-thirds of the gene complement are shared by all plastid lineages. The organization of plastid chromosomes is well conserved in vascular plants, while quite extensive rearrangements may be noted with that in algae. The plastid chromosomes of the glaucocystophycean, rhodophycean and chromophytic algae encode a significant number (40 - 80) additional genes compared to those of the chlorophyll *ab* lineage of plants [4]. The plastid chromosome of the rhodophyte *Porphyra purpurea* for instance, which is 191,028 bp in length (compared to 155,939 bp of that of tobacco), contains approximately 250 genes, about twice the number of those of higher plants, including additional information for ribosomal and thylakoid components, and various genes involved in plastid gene expression. On the other hand, the coding capacity of the plastid chromosome of *Euglena gracilis* is the lowest known for a plastome [e.g. 5]. Due to the overall similarity of plastid chromosomes from individual lineages, differences in gene reduction are diagnostic. They reflect different stages of an evolutionary pathway in which central organelle functions are gradually ceded to the nucleus [3-5].

The initial endocytobiotic cell conglomerate developed into the autotrophic version of the eukaryotic cell with three major lineages: Glaucocystophyta, Rhodophyta and Prasinophyceae/Chlorophyta/Charophyceae. Their plastids are surrounded by two envelope membranes. However, the evolution of plants, their plastids and genomes was by far more complex due to the existence of various plastid modifications appearing in multicellular plants and of plastids (but not of mitochondria) resulting from secondary, and even tertiary endosymbioses. The majority of the photoautotrophic organisms, such as Cryptophyta, Chlorarachniophyceae, Euglenophyta, Dinophyta, Haptophyta, and Chrysophyta, result from secondary, i.e. eukaryotic/eukaryotic, in the case of some fucoxanthin-containing Dinophyta even from tertiary (eukaryotic/secondary eukaryotic) endosymbioses. Their plastids are surrounded by three or four envelope membranes, of which each possesses a different origin and history. This has significant, not yet explored, functional consequences. Support for this concept is found in intermediate stages that exist in the Chlorarachniophyceae and Cryptophyta. Their multienveloped plastids contain a residual nucleus of the respective endocytobiotic eukaryotic microalga, the nucleomorph, and a residual cytosol with an additional category of 80S ribosomes in the space between the second and the third envelope layer (see e.g. [3,4,8]). Different from primary eukaryotes, these plants arose *polyphyletically* and hence evolved independently of each other, with both different hosts and different engulfed microalgae [3-5]. This is in line with the occurrence of different antenna systems and with differences in the degree of gene transfer between the individual lineages.

3. Multicellularity, Plastid Modifications

The „endosymbiotic" cell type had an enormous selective advantage when compared with the prokaryotic cell type. Apart from the fact that it appeared and survived polyphyletically, it was only the eukaryotic cell that developed advanced, true multicellular forms of life, first with plants, in combination with oxygenic photosynthesis also of terrestrial life. Multicellularity arose approximately 1.3 x 10^9 years ago, probably first with Rhodophyta. This step required new qualities of genetic information. Besides information for regulation in space, especially the step to land marks an interval of unparalleld biochemical innovation in the history of plant life. Novel syntheses as well as structures seem to have been the major responses to manage with new physiological conditions due to the transition from the aqueous to gaseous environment. As few other cellular structures, this includes the chloroplast. The biochemical and morphogenetic responses including plant diversification appeared gradually. Fossile records along with molecular, cytological and morphological phylogenetic work favour a single origin of land plants from charophycean thallophytes, and indicate that several major lineages of vascular plants had evolved more than 400 million years ago.

The biogenetic and morphological changes had to be necessarily paralleled by genome changes. These include the establishment of metabolic pathways, such as sporopollenin, lignin, flavonoid, wax and hormone synthesis, and the development of an enormous morphogenetic potential. Their evolution is poorly understood, although for instance metabolic sources for biochemical potentials have been detected in lower organisms. The morphological achievements, in turn, are apparent, spatially and temporally, in the three major organismic categories of plants, from protophytes (protists), through diverse forms of thallophytes (algae) to kormophytes (vascular plants), with their elaborate life cycles, the latter also with an extrordinary array of cells, tissues and organs. Spatially, they include the development of essential functions and organs (sexual organs, gas exchange surfaces, stomata etc.), of structurally and chemically protecting surfaces, of fluid transportation systems (vascular tissue), or of cell and tissue systems for stabilization (collenchym, sclerechym etc). Temporally, innovations are evident from the the two-phase gametophytic/sporophytic life cycles with meiotic and sexual nuclear ploidy changes, and the diverse morphological differentiations at both stages. The novel structural potential resulted in significantly more highly differentiated plants.

There is ample evidence to suggest that living plants are „genetic relics“. The varied ontogenies appear to be under similar general principles of developmental control in all main groups. Generation of morphological diversity seems to be based on a gradual appearance of novel genes or gene combinations, and their integration into regulatory networks. Also, much of it can be interpreted in terms of modification of basic structural and metabolic units, such as (often hormone-induced) dormancy/activation and abortion of meristematic, branching or cell expansion systems. Relevant examples are found in the organ „leaf“ which developed into functionally different photosynthetic-, floral- and storage-type structures or, at the cellular level, in a wide range of structurally and functionally diverse plastid forms (*leucoplasts*, *amyloplasts*, *elaioplasts* or *chromoplasts*) developed in advanced thallophytes and higher plants. The different achlorophyllous and physiologically specialized modifications of the organelle develop in distinct cells and tissues, and are involved in various fundamental biological processes, such as carbohydrate and lipid storage, and plant propagation. These features are brought about by specific, spatiotemporally regulated genes and genetic programmes. Their nature, origin and appearance remain to be deciphered.

4. Aspects of Chloroplast and Plant Genome Evolution

The development of the cyanobacterial or eukaryotic endosymbionts to photosynthetic organelles, and implicitly of plant cells, is characterized by an enormous intracellular rearrangement of genetic material. It is now known that *genome restructuration* is one of the most crucial features in the evolution of plant cells, and that the rearrangements were substantially more complex than generally assumed. They included the well established intracellular transfer, but also loss and gain of genetic information [3,9]. DNA rearragements must have been relatively frequent. They occurred probably between all genetic compartments, although the DNA net flow took place preferentially from the organelles to the nucleus with an establishment of nuclear regulatory dominance.

Compelling evidence for this scenario is found (i) in the fact that plastid chromosomes encode only a fraction, generally in the order of 5% or less, of the ancestral genome that may have been in the order of 2.000 - 4.000 genes based on the genome size of the cyanobacterium *Synechocystis* [7], (ii) in substantial amounts of organelle-derived (functional or promiscuous) DNA that have been detected in the nucleus, or of plastid (and nuclear) DNA sequences in mitochondria [e.g. 3,10], (iii) in the differences in the pattern of gene losses between different lineages, and (iv) in repeated functional gene losses and transfers of genes in the different photoautotrophic lineages [5]. For example, the relatively large plastid chromosome of *Prophyra* (see above) does not contain any *ndh* genes, encoding proteins which share substantial homology to subunits of complex I of the mitochondrial respiratory chain, different from those of plants of the chlorophyll *ab* lineage, however except those of gymnosperms. The elimination of redundant genes to avoid competition between the basic metabolisms of the symbiotic partners (or of otherwise dispensible genes) illustrates gene losses. It is both indicative of genetic integration and instructive with regard to the cell status and the complexity of changes.

Four points of general interest emerge from the outlined work: (1) The similarity of operon design in plastid chromosomes and, on the other hand, detail and degree of gene losses and transfers between plants with primary and secondary plastids reinforce that the major lineages have separated early [5]. -- (2) Most genome restructuration occurred at the unicellular level [5]. This and the temporal distance between the generation of plant cells, approximately 2,200 million years ago, and the appearance of multicellular forms of plant life suggest that the overall evolution of eukaryotes occurred in two principal steps, characterized by the unicellular and multicellular stage, respectively [3]. This implies that much information for regulation in space had to be generated *after* gene translocation (see below). -- (3) Different from what one would expect, the gene losses did not occur predominantly from the endosymbionts but from the host cell. The principal flux of genetic information from the organelles to the nucleus and the establishment of genetic nuclear regulatory dominance do not appear to reflect a metabolic dominance of the ancient host cell. Since it seems that only few (archaebacterial) elements from the basic carbohydrate metabolism of the host cell have survived [6], much more of the endosymbiont (plastid and mitochondrial) biochemistry than is reflected in organelle DNAs has been retained in the plant cell. Furthermore, a significant fraction of the enzymes of the chloroplast stroma or in the cytosol that are nuclear-coded is of α-proteobacterial origin. Also, functionally related isozymes located in different cellular spaces, e.g. of the Calvin cycle or glycolysis, may be of different phylogenetic origin [6]. This is indicative of substantial compartmental product detour [6,9], which required the acquisition of appropriate transit peptides, and of a massive intermixing of genetic material from the initial endosymbiotic cell association. The selective advantage of the eubacterial metabolic machinery, which dominates functionally in the plant cell, is not understood. -- (4) The comparison of the intracellular gene distribution for individual plastid types and of related products in plastids, mitochondria, and hydrogenosomes suggests that plastids and mitochondria have been integrated to different stages into the cellular entity [3]. Although the rearragements of genetic material occurred probably more or less stochastically, the individual translocation patterns can, in fact, be reconciled with a *directed functional gene transfer*. Obviously, genes for single or simple chain enzymes were transferred first. However, due to the complexity of biogenetic regulation, gene transfer for multisubunit structures was necessarily a gradual process. This, of course, caused the dual genetic origin of complex organelle structure. In such instances, regulatory subunits, in the photosynthetic machinery also peripheral components, were usually translocated first [3]. The extent of genetically hybrid structures in plastids and mitochondria (Table 1) and the different degree of organelle- and nuclear-coded components in corresponding structures of both organelles illustrate that the intercompartmental restructuration process of genetic information is more advanced for mitochondria than for plastids, and terminal for the majority of hydrogenosomes. The latter are usually considered to respresent mitochondrial derivatives [11] in protists growing under obligate anaeobic conditions. They lack generally DNA, probably since both the respiratory membrane and the organelle-own ribosomes for their synthesis, the only components of dual genetic origin in advanced mitochondria (Table 1), are dispensible in this environment [3].

Table 1: Structures of dual genetic origin in chloroplasts and plant mitochondria

thylakoid membrane - respiratory membrane	[2,3]
organelle stroma	
ribulose bisphosphate carboxylase/oxygenase	
70S ribosome - 70S/60S ribosome	
acteyl-CoA carboxylase	[12]
Clp protease	[13]
eubacterial-type RNA polymerase	[14]
transcript processing machinery	[15]
inner envelope membrane	[e.g. 16]

5. Functional consequences of gene rearrangements

The transfer of much of the genetic information from the original endosymbiotic cyanobacterium to the host (nuclear) genome and the return of the products of the translocated genes to the organelle [17] required the establishment of novel, intercompartmental regulatory circuitries, and caused the dual genetic origin of complex organelle structures, which both are fundamental features of the eukaryotic genetic system. In terms of biogenesis, the dispersal of genes for complex chloroplast structures among organelle and nucleus implies that the delivery of the components from two sources of protein synthesis in the cytosol and plastid respectively, has to be coordinated. This regulation is highly sophisticated since the regulatory schemes have to serve to coordinate the integrated genetic system, in *time*, *quantity*, inducibly, in multicellular organisms also in *space*, while allowing cellular input into the functional and developmental decision. Regulation occurs at all levels and with different mechanisms over a wide range of time scales. A few examples (i - iv) may illustrate this.

(i) *Gene expression in plastids:* The coding potential of plastid DNA is generally organized in operons, probably economically in a not yet intelligible way. Typical elements of genome compaction are found when plastid chromosomes are compared with their prokaryotic pendants [18,19]. Also, operons are frequently heterogenic. They harbor loci for more than one organelle structure. That genes for complex structure are dispersed throughout the chromosome and often cotranscribed with genes for other multisubunit assemblies has important consequences for gene expression.

Previous opinion that the major level of regulation in chloroplasts is not transcriptional but posttranscriptional [e.g. 20] has to be revised. Transcription is substantially more complex than previously assumed and unique, since it has now been proven that the chloroplast of higher plants operates with more than one RNA polymerase. These use different promoter types. Besides the ancient, organelle-coded eubacterial multisubunit core enzyme the existence of a nuclear-coded, second plastid-located RNA polymerase resembling phage-type and mitochondrial enzymes has recently been demonstrated (e.g. Börner et al., this Volume, [21]). In addition, in higher plant plastid chromosomes promoters are complex due to multiple transcriptional initiation sites. Some operons seem to be exclusively transcribed by either the plastid-coded or the nuclear-coded RNA-polymerase, the majority of them is transcribed by both, with different basic *cis* elements for each polymerase type [22,23].

The nuclear-coded enzyme appears to be the result of an internal duplication of the gene for the corresponding mitochondrial enzyme [21]. The appearance of this additional RNA polymerase in plastids that became the sole kind of enzyme in mitochondria [but see 24] has caused promoter changes in the respective organelle chromosomes. These changes are more or less terminal in mitochondria and probably at an intermediate stage in plastids [3]. The interaction of the two plastid RNA polymerases in tissue- and development-specific expression of plastid genes is not understood nor is the phylogenetic origin of the nuclear-coded enzyme, its appearance, and the reasons why a second enzyme has been established. Recent findings of our laboratory on the off-white *rpo*-deficient plastome mutants exclude simple assumptions such as that the plastid-encoded core polymerase were involved in the expression of photosynthetic genes, while the enzyme of nuclear origin in that of the housekeeping genes of the organelle.

As transcriptional control, modification of RNA molecules and translational control represent key levels of regulation in the chloroplast. The known polycistronic transcripts of plastid chromosomes often give rise to complex sets of overlapping RNA species, usually through a series of processing steps. Posttranscriptional RNA changes are complex. The multiple mechanisms known are interwoven into a regulatory network. At least ten different enzymatic machineries have been deduced to modify plastid RNA, such as activities involved in group I and group II intron splicing, 5' and 3' terminal shortage by endo- and exonucleolytic activities, 3' poly(A) addition to mRNA fragments, 3' CCA addition to tRNAs or nucleotide substitutions and nucleotide modifications. C-to-U conversions and the reverse process, designated RNA editing, are often essential for the decoding and/or for the structural and functional fidelity of the respective protein, or change the expression patterns and/or RNA stability [25]. The edited sites in plastid transcripts can be found in reading frames and in nontranslated RNA segments. With the exception of some components involved in the 3' processing of plastid encoded mRNAs and RNA stability [26], the enzymes and factors involved in posttranscriptional processing and transcript stability are not known. It is relevant to mention that the modification machinery is of dual genetic origin (Table 1, [15]) and that the extensive transcript processing in chloroplasts and some of its mechanisms are phylogenetically derived rather than original traits. They are not or to a much lesser extent found in prokaryotes or in thallophytes, and hence appear to be phylogenetic innovations [e.g. 27]. Circumstantial evidence suggests also that editing may have been acquired from the nuclear and mitochondrial compartments. If correct, this would be an interesting analogy to the nuclear-coded, second plastid RNA polymerase (see above).

(ii) *Promoter design and transcription of nuclear genes coding for chloroplast proteins*: The success of a chloroplast gene that has been translocated to the nucleus depends on the acquisition of an appropriate sequence context for expression and for protein import either during the translocation process or by rearrangement after translocation. Consistent with phylogenetically individual transfer events of originally organelle-encoded genes are the dispersal of genes in the nuclear genome [28], and that the generally coordinated, often induced (light, hormones) regulation or organ-specific expression of nuclear genes encoding thylakoid proteins is *not* brought about by common *cis*-elements or common *trans*-acting factors as is known from other collectively expressed groups of genes (e.g., heat shock genes) [3]. Even genes that code for different subunits of the same membrane complex usually possess entirely different promoters, unlike homologous genes in different higher plants. Moreover, comparative studies of promoters from photosynthetic genes, including those from C3 and C4 plants, indicate that at least some information to ensure (transcriptional) regulation in space must have been acquired during evolution, since the majority of such gene translocations have occurred at the unicellular level (see above). Important aspects of organelle biology, notably the streamlining and integration of acquired promoter sequences into the signal transduction chains, the principles of transcriptional changes as a means of physiological adaptation, and the link to the complex network of interacting signaling components and cascades remains to be explored.

(iii) An equivalent situation exists for later biogenetic processes. A functional intracellular gene transfer to the nucleus requires the return of the respective protein into the organelle. Import, transport to and assembly at various locations within the chloroplast of proteins depend on specific targeting and sorting signals, translocation machineries and protein folding/unfolding catalysts (chaperones) [29]. For chloroplasts, targeting signals show no obvious sequence homology, but fall into two principal classes, stroma- and stroma-/tylakoid-targeting presequences. The latter operate in two phases, with two distinct translocation and processing systems located in the envelope/stroma and thylakoids, and with a remarkable variety (at least four) of non-competable thylakoid integration/translocation routes, the Sec-, SRP-like, ΔpH- and spontaneous pathways. This complexity is only partly understood functionally and phylogenetically. With the exception of import sequences, of the spontaneous mechanism and of some of the second domains of bipartite transit peptides [3], most of this appears to be inherited from the prokaryotic progenitor of chloroplasts. The Rieske FeS protein of the cytochome b_6f complex is a first example for an ancient as well as for an integral protein using the ΔpH route (I. Karnauchov, personal communication). Different from all other compounds known to use this route, it operates with an uncleaved thylakoid targeting signal, and integration of its FeS center appears to be required for transport, reminiscent of a recently found specific group of bacterial proteins which share a cofactor-mediated folded state during membrane translocation [see 29]. It also lacks the twin arginine motif characteristic of all other ΔpH-dependent transit peptides, and interacts with several chaperones, successively with cpn60, Rubisco activase, and with the stromal Hsp100 (ClpC). On the other hand, the study of the related ATP synthase subunits CFo-I and -II (b, b′), which are favourably encoded in different cellular compartments, has shown that gene translocation can change late biogenetic processes [see 3].

(iv) Regulatory complexity emerges also for components and mechanisms contributing to the modification (maintenance, optimization, protection, senescence) of existing biological structure, which often includes components different from those regulating input synthesis [e.g. 2]. Examples for long-term changes are found in the spectrum of more than two dozen biochemically discernible proteolytic activities in the chloroplast [13,30,31] that include two energy-dependent proteases with homologs known from prokaryotic ancestors of the organelles, the metalloprotease FtsH and the multisubunit Clp protease. Complexity of short term adaption is illustrated by the recently discovered multifunctional complex 40 kDa immunophilin-like peptidyl-prolyl *cis-trans* isomerase TLP40 in the thylakoid lumen. The protein folding catalyst may be involved in the redox-controlled kinase/phosphatase system which ensures a fast balance of the light-driven electron transfer between the two photosystems (state I/state II transitions), and link protein turnover, folding and phosphorylation, and repair processes [32].

Conclusion: For efficient propagation, plants integrate multiple signals from their environment to regulate biogenesis, acclimatization and maintenance of the chloroplast. Each distinct situation requires the coordination of information from nuclear and plastid genes and regulation at various levels. We are only now beginning to understand the outlines of the respective role of these levels, their multiple interactions, and the enormous role which the phylogenetic integration of organelle ancestors have played in this scenario.

Acknowledgements: This work was supported by the German Research Foundation (SFB 184) and the Human Frontier Science Program (HFSP).

References

1. Hankamer, B., Barber, J. and Boekema, E.J. (1997) Structure and membrane organization of photosystem II in green plants, *Annu. Rev. Plant Physiol.. Plant Mol. Biol.* **48**, 641-671
2. Herrmann, R.G. (1996) Photosynthesis research: aspects and perspectives, in B. Andersson, H.A. Salter and J. Barber (eds.), *Frontiers of Molecular Biology. Molecular Genetics in Photosynthesis*, IRL Press, Oxford, pp. 1-44.
3. Herrmann, R.G. (1997) Eukaryotism, towards a new interpretation, in H.E.A. Schenk, R.G. Herrmann, K.W. Jeon, N.E. Müller, and W. Schwemmler (eds.), *Eukaryotism and Symbiosis*, Springer, Heidelberg, New York, pp. 73-118.
4. Kowallik, K. (1997) Origin and evolution of chloroplasts: Current status and future perspectives, in H.E.A. Schenk, R.G. Herrmann, K.W. Jeon, N.E. Müller and W. Schwemmler (eds.), *Eukaryotism and Symbiosis*, Springer, Heidelberg, New York, pp. 3-23.
5. Martin, W., Stoebe, B., Goremykin, V., Hansmann, S., Hasegawa, M. and Kowallik, K.V. (1998) Gene transfer to the nucleus and the evolution of chloroplasts, *Nature* **393**, 162-165.
6. Martin, W., and Schnarrenberger, C. (1997) The evolution of the Calvin cycle from prokaryotic to eukaryotic chromosomes: A case study of functional redundancy in ancient pathways through endosymbiosis, *Curr. Genet.* **32**, 1-8.

7. Kaneko, T. et al. (1996) Sequence analysis of the genome of the unicellular cyanobacterium *Synechocystis* sp. strain PCC6803. II. Sequence determination of the entire genome and assignment of potential protein-coding regions, *DNA Res.* **3**, 109-136.
8. McFadden, G.I. and Gilson, P. (1997) Whats eating Eu? The role of eukaryote/eukaryote endosymbioses in plastid origins, in H.E.A. Schenk, R.G. Herrmann, K.W. Jeon, N.E. Müller and W. Schwemmler (eds.), *Eukaryotism and Symbiosis*, Springer, Heidelberg, New York, pp. 24-39.
9. Martin, W. and Herrmann, R.G. (1998) Gene transfer from organelles to the nucleus: how much, what happens, and why?, *Plant Physiol.* **118**, 9-17.
10. Brennicke, A., Grohmann, L., Hiesel, R., Knoop, V., and Schuster, W. (1993). The mitochondrial genome on its way to the nucleus: different stages of gene transfer in higher plants, *FEBS Lett.* **325**, 140-145.
11. Bui, E.T., Bradley, P.J. and Johnson P.J. (1996) A common evolutionary origin for mitochondria and hydrogenosomes, *Proc. Natl. Acad. Sci. USA* **93**, 9651-9656.
12. Sasaki, Y., Sekiguchi, K., Nagano, Y. and Matsuno R. (1993) Chloroplast envelope protein encoded by chloroplast genome, *FEBS Lett.* **316**, 93-98.
13. Sokolenko. A., Lerbs-Mache, S., Altschmied, L., and Herrmann, R.G. (1998). Clp protease complexes and their diversity in chloroplasts, *Planta*, in press
14. Link, G. (1996) Green life: control of chloroplast gene transcription, *BioEssays* **18**, 465-471.
15. Rochaix, J.D. (1996) Post-transcriptional regulation of chloroplast gene expression in *Chlamydomonas reinhardtii*, *Plant Mol. Biol.* **32**, 327-341.
16. Rolland, N., Dorne, A.J., Amoroso, G., Sultemeyer, D.F., Joyard, J., and Rochaix, J.D. (1997) Disruption of the plastid *ycf*10 open reading frame affects uptake of inorganic carbon in the chloroplast of *Chlamydomonas*, *EMBO J.* **16**, 6713-6726.
17. Weeden, N.W. (1981) Genetic and biochemical implications of the endosymbiotic origin of the chloroplast, *J. Mol. Evol.* **17**, 133-139.
18. Herrmann, R.G., Westhoff, P. and Link G. (1992) Biogenesis of plastids in higher plants, in R.G. Herrmann (ed), *Plant gene research*, vol 6, Springer, Wien, New York, pp 275-349
19. Sugita, M. and Sugiura, M. (1996) Regulation of gene expression in chloroplasts of higher plants, *Plant Mol. Biol.* **32**, 315-326.
20. Gruissem, W. (1989) Chloroplast gene expression: how plants turn their plastids on?, *Cell* **56**, 161-170.
21. Hedtke, B., Börner, T. and Weihe, A. (1997) Mitochondrial and chloroplast phage-type RNA polymerases in *Arabidopsis*, *Science* **277**, 809-811.
22. Hajdukiewicz, P.T.J., Allison, L.A., and Maliga, P. (1997) The two RNA polymerases encoded by the nuclear and plastid compartments transcribe distinct groups of genes in tobacco plastids, *EMBO J.* **16**, 4041-4048.
23. Kapoor, S., Suzuki, Y.-Y., and Sugiura, M. (1997) Identification and functional significance of a new class of non-consensus-type plastid promoters, *Plant J.* **11**, 327-337.
24. Lang, B.F., Burger, G., O'Kelly, C.J., Cedergren, R., Golding, G.B., Lemieux, C., Sankoff, D., Turmel, M, and Gray M.W. (1997) An ancestral mitochondrial DNA resembling a eubacterial genome in miniature, *Nature* **387**, 493-497.
25. Maier, R.M., Zeltz, P., Kössel, H., Bonnard, G., Gualberto, J.M. and Grienenberger, J.M. (1996) RNA editing in plant mitochondria and chloroplasts, *Plant Mol. Biol.* **32**, 343-365.
26. Hayes, R., Kudla, J., Schuster, G., Gabay, L., Maliga, P. and Gruissem, W. (1996) Chloroplast mRNA 3'-end processing by a high molecular weight protein complex is regulated by nuclear encoded RNA binding proteins, *EMBO J.* **15**,1132-1141.
27. Malek, O., Lattig, K., Hiesel, R., Brennicke, A. and Knoop, V. (1996) RNA editing in bryophytes and a molecular phylogeny of land plants, *EMBO J.* **5**, 1403-1411.
28. Pillen, K., Jung, C. and Herrmann, R.G. (1996) Genetic mapping of loci for thirteen nuclear-encoded polypeptides associated with the thylakoid membrane in *Beta vulgaris* L, *FEBS Lett.* **395**, 58-62.
29. Robinson, C., Hynds, P.J., Robinson, D. and Mant A. (1998) Multiple pathways for the targeting of thylakoid proteins in chloroplasts, *Plant Mol. Biol.* **38**, 209-221.
30. Adam, Z. (1996) Protein stability and degradation in chloroplasts, *Plant Mol. Biol.* **32**, 773-783.
31. Sokolenko, A., Altschmied, L. and Herrmann, R.G.(1997) Sodium dodecylsulfate-stable proteases in chloroplasts, *Plant Physiol.* **115**, 827-832.
32. Fulgosi, H., Vener, A.V., Altschmied, L., Herrmann, R.G. and Andersson, B. (1998) A novel multifunctional chloroplast protein: identification of a 40 kDa immunophilin-like protein located in the thylakoid lumen, *EMBO J.* **17**, 101-111.

PHAGE-TYPE RNA POLYMERASES IN HIGHER PLANTS

T. BÖRNER[1], B. HEDTKE[1,2], W. R. HESS[1], J. LEGEN[1,2],
R.G. HERRMANN[2], A. WEIHE[1]

[1]*Institute of Biology, Humboldt University, D-10115 Berlin, and*
[2]*Botanical Institute, University of Munich, D-80638 Munich, Germany*

1. Introduction

Plastids harbor a multi-subunit RNA polymerase (RNAP) highly similar to eubacterial RNA polymerase, encoded by the plastid *rpoA*, *rpoB* and *rpoC* genes. The *rpo* genes have been found in all plant species so far investigated, with the exception of the parasite *Epifagus virginiana*. The eubacterial-like plastid RNA polymerase has been designated PEP (= plastid encoded polymerase). Several lines of evidence implicate the existence of a second RNA polymerase in the plastids of higher plants which is encoded in the nucleus (NEP = nuclear encoded polymerase. The plastid enzyme, NEP, functions in a coordinated manner, which can be described as „division of labour", together with the plastid encoded PEP [4, 10, 14, 17].

Earlier biochemical data, based on partial purification and immmunological analyses, suggested that the NEP might be a single-subunit phage-type RNA polymerase [13]. Further evidence for the phage-type nature of NEP can be provided by inhibitor studies: Tagetitoxin, a potent inhibitor of PEP activity, but not of the T7/T3 polymerases, does not affect NEP related run-on transcription in the ribosome-free plastids of *albostrians* barley (Rajashekar and Börner, unpubl.). Distinct promoters for PEP and NEP have been characterized in chloroplast DNA [6, 12, reviewed in 10]. While PEP uses typical sigma-type promoters, different non-consensus type promoter sequences are recognized by NEP. Interestingly, a YRTA motif - conserved in plant mitochondrial promoters - is also detectable in a similar position relative to the transcription initiation site in most plastid NEP promoters [10, 11]. The mitochondrial RNA polymerase of fungi and vertebrates is a T7/T3-type enzyme [3, 15, 20]. The recent identification of single-subunit phage-type RNA polymerase genes in *Chenopodium album* and *Arabidopsis thaliana* [8, 21] and *in vitro* import data suggest that the plant mitochondrial RNA polymerase and the plastid NEP are also phage-related enzymes. Thus, transcription in both mitochondria and plastids seems to be dependent on similar nuclear-encoded RNA polymerases. We report here on the existence of small gene families encoding phage-type RNA polymerases in the nuclear genomes of *A. thaliana* and *Nicotiana tabacum*. Sequence comparison reveals a high conservation of the intron-exon structure among plant genes encoding phage-type RNA polyerase including also genes from moncots.

J. H. Argyroudi-Akoyunoglou and H. Senger (eds.),
The Chloroplast: From Molecular Biology to Biotechnology, 73–78.

2. Material and Methods

RNA and DNA was isolated from young leaves of *A. thaliana* (ecotype Columbia), *N. tabacum* (cv.SNN) and the *albostrians* mutant of *Hordeum vulgare*. cDNAs and genomic clones encoding phage-type RNA polymerases were obtained as described [8, 21]. Alignments and phylogenetic analysis were performed using the programs CLUSTALW [19] and PUZZLE [18]. For *in vivo* targeting studies, cDNA sequences corresponding to the putative transit peptides of the *Arabidopsis* RNAPs were fused to a modified green fluorescent protein (GFP) sequence [7] and set under control of the CaMV35S promotor in the vector pGPTV-bar [2]. The constructs were used to transform *A. thaliana* by an *Agrobacterium*-mediated *in planta* infiltration method [1]. Plants expressing the transit peptide-GFP fusion proteins were examined by epifluorescence and confocal laser scanning microscopy [9].

3. Results and Discussion

The identification of two rice EST sequences (Acc. No. D23514 and D24565) highly similar to the mitochondrial RNAP from fungi enabled us to isolate homologous sequences from a number of higher plants. Using RT-PCR and adaptor-ligated RACE techniques, a full-length cDNA sequence was obtained from the dicotyledoneous plant *C. album* [21]. This clone was employed to screen cDNA and genomic libraries from *A. thaliana, N. tabacum* and *H. vulgare* (*albostrians* mutant). In Arabidopsis, three related genes - designated *RpoT;1* [8], *RpoT;2* (Schuster et al., unpublished) and *RpoT;3* [8] - with striking similarity to sequences encoding single-subunit RNAPs of the phage type were found.

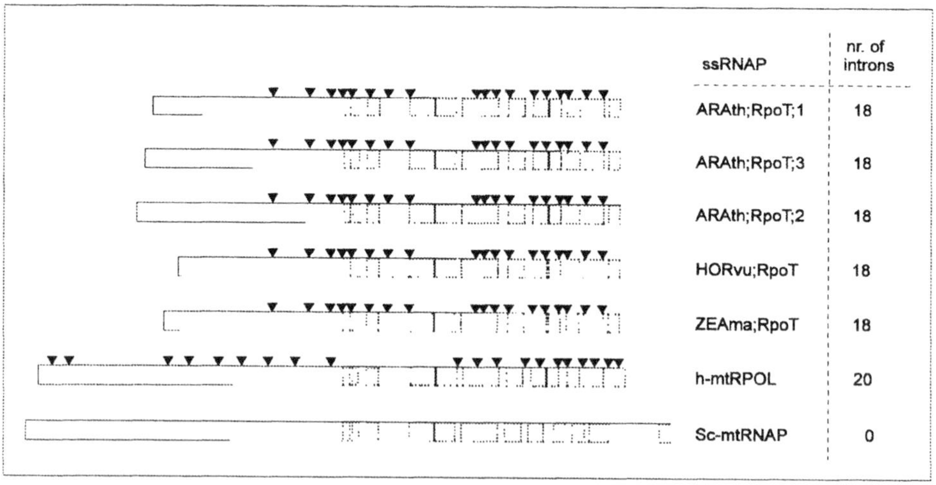

Figure 1. Schematic representation of the amino acid sequences of nuclear-encoded single-subunit RNA polymerases from *A. thaliana*, *H. vulgare* and *Z. mays* (ARAth, HORvu, ZEAma), human (h-mtRPOL) and yeast (Sc-mtRNAP). Shaded areas indicate conserved sequence blocks, intron positions are designated by black arrow heads.

The putative translation products of the three genes possess all amino acids reported to be essential for the function of the T7 RNAP [16]. Highly similar genes have recently been isolated from *Zea mays* [22] and barley (Hess et al., in preparation). Sequencing of genomic clones from tobacco is in progress in our laboratories. In contrast to the intronless mitochondrial RNAP genes from *Saccharomyces cerevisiae* and *Neurospora crassa* [3, 15], the three *RpoT* genes from *A. thaliana* all contain 18 introns at exactly the same positions in the amino acid sequence. The number and position of the introns in the single-subunit RNAP genes from maize and barley are identical to those of the *Arabidopsis* genes. Moreover, identical intron positions have been found in those parts of the tobacco *RpoT* genes sequenced so far (Hedtke et al., unpublished). In contrast, the human *RpoT* gene coding for the mitochondrial RNAP exhibits a completely different exon-intron-structure [20; Fig. 1]. The conservation of the exon-intron structure of the phage-type RNAP genes of higher plants as well as their relatively high sequence similarity suggest that the common ancestor of monocots and dicots exhibited already the same structural *RpoT* organization and that the small gene family of nuclear encoded phage-type RNAPs of higher plants arose by duplication events.

Gene	Species	Localization	Data	References
ARAth;RpoT;1	*A. thaliana*	mitochondria	*in vitro, in vivo*	[11, 12]
ARAth;RpoT;2	*A. thaliana*	mitochondria	*in vitro, in vivo*	[see text]
ARAth;RpoT;3	*A. thaliana*	plastids	*in vitro, in vivo*	[11, 12]
CHEal;RpoT;1	*C. album*	mitochondria	*in vivo*	[12]

Table 1. Subcellular localization of plant phage-type RNAPs as determined by *in vitro* and *in vivo* targeting experiments

The monocot phage-type RNAPs analyzed so far seem to share one peculiar feature which is missing among the dicot polymerases: Both the maize *RpoT* and an orthologous partial sequence from wheat show alternative splicing of intron-12 which would lead to a truncated protein lacking most of its functionally essential domains [22]. In an *albostrians* barley cDNA library also several clones were identified showing alternative splicing at exactly the same position in intron 12 (Hess et al., unpublished). The conservation of this feature between the monocot RNAPs suggests a functional role for the truncated polypeptide resulting from alternative splicing. However, further investigations will be necessary to verify this hypothesis.

Targeting of four of the plant *RpoT* genes has been experimentally addressed by *in vitro* import experiments with purified organelles and *in vivo* targeting using GFP as a marker fused to the putative transit peptides of the RNAPs. Data are summarized in Table 1.

The transit sequences of Rpo;T1 and Rpo;T3 from *A. thaliana* were found to target GFP into purified mitochondria and chloroplasts, respectively [8]. Identical results have been obtained *in planta* with transgenic *Arabidopsis* plants expressing GFP fused to the respective transit peptide [9]. Preliminary results suggest that the transit peptide of the third phage-type RNAP from *Arabidopsis*, Rpo;T2, confers targeting to mitochondria, both *in vitro* and *in vivo* (Schuster et al., pers. commun.; Weihe et al., unpublished). In the same way we could demonstrate that the previously identified cDNA from *C. album* [21] codes for a mitochondrial enzyme [9].

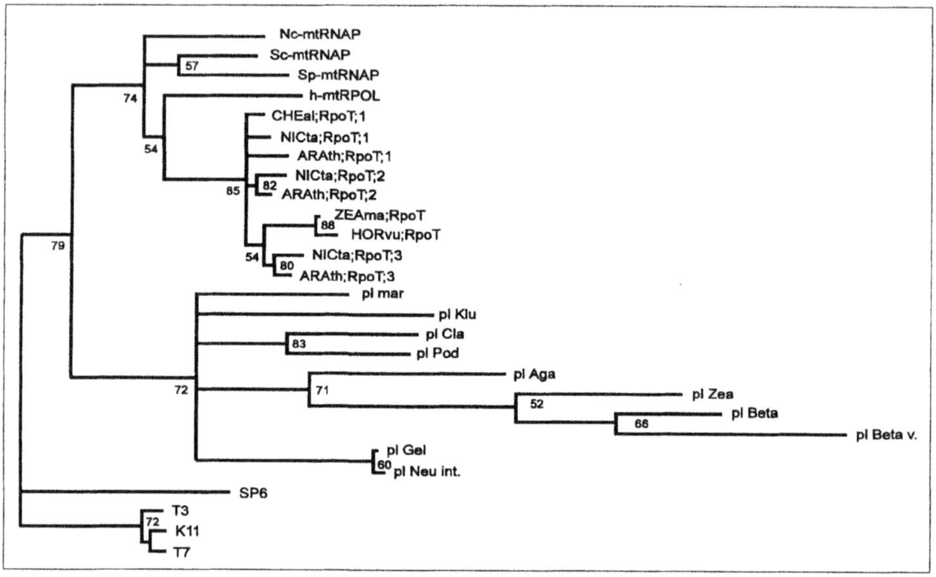

Figure 2. Phylogenetic tree of bacteriophage-type RNAPs based on maximum likelihood distances calculated with the PUZZLE programme. Branch support values are shown at the nodes. Distances are proportional to relative sequence divergence. Bacteriophage RNAPs: T7 [M38308], T3 [P07659], K11 [P18147], SP6 [P06221]. Fungal and human mitochondrial RNAPs: Nc-mtRNAP [P38671], *Neurospora crassa*, Sc-mtRNAP [P13433], *Saccharomyces cerevisiae*; Sp-mtRNAP [Z99126], *Schizosaccharomyces pombe*; h-mtRPOL [U75370], human mtRNAP. Plant polymerases: CHEal;RpoT [Y08067], *Chenopodium album*; NICta;RpoT;1, NICta;RpoT;2, NICta;Rpo;T3, *Nicotiana tabacum*; ARAth;RpoT;1 [Y08137], ARAth;RpoT;3 [Y08722], ARAth;RpoT;2 [AJ001037], *Arabidopsis thaliana*; ZEAma;RpoT [AJ005343], *Zea mays*; HORvu;RpoT, *Hordeum vulgare*. Plasmid-encoded mitochondrial RNAPs of unknown function from: pl mar [P33540], *Neurospora crassa*; pl Pic [Y11606], *Pichia kluyveri*; pl Cla [P22372], *Claviceps purpurea*; pl Pod [Q01521], *Podospora anserina*; pl Aga [P33539], *Agaricus bitorquis*; pl_Gel [S62751], *Gelasinospora* sp.; pl Neuin [P33541], *Neurospora intermedia*; pl Zea [P10581], *Zea mays*; pl Beta [Y10854], *Beta maritima*; pl Betav [Z34298]; *Beta vulgaris*. Database accession numbers are given in brackets.

According to our data, the nuclear-encoded RNA polymerase in the plastids of *A. thaliana* (the postulated NEP which has long been searched for) is most likely the product *of Rpo;T3* , i.e. an enzyme belonging to the family of single-subunit RNA polymerases. The presence of three different *RpoT* genes in *A. thaliana* raises the question whether other species also contain small families of phage-type RNAPs. Indeed, we have identified several *RpoT* genes in *Nicotiana tabacum* (sequencing in progress). At least one gene in addition to *ZEAma;Rpo;T1* (see above) has been found in maize (Stern et al., pers. commun.) and two genes were identified so far in wheat (Ikeda et al., pers. commun.).

We have performed phylogenetic analyses of the phage-type RNA polymerases (Fig. 2). To include all available data, only partial sequences corresponding to a conserved domain in exon 9 were used for tree reconstruction. From the topology of the tree, three well separated branches can be distinguished: 1) the RNAPs of phages, 2) single-subunit RNAPs of still unknown function encoded on mitochondrial plasmids of fungi and some plants [7], and 3) organellar RNA polymerases encoded by nuclear *RpoT* genes. The latter group comprises the mitochondrial RNA polymerases from three fungi (*S. cerevisiae, S. pombe, N. crassa*) and from *Homo sapiens*, and the plant RNAPs. Interestingly, the plant polymerases form, with a relatively high branch support value, two subgroups: one consisting of ARAth;Rpo;T1 and T2, CHEal;Rpo;T, and two of the tobacco polymerases NICta;Rpo;T1 and T2, while ZEAma;T and ARAth;Rpo;T3 cluster together with the barley sequence HORvu;Rpo;T and the third polymerase from *N. tabacum*, NICta;Rpo;T3. The overall topology of the tree was essentially the same, when the phylogenetic analyses were based on all 12 sequence blocks conserved between single-subunit RNAPs. For ARAth;Rpo;T3 plastid targeting has been experimentally demonstrated, and preliminary data suggest that also the putative transit peptide of ZEAam;RpoT confers targeting to plastids (D. Stern, pers. commun.). We, therefore, propose that all members of this clade represent plastid RNAPs, i.e. NEP enzymes. Accordingly, the other branch would consist of mitochondrial RNAPs, an assumption supported by experimental data for three of its members. Our data implicate that *Arabidopsis* is not the only species harboring several *RpoT* genes. The same situation is met in tobacco, where we have identified, in addition to the three genes included in Fig. 2, even two further copies of *RpoT* genes. This might be a consequence of the allotetraploid nature of *N. tabacum*, and therefore a sixth gene might be expected. Thus, our data suggest that the presence of two phage-type RNA polymerases in mitochondria, and of a phage-type as well as a eubacterial-type RNA polymerase in plastids, might be a general phenomenon among higher plants.

Acknowlegement

We are thankful to C. Stock for excellent technical assistance and to the Deutsche Forschungsgemeinschaft, Bonn, for financial support.

4. References

1. Bechtold, N., Ellis, J. and Pelletier, G. (1993) *In planta* Agrobacterium mediated gene transfer by infiltration of adult *Arabidopsis thaliana* plants, *Comt. Rend. Acad. Sci., Life Sci.* **316**, 1194-1199
2. Becker, D, Kemper, E., Schell, J. and Masterson, R. (1992) New plant binary vectors with selectable markers located proximal to the left T-DNA border, *Plant Mol. Biol.* **20**, 1195-1197

3. Chen, B., Kubelik, A.R., Mohr, S. and Breitenberger, C.A. (1996) Cloning and characterization of the *Neurospora crassa cyt-5* gene. A nuclear-coded mitochondrial RNA polymerase with a polyglutamine repeat, *J. Biol. Chem.* **271**, 6537-6544
4. Gray, M. W., Lang, B. F. (1998) Transcription in chloroplasts and mitochondria: a tale of two polymerases, *Trends Microbiol.* **6**, 1-3
5. Griffiths, A.J.F. (1995) Natural plasmids of filamentous fungi, *Microbiol. Rev.* **59**, 673-685
6. Hajdukiewicz, P.T.J., Allison, L.A. and Maliga, P. (1997) The two RNA polymerases encoded by the nuclear and plastid compartments transcribe distinct groups of genes in tobacco plastids, *EMBO J.* **13**, 4041-4048
7. Haseloff, J., Siemering, K. R., Prasher, D. C. and Hodge, S. (1997) Removal of a cryptic intron and subcellular localization of green fluorescent protein are required to mark transgenic Arabidopsis plants brightly, *Proc. Natl. Acad. Sci. USA* **94**, 2122-2127
8. Hedtke, B., Börner, T. and Weihe, A. (1997) Mitochondrial and chloroplast phage-type RNA polymerases in Arabidopsis, *Science* **277**, 809-811
9. Hedtke, B., Meixner, M., Gillamdt, S., Börner, T. and Weihe, A. (1998) Green fluorescent protein as a marker to investigate targeting of organellar RNA polymerases of higher plants *in vivo*, (submitted)
10. Hess, W. R. and Börner, T. (1998) Organellar RNA polymerases, *Int Rev. Cytol.* (in press)
11. Hübschmann, T., Hess, W.R. and Börner, T. (1998) Characterisation of transcript initiation sites in ribosome-deficient barley plastids, *Plant Mol. Biol.* **36**, 493-496
12. Kapoor, S., Suzuki, J.Y. and Sugiura, M. (1997) Identification and functional significance of a new class of non-consensus-type plastid promoters, *Plant J.* **11**, 327-337
13. Lerbs-Mache, S. (1993) The 110-kDa polypeptide of spinach plastid DNA-dependent RNA polymerase: Single-subunit enzyme or catalytic core of multimeric enzyme complexes?, *Proc. Natl. Acad. Sci. USA* **90**, 5509-513
14. Maliga, P. (1998) Two plastid RNA polymerases of higher plants: an evolving story, *Trends Plant Sci.* **3**, 4-6
15. Masters, B.S., Stohl, L.L. and Clayton, D.A. (1987) Yeast mitochondrial RNA polymerase is homologous to those encoded by bacteriophages T3 and T7, *Cell* **51**, 89-99
16. Sousa, R., Chung, Y.J., Rose, J.P. and Wang, B.C. (1993) Crystal structure of bacteriophage T7 RNA polymerase at 3.3 A resolution, *Nature* **364**, 593-599
17. Stern, D. B., Higgs, D. C. and Yang, J. (1997) Transcription and translation in chloroplasts, *Trends Plant Sci.* **2**, 308-315
18. Strimmer, K. and von Haeseler, A. (1997) Likelihood-mapping: a simple method to visualize phylogenetic content of a sequence alignment, *Proc. Natl. Acad. Sci. USA* **94**, 6815-6819
19. Thompson, J.D., Higgens, D.G. and Gibson, T.J. (1994) CLUSTAL W: improving the sensivity of progressive multiple sequence alignment through sequence weighting, positions-specific gap penalties and weight matrix choice, *Nucl. Acids Res.* **22**, 4673-4680
20. Tiranti, V., Savoia, A., Forti, F., D'Apolito, M.F., Centra, M., Rocchi, M. and Zeviani, M. (1997) Identification of the gene encoding the human mitochondrial RNA polymerase (h-mtRPOL) by cyberscreening of the Expressed Sequence Tags database, *Hum. Mol. Genet.* **6**, 615-625
21. Weihe, A., Hedtke, B. and Börner, T. (1997) Cloning and characterization of a cDNA encoding a bacteriophage-type RNA polymerase from the higher plant *Chenopodium album*, *Nucl. Acids Res.* **25**, 2319-2325
22. Young, D.A., Allen, R.L., Harvey, A.J. and Lonsdale, D. M. (1998) Characterization of a gene encoding a single-subunit bacteriophage-type RNA polymerase from maize which is alternatively spliced, *Mol. Gen. Genet.* (in press)

NOVEL *IN VITRO* TRANSCRIPTION ASSAY INDICATES THAT THE *ACCD* NEP PROMOTER IS CONTAINED IN A 19 BP FRAGMENT

K. LIERE AND P. MALIGA
Waksman Institute, Rutgers, The State University of New Jersey
190 Frelinghuysen Rd., Piscataway, New Jersey, 08854-8020

In order to characterize promoters for the nuclear-encoded plastid RNA polymerase (NEP), we developed a novel *in vitro* transcription system. The assay is based on extracts from plants lacking the plastid-encoded RNA polymerase (PEP) activity due to targeted deletion of the PEP α subunit gene. We report here the characterization of P*accD*-129, the promoter for *accD*, encoding a subunit of the acetyl-CoA carboxylase. Dissection of a ~0.8 kb fragment determined that the *accD* promoter is contained within a 19-bp fragment (-17/+2).

1. Introduction

The plastid genome contains functional *rpo* genes encoding the homologues of the eubacterial RNA polymerase α, β and β' subunits, which form the core of the plastid-encoded RNA polymerase [PEP; 10, 25]. A set of sigma-like factors interacts with PEP conferring promoter-specific binding and transcription specificity [11, 13, 26-28]. A second, plastid-localized, nuclear-encoded RNA polymerase (NEP) has been proposed by several researchers [8, 14, 17]. This has been recently established by showing transcription from novel promoters in plants lacking PEP due to targeted deletion of plastid *rpo* subunits [1, 6, 23]. A gene for the putative NEP catalytic subunit has been cloned, and its product was found to be similar to mitochondrial and phage RNA polymerases [7].

The identification of transcription initiation sites for NEP was possible in plants, which lack PEP, or in tissue-culture cells, both having proplastid-like non-photosynthetic plastids [6, 9, 12, 16, reviewed in 15]. These studies revealed that only few genes are transcribed exclusively from a NEP promoter: *accD,* encoding a subunit of the acetyl-CoA carboxylase in dicots [6]; *clpP* encoding the proteolytic subunit of the Clp ATP-dependent protease in monocots; and the *rpoB* operon in all higher plants [24]. The sequences flanking the *accD* transcription initiation site fit into the loose 10-nucleotide consensus for Type I NEP promoters [6].

Our long-term interest is to understand the mechanism by which NEP transcription rates are regulated. In this study we focused on identifying the *accD* promoter and its position relative to the transcription start site. Using our novel *in vitro* transcription system we demonstrate here that NEP, with respect to its promoter position, is indeed

J. H. Argyroudi-Akoyunoglou and H. Senger (eds.),
The Chloroplast: From Molecular Biology to Biotechnology, 79–84.

similar to phage-type polymerases. Our data indicate that the *accD* promoter is comprised within a 19-nucleotide DNA segment extending from -17 to +2 relative to the transcription initiation site (+1).

2. Results and Discussion

Definition of the *accD* promoter (P*accD*-129) was carried out in a novel *in vitro* transcription system utilizing a unique source of NEP activity, plants lacking the *rpoA* gene [23]. Since the extracts entirely lack PEP, a comprehensive characterization of NEP transcription activity could be accomplished without the biochemical separation of NEP and PEP activities.

In order to test transcripts initiating from the *accD*-129 NEP promoter *in vitro*, a promoter-terminator cassette was constructed consisting of a 374 bp fragment containing the *accD*-129 promoter (-268/+96) and a pair of transcription terminators. Termination at the histidine-attenuator [*hisa*; 2] is expected to add 145 nt. Termination at the more downstream located threonine-attenuator [*thra*; 4] would extend transcripts by additional 72 nt (Figure 1C). RNA polymerase activity was detected in run-off type experiments by incubation of DNA templates (the promoter-terminator cassettes) in transcriptionally active plastid extracts in the presence of labeled nucleotides [5, 19].

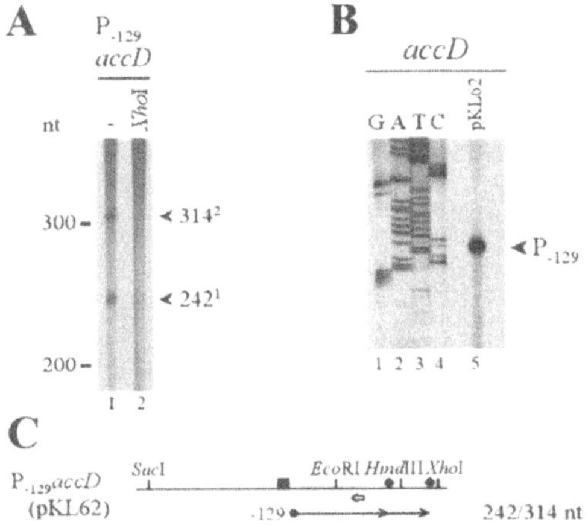

Figure 1. NEP transcription *in vitro* is dependent on supercoiled DNA template and initiates accurately. (**A**) *In vitro* transcription products from supercoiled and linear DNA templates. Plastid extracts of Δ*rpoA* plants were incubated with supercoiled (lane 1) and *Xho*I-linearized (lane 2) pKL62 in the presence of [α-^{32}P]UTP for 45 min at 30°C. Products were analyzed on 6% sequencing gels. Sizes of the reaction products are indicated, with 1 and 2 in superscript marking the position of products of the first and second termination event, respectively. (**B**) Primer extension analysis to map the 5'-end of *accD in vitro* transcripts. For reference, the same end-labeled primer was used to generate a DNA sequence ladder. (**C**) Map of promoter-terminator cassette. Site of transcription initiation for NEP (■) and corresponding RNA 5'-end (●) is marked. The site of transcription terminators is also denoted (◆). Arrows with solid lines symbolize NEP transcripts; the unfilled arrow marks the location of the *hisa*PE primer.

As shown in Figure 1A, two transcripts were obtained from supercoiled DNA templates as expected if the NEP initiated accurately from the *accD* promoter (lane 1). From linearized DNA templates only negligible amounts of transcripts could be detected (lane 2). Therefore, we conclude that NEP transcription activity from the *accD* promoter is depending on supercoiled DNA as a template *in vitro*. Primer extension analysis confirmed that transcription from P*accD*-129 NEP initiates accurately to the nucleotide mapped *in vivo* (Figure 1B; [6]). These data indicate that the *in vitro* transcription system contains all essential components required for accurate transcription initiation from the *accD* promoter.

The availability of a faithful *accD in vitro* transcription system allowed the testing of sequences required for *accD* promoter function. To define the *accD* promoter boundaries, a series of constructs containing promoter deletions was tested in *in vitro* transcription assays (Figure 2). Sequences upstream of the transcription initiation site have been deleted in seven steps between nucleotide -735 and -6.

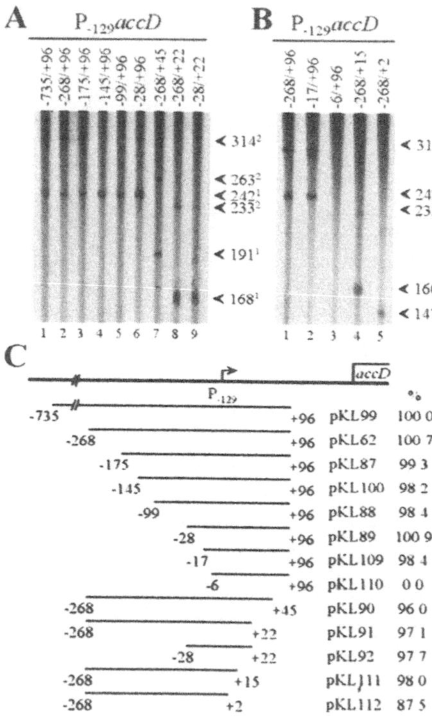

Figure 2. In vitro definition of the tobacco P*accD*-129 promoter. (**A, B**) *In vitro* transcription products separated on sequencing gels. Sizes of the reaction products are indicated, with 1 and 2 in superscript marking the position of products of the first and second termination event, respectively. (**C**) Map of the *accD* promoter region and its deletion derivatives. Numbers indicate deletion endpoints in nucleotides relative to transcription initiation site. Plasmid carrying the constructs and transcription activities (% of largest fragment) are also listed. The % values represent the average of at least three experiments and thus may be slightly different than reflected by the gel.

Deletions between -735 and -17 had no significant effect on the *in vitro* transcription activity whereas removal of sequences between -17 and -6 practically eliminated transcription activity. Sequences downstream of the transcription initiation site were deleted in three steps between position +45 and +2. Deletions down to position +2 had no significant effect. We conclude from these data, that the *accD* promoter is contained within the -17 to +2 fragment relative to the transcription initiation site. This suggests that up- or downstream elements are not required for regulation and efficient *in vitro* transcription from P*accD*-129. However, we cannot rule out the possibility that sequence elements may influence the *accD* transcription *in vivo* through interacting factor(s) that are not present in our plastid extracts.

Our data confirm the classification of P*accD*-129 as a Type I NEP promoter based on sequence conservation [6] and suggests an important roll of the conserved sequences in NEP promoter function. With most of the transcriptional important sequences upstream of the transcription initiation site, the P*accD*-129 Type I architecture resembles the structure of plant mitochondrial promoters [3, 20, 21]. The similar promoter architecture supports the postulated common origin of the nuclear-encoded phage-type transcription apparatus of plant mitochondria and chloroplasts [7].

3. Material and methods

3.1. EXTRACT PREPARATION

Δ*rpoA* plants [23] were grafted onto wild-type tobacco plants and grown in the greenhouse for 6 - 8 weeks prior to harvesting the leaves. Plastids were isolated following the protocol of Gruissem *et al.* [5] and Nickelsen and Link [18]. After lysis, plastid proteins were precipitated with solid ammonium sulfate to 55% saturation and stored at -80°C after quick-freezing in liquid nitrogen.

3.2. *IN VITRO* TRANSCRIPTION ASSAYS

NEP *in vitro* transcription assays were carried out for 45 min at 30°C in 20 µl reaction mixture containing 40 mM Tris/HCl pH 7.9, 10 mM KCl, 6 mM $MgCl_2$, 5 mM DTT, 2 mM spermidine, 500 µM each ATP, GTP, CTP, 25 µM unlabelled UTP, and 40 µCi of [α-^{32}P]UTP (6000 Ci/mmol), 30 U RNAguard (Pharmacia), 0.5 µg template DNA and 8 µl plastid extract. The reaction products were Phenol/Chloroform extracted, isopropanol precipitated, and analyzed on 6% sequencing gels. For primer extension analysis, *in vitro* transcription assays were carried out in the presence of 500 µM each ATP, GTP, CTP, and UTP.

3.3. PRIMER EXTENSION ANALYSIS

Primer extension reactions were carried out with transcripts derived by NEP *in vitro* transcription according to standard protocols [22] using primer *hisa*PE (#844; CAGCAGACATCGCCTGAAAGACTAGTCAGG), which anneals to the *hisa* region between +103 and +74 referring to the native transcription start site [2].

4. References

1. Allison L.A., Simon L.D., Maliga P. (1996) Deletion of *rpoB* reveals a second distinct transcription system in plastids of higher plants, *EMBO J.* **15**, 2802-2809.
2. Barnes W.M., Tuley E. (1983) DNA sequence changes of mutations in the histidine operon control region that decrease attenuation, *J. Mol. Biol.* **165**, 443-59.
3. Caoile A.G.F.S., Stern D.B. (1997) A conserved core element is functionally important for maize mitochondrial promoter activity *in vitro*, *Nucleic Acids Res.* **25**, 4055-4060.
4. Gardner J.F. (1982) Initiation, pausing, and termination of transcription in the threonine operon regulatory region of Escherichia coli, *J. Biol. Chem.* **257**, 3896-904.
5. Gruissem W., Greenberg B.M., Zurawski G., Hallick R.B. (1986) Chloroplast gene expression and promoter identification in chloroplast extracts, *Meth. Enzymol.* **118**, 253-70.
6. Hajdukiewicz P.T.J., Allison L.A., Maliga P. (1997) The two RNA polymerases encoded by the nuclear and the plastid compartments transcribe distinct groups of genes in tobacco plastids, *EMBO J.* **16**, 4041-4048.
7. Hedtke B., Börner T., Weihe A. (1997) Mitochondrial and chloroplast phage-type RNA polymerases in *Arabidopsis*, *Science* **277**, 809-811.
8. Hess W.R., Prombona A., Fieder B., Subramanian A.R., Börner T. (1993) Chloroplast *rps15* and the *rpoB/C1/C2* gene cluster are strongly transcribed in ribosome-deficient plastids: evidence for a functioning non-chloroplast-encoded RNA polymerase, *EMBO J.* **12**, 563-71.
9. Hübschmann T., Börner T. (1998) Characterization of transcript initiation sites in ribosome-deficient barley plastids, *Plant Mol. Biol.* **36**, 493-496.
10. Igloi G.L., Kössel H. (1992) The transcriptional aparatus of chloroplasts, *Crit. Rev. Plant Sci.* **10**, 525-558.
11. Isono K., Niwa Y., Satoh K., Kobayashi H. (1997) Evidence for transcriptional regulation of plastid photosynthesis genes in *Arabidopsis thaliana* roots, *Plant Physiol.* **114**, 623-630.
12. Kapoor S., Suzuki J.Y., Sugiura M. (1997) Identification and functional significance of a new class of non-consensus-type plastid promoters, *Plant J.* **11**, 327-37.
13. Kestermann M., Neukirchen S., Kloppstech K., Link G. (1998) Sequence and expression characteristics of a nuclear-encoded chloroplast sigma factor from mustard (*Sinapis alba*), *Nucleic Acids Res.* **26**, 2747-2753.
14. Lerbs-Mache S. (1993) The 110-kDa polypeptide of spinach plastid DNA-dependent RNA polymerase: single-subunit enzyme or catalytic core of multimeric enzyme complexes?, *Proc. Natl. Acad. Sci. USA* **90**, 5509-13.
15. Maliga P. (1998) Two plastid polymerases of higher plants: an evolving story, *Trends Plant Sci.* **3**, 4-6.
16. Miyagi T., Kapoor M., Sugita M., Sugiura M. (1998) Transcript analysis of the tobacco plastid operon *rps2/atpI/H/F/A* reveals the existance of a non-consensus type II (NCII) promoter upstream of the *atpI* coding sequence, *Mol. Gen. Genet.* **257**, 299-307.
17. Morden C.W., Wolfe K.H., dePamphilis C.W., Palmer J.D. (1991) Plastid translation and transcription genes in a non-photosynthetic plant: intact, missing and pseudo genes, *EMBO J.* **10**, 3281-8.
18. Nickelsen J., Link G. (1993) The 54 kDa RNA-binding protein from mustard chloroplasts mediates endonucleolytic transcript 3' end formation in vitro, *Plant J.* **3**, 537-44.
19. Orozco E.M., Jr., Mullet J.E., Hanley-Bowdoin L., Chua N.H. (1986) *In vitro* transcription of chloroplast protein genes, *Meth. Enzymol.* **118**, 232-53.
20. Rapp W.D., Lupold D.S., Mack S., Stern D.B. (1993) Architecture of the maize mitochondrial *atp1* promoter as determined by linker-scanning and point mutagenesis, *Mol. Cell. Biol.* **13**, 7232-7238.
21. Rapp W.D., Stern D.B. (1992) A conserved 11 nucleotide sequence contains an essential promoter element of the maize mitochondrial *atp1* gene, *EMBO J.* **11**, 1065-1073.
22. Sambrook J., Fritsch E.F., Maniatis T. (1989) *Molecular Cloning. A Laboratory Manual.*, Cold Spring Harbor Laboratory Press, Cold Spring Harbor, NY.
23. Serino G., Maliga P. (1998) RNA polymerase subunits encoded by the plastid *rpo* genes are not shared with the nucleus-encoded plastid enzyme, *Plant Physiol.* **117**, 1165-1170.
24. Silhavy D., Maliga P. (1998) Mapping of the promoters for the nucleus-encoded plastid RNA polymerase (NEP) in the *iojap* maize mutant, *Curr. Genet.* **33**, 340-344.
25. Sugiura M. (1992) The Chloroplast Genome, *Plant Mol. Bio.* **19**, 149-168.
26. Tanaka K., Kosuke O., Ohata N., Kuroiwa H., Kuroiwa T., Takahshi H. (1996) Nuclear encoding of a chloroplast RNA polymerase sigma subunit in a red alga, *Science* **272**, 1932-1935.

27. Tanaka K., Tozawa Y., Mochizuki N., Shinozaki K., Nagatani A., Wakasa K., Takahashi H. (1997) Characterization of three cDNA species encoding plastid RNA polymerase sigma factors in *Arabidopsis thaliana* : evidence for the sigma factor heterogeneity in higher plant plastids, *FEBS Letters* **413**, 309-313.
28. Tozawa Y., Tanaka K., Takahashi H., Wakasa K. (1998) Nuclear encoding of a plastid sigma factor in rice and its tissue- and light-dependent expression, *Nucleic Acids Res* **26**, 415-419.

POLYADENYLATION AND DEGRADATION OF mRNA IN THE CHLOROPLAST

I. LISITSKY and G. SCHUSTER
Department of Biology, Technion - Israel Institute of Technology, Haifa 32000, Israel

1. Introduction

In the chloroplasts of higher plant and green algae mRNAs are transcribed as precursor RNAs which undergo a variety of maturation events, including *cis-* and *trans*-splicing, cleavage of polycistronic messages, processing of 5' and 3' ends, and RNA editing [1-4]. These post-transcriptional processes play an important role in regulating gene expression and in the chloroplast, they may be controlled in part by nuclear encoded-genes. For example, mutants of nuclear-encoded genes which affect the accumulation of specific chloroplast transcripts have been characterized in maize [5], *Arabidopsis* [6] and the green alga *Chlamydomonas reinhardtii,* reviewed in [4, 7]. A general characteristic of the plastid protein-coding region is the presence of inverted-repeat sequences in the 3' untranslated region (UTR), which form a stem-loop structure when transcribed to RNA. The 3' end of mature chloroplast mRNAs is located several nucleotides 3' to this stem-loop structure. In contrast to similar structures found in bacterial mRNA, in chloroplasts these elements do not function as transcriptional terminators [8-10]. Instead, they serve as RNA-processing elements *in vitro* and are capable of stabilizing upstream RNA fragments *in vivo* and *in vitro* [8, 9, 11-14]. Here we focus on recent discoveries concerning the molecular mechanism of mRNA degradation in the chloroplast and the proteins involved.

2. Polyadenylation of mRNA in the chloroplast

Post-transcriptional addition of poly(A) tail to the 3' end of mRNA was first identified and characterized in eukaryotic cells for viral and nuclear-encoded mRNAs [15, 16]. The long poly(A) tail of eukaryotic nuclear-encoded mRNAs is an important determinant of their maturation and of the initiation of translation. In addition, deadenylation may be a major step in the degradation pathway of nuclear-encoded mRNAs.

J. H. Argyroudi-Akoyunoglou and H. Senger (eds.),
The Chloroplast: From Molecular Biology to Biotechnology, 85–90.

Poly(A) tails have also been described in bacteria for several mRNA 3' ends [17]. In the nucleus and cytoplasm of eukaryotic cells, deadenylation of the long poly(A) tail is part of the mRNA degradation pathway. In bacteria, the addition of poly(A) tails to several mRNAs promotes their degradation.

Polyadenylation has commonly been regarded as a feature of eukaryotic nuclear and viral mRNAs. Nevertheless, polyadenylation of prokaryotic and organellar mRNAs has recently returned to the focus of research as part of the mRNA degradation mechanism. It is interesting to note that poly(A) RNA was detected in the chloroplast by Bogorad and others more than 20 years ago [18, 19].

The post-transcriptional addition of poly(A)-rich sequences to the plastid-encoded *psbA* mRNA, as well as to several other transcripts, has recently been described using RT-PCR of oligo-dT-primed RNA of spinach chloroplasts [20, 21]. Unlike in eukaryotic nuclear-encoded and bacterial RNAs, the poly(A) moiety in the chloroplast was not found to be a ribohomopolymer of adenosine residues but rather clusters of adenosines bound mostly by guanosines and on rare occasion by cytidines and uridines. The tails could be several hundred nucleotides long [20]. Most of the polyadenylation sites that were found by RT-PCR of oligo-dT-primed chloroplast RNA were localized within the amino-acid coding region or the 3' UTR of the *psbA* mRNA. This result indicated that mostly truncated mRNAs are polyadenylated, while the truncation may originate from either early transcription termination or by cleavage of full-length transcript. Our results suggested that *in vivo* polyadenylation occurs subsequent to the cleavage of an mRNA as part of the specific degradation pathway [20].

3. Polyadenylation is required for the degradation of chloroplast mRNA

The question of whether polyadenylation is required for mRNA degradation or whether other decay pathways also exist for chloroplast mRNAs was approached via studies using the lysed-chloroplast system and the polyadenylation inhibitor 3'-dATP (cordycepin triphosphate). Blocking the polyadenylation of RNA inhibits RNA degradation and has similar, if not identical, effects as direct blocking of the exonucleases by the addition of excess yeast-tRNA [22]. In both treatments, the full-length mRNA was endonucleolytically cleaved to distinct degradation products that accumulated instead of being exonucleolytically degraded. Therefore, the addition of poly(A)-rich sequences to endonucleolytic cleavage products of mRNA is required to target these molecules for rapid exonucleolytic degradation in the chloroplast.

4. *In vitro* polyadenylation of chloroplast mRNAs

An *in vitro* polyadenylation system was used to elucidate the biochemistry of mRNA polyadenylation and degradation activities, to isolate the proteins involved, and to reconstitute their activities. *In vitro* transcribed RNAs corresponding to the chloroplast RNAs could be polyadenylated at their 3' end using a soluble chloroplast protein extract complemented by the addition of ATP [20]. *In vitro* analysis of chloroplast polyadenylation activity revealed specificity to ATP and GTP, reflecting the composition of the poly(A)-rich tails *in vivo*. Furthermore, the *in vitro* polyadenylation activity is dependent on the substrate structure. 3' end unstructured RNAs were polyadenylated with high efficiency as compared to those molecules forming the stem-loop structure characteristic of the mature plastid mRNA [20].

In vitro transcribed synthetic polyadenylated RNA was rapidly degraded as compared to the same non-polyadenylated RNA when incubated in a soluble chloroplast protein extract. Competition experiments revealed that polyadenylated RNA molecules are more efficient competitors for the degradation machinery than their non-polyadenylated counterparts [23]. These results suggested that poly(A)-rich tails play a major role in the rapid degradation of intermediate products of mRNA decay in the chloroplast, by targeting the cleavage products for rapid degradation due to their high affinity to chloroplast exonuclease(s).

The chloroplast exoribonuclese 100RNP/PNPase, like it's bacterial homologue PNPase, is a processive exoribonuclease that binds to the 3' end and digests the RNA nucleotide by nucleotide, without dissociating from the molecule [12]. In competition experiments using isolated, purified 100RNP/PNPase, the polyadenylated RNA competed with the non-polyadenylated RNA for the exonuclease for the soluble protein extract [23]. Binding assays of the 100RNP/PNPase to poly(A) as well as to other RNA molecules displayed higher affinity of this protein to poly(A) than to other RNA molecules. These results suggested that the preferential degradation of polyadenylated RNA in the chloroplast is based on the exoribonuclease 100RNP/PNPase's high binding affinity to poly(A) sequence. However, the possibility that another, as yet unidentified chloroplast exoribonuclease(s) also binds polyadenylated RNA with higher affinity than non-polyadenylated RNA can not be ruled out.

5. The molecular mechanism of mRNA degradation in the chloroplast

Our recent model of the mRNA degradation pathway in the chloroplast is presented in Figure 1. The initial event is endonucleolytic cleavage(s) that produces RNA molecules with no stem-loop at the 3' end (B in Figure 1). RNAs that terminate with a stem-loop structure were poorly polyadenylated *in vitro*, and RT-PCR clones of poly(A)-rich sequences at the end of the mRNA molecule (characterized by a stem-loop structure) were obtained with much lower frequency than those having the additional site inside the coding region. Therefore, we suggest that the stem-loop structures characterizing most of the chloroplast mRNAs and shown to be effective 3' end processing signals also

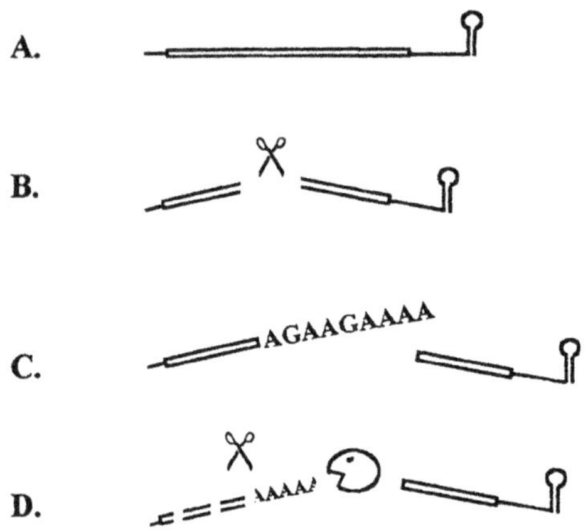

Fig. 1. A model for the degradation pathway of mRNA in the chloroplast.
A schematic representation of the *psbA* mRNA molecule. The open box represents the amino-acid coding region, and stem-loop structures represent inverted repeats in the 3' untranslated region which potentially form stem-loop structure. **B**. The initial step in the mRNA degradation process is suggested to be endonucleolytic cleavage by an as yet unidentified endonuclease(s). The endonuclease is schematically symbolized by scissors. **C**. A poly(A)-rich tail, which can be up to several hundred nucleotides in length, is then added to the 3' end of the 5' endonucleolytic cleavage product. **D**. The polyadenylated RNA molecule is rapidly degraded by an exonuclease(s), possibly the 100RNP/PNPase.

serve as poor polyadenylation sites in order to prevent exonucleolytic degradation of the functional molecule. Following the endonucleolytic cleavage(s), the proximal fragments are polyadenylated by the addition of poly(A)-rich sequences (C in Figure 1). This stage is inhibited by 3'-dATP (cordycepin) and is required for the continuation of mRNA degradation in the chloroplast, because only following polyadenylation the RNAs are rapidly digested by exonuclease(s) due to the higher affinity of this enzyme(s) to the poly(A)-rich sequence (D in Figure 1).

The model suggests two mechanisms to modulate the half-life of a particular RNA molecule. The first implies that once the mRNA molecule is endonucleolytically cleaved, it is targeted to the degradation process. In this model, the rate-limiting step is the initial endonucleolytic cleavage; once this has occurred, cleaved mRNA will be rapidly polyadenylated and exonucleolytically degraded. In such a mechanism, the nature, specificity and modulation of activity and/or expression of the endonuclease(s) determine the half-life of a particular mRNA molecule, and the activities of the poly(A) polymerase and exonuclease do not figure the rate limiting step. The second model suggests that other steps in the degradation pathway could limit the degradation rate. We do not yet know if the length of the poly(A) tail, which can amount to several hundred nucleotides [20], influences the rate of degradation. Furthermore, the guanosine residues that are characteristic of the poly(A) tails of chloroplast *psbA* mRNA may be involved in modulating the activity of the exonucleolytic enzyme. On the other hand, they may simply reflect the specificity of the poly(A) polymerase(s) [18, 20].

So far, our model does not explain how the distal endonucleolytic cleavage product is degraded. One possibility is that many endonucleolytic cleavages along the mRNA occur until the small RNA fragment representing the 3' UTR is degraded by the polyadenylation dependent pathway, possibly by demolishing the stem-loop structure, thereby enabling polyadenylation. A low degree of polyadenylation of the mature 3' end has already been obtained *in vivo* and *in vitro* [20]. Moreover, an additional exonucleolytic degradation pathway in the chloroplast that is not dependent on polyadenylation may exist. Whether or not there is an additional degradation pathway, the results of the experiments using the polyadenylation inhibitor 3'-dATP (cordycepin) indicated that the *rbcL* and *psbA* mRNAs are exonucleolytically degraded only in the polyadenylation dependent degradation pathway. To understand the role of this degradation pathway in the developmental regulation of plastid mRNA stability, the degradation pathway and the necessity for polyadenylation therein in etioplasts and root amyloplasts has to be determined. In these developmental stages, the *psbA* and *rbcL* RNAs have short half-lives and are rapidly degraded.

***

In the last few years, our understanding of the chloroplast mRNA degradation pathway has already progressed. For certain mRNAs, such as *psbA,* the different steps of its specific decay have been disclosed. The mRNA degradation events are similar to those described for bacteria. This is supported by the observation that one of the enzymes involved (100RNP/PNPase) has been identified as sharing structural and functional homology with its prokaryotic counterpart. Gene expression in the chloroplast, however, is in part regulated by the nucleus, as well as by external stimuli such as light. Moreover, the longevity of its mRNA better reflects the properties of plants than of bacteria. Therefore, the chloroplast probably adopted the bacteria features of the mRNA degradation mechanism and developed additional novel characteristics. Following the biochemical pathway of plastid mRNA degradation, research will continue on the regulation of mRNA stability as part of the regulatory network which determines leaf development and adaptation to environmental conditions.

Acknowledgement

This work was supported by the Israel Science Foundation, administered by the Israel Academy of Science and Humanities, by a grant from the Niedersachsen Foundation for Science and Art and by a grant from the Israel-Japan Scientific Cooperation Research Foundation.

References

1. Gruissem, W., Schuster, G. (1993) Control of mRNA stability and degradation in organelles. In G. Brawerman, J. Belasco (eds) *Control of mRNA degradation in organelles*, pp. 329-365. Orlando, FL: Academic Press.
2. Schuster, G., Lisitsky, I., Klaff, P. (1997) Polyadenylation and degradation of mRNA in the chloroplast. In K. Bansal (ed.) *Recent Advances in Plant Molecular Biology* Oxford and IBH Publishers, New Delhi. (in press)
3. Drager, R. G., Stern, D. B. (1998) Chloroplast RNA Synthesis and Processing. In *Molecular Biology of Chlamydomonas: Chloroplasts and Mitochondria* J.-D. Rochaix, Goldschmidt-Clermont, M. and Merchant, S. (eds.) Kluwer Academic Publishers, Dordrecht, (in press).
4. Rochaix, J.-D. (1996) Post-transcriptional regulation of chloroplast gene expression in *Chlamydomonas reinhardtii*. *Plant Mol. Biol.* **32**, 327-341.
5. Barkan, A., Walker, M., Nolasco, M., Johnson, D. (1994) A nuclear mutation in maize blocks the processing and translation of several chloroplast mRNAs and provides evidence for the differential translation of alternative mRNA forms. *EMBO J* **13**, 3170-3181.
6. Meurer, J., Berger, A., Westhoff, P. (1996) A nuclear mutant of Arabidopsis with impaired stability on distinct transcripts of the plastid *psbB*, *psbD/C*, *ndhH*, and *ndhC* operons. *Plant Cell* **8**, 1193-1207.
7. Rochaix, J.-D. (1992) Post-transcriptional steps in the expression of chloroplast genes. *Annu. Rev. Cell Biol.* **8**, 1-28.
8. Stern, D. B., Gruissem, W. (1987) Control of plastid gene expression: 3' inverted repeats act as mRNA processing and stabilizing elements, but do not terminate transcription. *Cell* **51**, 1145-1157.
9. Stern, D. B., Radwanski, E. R., Kindle, K. L. (1991) A 3' stem/loop structure of the *Chlamydomonas* chloroplast *atpB* gene regulates mRNA accumulation *in vivo*. *Plant Cell* **3**, 285-297.
10. Rott, R., Drager, R. G., Stern, D. B., Schuster, G. (1996) The 3' untranslated regions of chloroplast genes in *Chlamydomonas reinhardtii* do not serve as efficient transcriptional terminators. *Mol. Gen. Genet.* **252**, 676-683.
11. Rott, R., Liveanu, V., Drager, R. G., Stern, D. B., Schuster, G. (1998) The sequence and structure of the 3' untranlated regions of chloroplast transcripts are important determinants of mRNA accumulation and stability. *Plant. Mol. Biol.* **36**, 307-314.
12. Hayes, R., Kudla, J., Schuster, G., Gabay, L., Maliga, P., Gruissem, W. (1996) Chloroplast mRNA 3'-end processing by a high molecular weight protein complex is regulated by nuclear encoded RNA binding proteins. *EMBO J.* **15**, 1132-1141.
13. Schuster, G., Gruissem, W. (1991) Chloroplast mRNA 3' end processing requires a nuclear-encoded RNA-binding protein. *EMBO J.* **10**, 1493-1502.
14. Blowers, A. D., Klein, U., Ellmore, G. S., Bogorad, L. (1993) Functional *in vivo* analyses of the 3' flanking sequences of the *Chlamydomonas* chloroplast *rbcL* and *psaB* genes. *Mol. Gen. Genet.* **238**, 339-349.
15. Jackson, R. J., Standart, N. (1990) Do the poly(A) tail and 3' untranslated region control mRNA translation? *Cell* **62**, 15-24.
16. Sachs, A. B., Sarnow, P., Hentze, M. W. (1997) Starting at the beginning, middle, and end: translation initiation in eukaryotes. *Cell* **89**, 831-838.
17. Sarkar, N. (1997) Polyadenylation of mRNA in prokaryotes. *Annu. Rev. Biochem.* **66**, 173-197.
18. Burkard, G., Keller, E. B. (1974) Poly(A) polimerase and poly(G) polymerase in wheat chloroplasts. *Proc. Natl. Acad. Sci. USA* **71**, 389-393.
19. Haff, L. A., Bogorad, L. (1976) Poly(adenylic acid) -containing RNA from plastids of maize. *Biochemistry* **15**, 4110-4115.
20. Lisitsky, I., Klaff, P., Schuster, G. (1996) Addition of poly(A)-rich sequences to endonucleolytic cleavage sites in the degradation of spinach chloroplast mRNA. *Proc. Natl. Acad. Sci. USA* **93**, 13398-13403.
21. Kudla, J., Hayes, R., Gruissem, W. (1996) Polyadenylation accelerates degradation of chloroplast mRNA. *EMBO J.* **15**, 7137-7146.
22. Lisitsky, I., Klaff, P., Schuster, G. 1997. Blocking polyadenylation of mRNA in the chloroplast inhibits its degradation. *Plant J.* **12**, 1173-1178.
23. Lisitsky, I., Kotler, A., Schuster, G. (1997) The mechanism of preferential degradation of polyadenylated RNA in the chloroplast: the exoribonuclease 100RNP/PNPase displays high binding affinity for poly(A) sequence. *J. Biol. Chem.* **272**, 17648-17653.

DNA-BINDING PROTEINS OF CHLOROPLAST NUCLEOIDS

N.P. YURINA, Y.P. OLESKINA, S.M. MELNIK,
M.S. ODINTSOVA
A.N. Bach Institute of Biochemistry RAS, Moscow 117071,
Russia

1. Introduction

Chloroplasts are the major sites of energy conversion in plant cell. These specialized organelles contain their own genetic system, consisting of multiple copies of circular double-stranded DNA and the machinery necessary for its replication, transcription and translation. In each organelle the plastid DNA is associated with proteins [1-3] and packaged into discrete, highly compact structures termed nucleoids, which rarely contain only one DNA molecule and can contain as many as 100 copies of the plastid genome. However, the arrangement of individual organellar DNA molecules within a nucleoid structure remains uncertain. The form of DNA packaging in plastid nucleoids and/or proteins in plastid nucleoids may be involved in transcriptional control of plastid gene expression. Since it is well known that, in addition to the structural role of the basic proteins and, namely, DNA-binding proteins, these proteins participate in the initiation of DNA replication and transcription, gene expression, cell division, etc., the analysis of basic proteins of chloroplast nucleoids is of great interest.

For investigation of proteins of chloroplast nucleoids we used two approaches: fixation of DNA-protein contacts in organelles and isolation and characterization of individual proteins of chloroplast nucleoids.

J. H. Argyroudi-Akoyunoglou and H. Senger (eds.),
The Chloroplast: From Molecular Biology to Biotechnology, 91–96.

2. Materials and Methods

Chloroplasts were isolated from 10-12-day-old pea seedlings (*Pisum sativum* L., cv. Izumrud) by differential centrifugation (3000g, 15 min), washed twice with homogenization buffer, and purified by centrifugation in a sucrose density gradient.

To identify proteins bound to chloroplast DNA (cpDNA) we fixed of DNA-protein contacts directly in chloroplasts by their chemical covalent crosslinking and measuring the molecular masses of crosslinked proteins. Experimental procedure for covalent DNA-protein crosslinking by the radical-producing system is described in detail [4].

DNA-protein crosslinked complexes were fractionated by diagonal two-dimentional gel retardation electrophoresis [5]. After electrophoresis in the first direction, DNA was released from protein by their digestion directly in a gel by pronase, and electrophoresed in the second direction. DNA was hybridized to pea chloroplast DNA probes (1,1 kbp 3'psbD and 5'psbC genes; 1,8 kbp, containing 5'atpB and 5' rbcL regions; 5,7 kbp, containing 5'-psbD region, trnE, trnY, trnD genes and 5'-psaA region; 0,7 kbp, containing a part of 23S rRNA gene).

The procedure to identify of crosslinked proteins included an exhaustive hydrolysis of crosslinked DNA in the DNA-protein complexes with micrococcal nuclease. Micrococcal nuclease digestion of the covalent adducts leaves a short nucleotide tail which was tagged by ^{32}P. Individual labeled proteins were fractionated by SDS-PAGE and autoradiographed.

The isolation of nucleoids from pea chloroplasts was performed according to the modified procedure of Briat et al. [6] . Briefly, purified chloroplasts were lysed with 1% Triton X-100, and the lysate obtained was subjected to gel-filtration on a Toyopearl HW-65 column in 0.05 M Tris-HCl, pH7.6, 0.04 M 2-mercaptoethanol, 25% glycerol, 4 mM EDTA, 1% Triton X-100 and 0.1 M $(NH_4)_2SO_4$.

Nucleoids were eluted in a void volume and well separated from ribosomes, chlorophyll-containing protein peak and chlorophyll.

Basic proteins were extracted from isolated nucleoids with 0,4 M H_2SO_4 and precipitated with acidified acetone.

For RP-HPLC chromatography the basic proteins of nucleoids were dissolved in 150 μl of 0.2 M Tris-HCl, pH 7.8, containing 5 M guanidine-HCl, 1 mM EDTA, and 0.1% trifluoroacetic acid (TFA) and applied to an Aquapore RP300 column (4 ,6 x 100 mm) (Applied Biosystems). Proteins were eluted by a linear gradient of solvent A (0,1% TFA in water) and solvent B (0,1% TFA in acetonitrile) at room temperature. The flow rate was 0.5 ml/min. Proteins were detected at 254 and 214 nm .

The purified nucleoid proteins (about 20 μg) were digested with trypsin (Sigma) at an enzyme/substrate ratio of 1 : 50 in 50 μl of 50mM Tris-HCl, pH 8.0 at 37 ^{o} C for 24 h. The reaction was terminated by the addition of TFA to the final concentration of 0.1%.

Tryptic peptides were fractionated on an Aquapore RP300 column (4.6 x 220 mm) (Applied Biosystems) with a linear acetonitrile gradient from 0%B to 60%B for 90 min at 40 ^{o} C. Solvents A and B were the same as for protein separation. The peptides were detected at 214 nm, and lyophilized using a SpeedVac concentrator (Savant).

The sequencing of protein and tryptic peptides was performed with a Procise protein sequencer (Perkin-Elmer/Applied Biosystems). The SwissProt data base was used for sequence comparison .

3. Results and Discussion

The method of DNA-protein crosslinking by the radical-producing system allowed us to investigate native structure of DNA-protein complexes in chloroplasts, determine and compare the protein content of chloroplast nucleoids in different conditions.

We identified proteins which are crosslinked to DNA using identification of polypeptides crosslinked to DNA by labeling mononucleotides covalently linked to protein after trimming of DNA-protein complexes by micrococcal nuclease. The identified labeled proteins after SDS-PAGE were compared with the molecular mass markers.

We observed five bands, corresponding to five crosslinked proteins with molecular masses from 14 kDa to 31 kDa, respectively (noncrosslinked chloroplasts were the control in experiments). Hybridization patterns of two-dimentional gels do not depend on the chosen probes of cpDNA, which means that these proteins interact with all investigated genome regions of chloroplast. This data suggest that revealed proteins take part in packing of cpDNA and are structural proteins. Apparently the cpDNA is organized with these proteins as a regular structure over whole chloroplast genome or at least over the regions studied.

Earlier by SDS-PAGE analysis the fraction of basic proteins of pea chloroplast nucleoids was separated into seven components with molecular masses from 17 to 37 kDa [3]. However, RP-HPLC separation of basic nucleoid proteins from pea chloroplasts yields 25 peaks with four major ones (fractions No. 11-14) (Fig. 1A). SDS-PAGE analysis of these protein fractions showed that they contain proteins

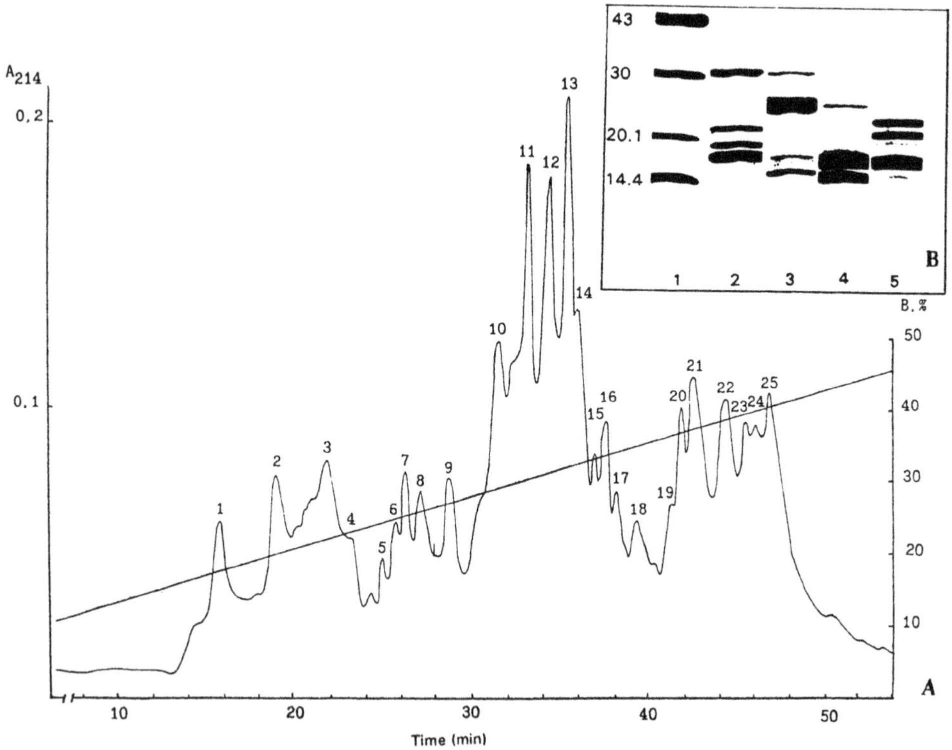

Fig. 1. Separation of nucleoid basic proteins from pea chloroplasts.

A. RP-HPLC on an Aquapore RP300 column (4.6 x 100 mm).

B. SDS-PAGE of RP-HPLC-fractions: 1 - markers; 2-5 - fractions no. 11-14.

with molecular masses ranging from 17 to 30 kDa (17, 18, 19, 22, 25, 28, and 30 kDa), namely, similar proteins as we discovered earlier. For further analysis the fraction No.12 was chosen as more pure one. SDS-PAGE of fraction No.12 on an overloaded gel revealed a major 28 kDa protein band and seven additional minor polypeptides. The amount of prevailing minor polypeptides does not exceed 10% from the major band after Coomassie Brilliant Blue R-250 staining (Fig.1B).

The major component of fraction No.12 was called P28. To identify P28, the fraction No. 12 was first subjected to N-terminal sequencing . However, no amino acids were detected, this indicated that the N-terminal amino acid in P28 was blocked. Therefore, either P28 and minor proteins revealed by electrophoresis also have blocked N-terminal amino acids, that seems unlikely, or their relative amount is low enough not to be detected by N-terminal sequencing.

After digestion of fraction No.12 with trypsin and RP-HPLC-separation of tryptic peptides, 21 fractions were obtained . The sequencing of three peptides (fractions No. 6, 17, and 19) gave the following sequences:

peptide T6 : NGTIVK,

peptide T17 : VWLDPNETSEIAQAN,

peptide T19 : LPSQVVW.

A single sequence was obtained for each peptide. The search in SwissProt data base showed that the primary structure of the isolated tryptic peptides completely coincides with the primary structure of the ribosomal protein L19 from Saccharomyces cerevisiae. Since all the sequences of three peptides studied (one of which consists of 15 amino acids) fall within the structure of one and the same protein, we conclude that the minor components in the fraction No.12 did not interfere with the results of sequencing. Summing up the above data it can be concluded that the protein of chloroplast nucleoids homologous to ribosomal protein L19 has been identified.

It should be noted that the procedure for isolation and purification of chloroplasts, and subsequent isolation of nucleoids completely excludes contamination of this fraction with cytoplasmic ribosomes. This is confirmed by our previous data. We did not find cytoplasmic ribosomes in our nucleoid preparations using sedimentation analysis in sucrose and CsCl density gradients [3]. The fact that P28 has no homology to the chloroplast encoded ribosomal proteins means that it is nuclear-encoded.

In chloroplast nucleoids the ribosomal protein was first detected in this research. Earlier it was shown that the histone-like protein H from E. coli nucleoids is a ribosomal protein S3. It was also reported that three more ribosomal proteins L2,S4, and S5 together with S3 are associated with bacterial DNA and involved in formation of E. coli nucleoid structure [6]. The association of several ribosomal proteins with DNA is not surprising due to their basic character and affinity to rRNA. The functional role of ribosomal proteins in cells is not completely elucidated. Abundant data indicate that these proteins may have non-ribosomal functions, namely, they may participate in DNA reparation, chromosome rearrangements, plastid differentiation, etc.

We suppose that the nucleoid ribosomal protein P28 studied in this work may participate in differentiation of proplastids into chloroplasts. However, the elucidation of its functions needs further investigation.

4. References

1. Briat, J.F., Letoffe, S., Mache, R. and Rouviere-Yaniv, J. (1984) Similarity between the bacterial histone-like protein HU and a protein from spinach hloroplasts, *FEBS Lett.* **72**, 75-80.
2. Sato, N., Albrieux, C., Joyard, J., Douce, R. and Kuroiwa, T. (1993) Detection and characterization of a plastid envelope DNA-binding protein which may ancor plastid nucleoids, *EMBO J.* **12**, 555-561.
3. Yurina, N.P., Belkina, G.G., Karapetyan, N.V. and Odintsova, M.S. (1995) Nucleoids of pea chloroplasts: microscopic and chemical characterization, occurrence of histone-like proteins, *Biochem. and Mol. Biol. Int.* **36**, 145-154.
4. Gavin, I.M., Melnik S.M., Yurina N.P., Bavykin S.G. Zero-length protein-nucleic acid crosslinking by radical-producing chemical agents as a probe for analysis of protein- DNA interactions in vitro and *in* vivo. in press.
5. Mirzabekov, A.D., Bavykin, S.G., Belyavsky, A.V., Karpov, V.L., Preobrazhenskaya, O.V., Shick, V.V. and Ebralidze, K.K. (1989) Mapping DNA-protein interactions by crosslinking, *Methods Enzymol.* **170**, 386-408.
6. Cockayne, D.J.H., Harris, J.R., Egerton, R.F. and Jap, B.K. (1995) Nucleoid proteins, *Micron* **26**, 461-480.

PIGMENT-PROTEIN COMPLEXES, PLASTID DEVELOPMENT AND PHOTOOXIDATIVE PROTECTION

The effects of PORA and PORB overexpression on Arabidopsis *seedlings shifted from far-red to white light*

Ulrich Sperling, Geneviève Frick, Barbara van Cleve, Klaus Apel and Gregory A. Armstrong*
Institute for Plant Sciences, Plant Genetics,
Swiss Federal Institute of Technology (ETH),
CH-8092 Zürich, Switzerland
*gregory.armstrong@ipw.biol.ethz.ch

1. Introduction

Etiolated seedlings of angiosperms, the most highly evolved and diverse group of higher plants, synthesize chlorophyll (Chl) only upon exposure to light (1). In contrast, most oxygenic photosynthetic organisms also green when grown initially in the dark (2). The biochemical basis of light-dependent Chl biosynthesis is the strictly light- and NADPH-dependent enzymatic reduction of protochlorophyllide (Pchlide), a late Chl biosynthetic precursor. Pchlide is converted to chlorophyllide (Chlide) by two structurally related but differentially regulated enzymes, NADPH:Pchlide oxidoreductases (POR; EC 1.3.1.33) A and B, in the angiosperms *Arabidopsis thaliana* (3) and barley (4). Although PORA and PORB are both nuclear-encoded, translated in the cytosol, and ultimately imported into plastids, the *PORA* and *PORB* genes display dramatic differences in their regulation by light and developmental state, and the two enzymes have different plastid import requirements (5).

We are currently using *Arabidopsis* as a model system for molecular-genetic studies of plastid differentiation and greening in higher plants (6). Major focuses of our recent research have been the roles of PORA and PORB in etioplast differentiation, photooxidative protection and light-dependent Chl biosynthesis, and the differential negative light regulation of the *POR* genes by phytochrome (PHY) photoreceptors (7-11). To address POR function we have used a transgenic approach to constitutively overexpress either PORA or PORB in seedlings depleted of endogenous POR. Such a POR-depleted background can be obtained by exploiting the developmental and the strong negative light regulation of *PORA*, and to a lesser extent *PORB*, in the *Arabidopsis* wild type grown in PHYA-activating continuous far-red light (cFR) (7-9, 11, 12), and in the dark-grown *cop1-18* (previously *det340*) pleiotropic photomorphogenic mutant (10, 13), one of a number of similar mutants in which many normally PHY-mediated processes occur constitutively (14). 4-d-old cFR-grown wildtype and dark-grown *cop1* mutant seedlings retain a morphology typical of plants

J. H. Argyroudi-Akoyunoglou and H. Senger (eds.),
The Chloroplast: From Molecular Biology to Biotechnology, 97–102.

germinated in white light (WL), but contain little or no Chl, respectively. Furthermore, such seedlings lack detectable *PORA* mRNA and PORA, retain only low amounts of *PORB* mRNA and PORB, and are almost entirely deficient in the prolamellar body (PLB), the characteristic lattice-like internal membrane of etioplasts, in photoactive (ie. enzymatically reducible) Pchlide:NADPH:POR ternary complexes with a 77 K fluorescence emission maximum at 655 nm (Pchlide-F655), and in the ability to green after subsequent transfer to WL (7, 8, 10, 12, 13). Finally, such POR-depleted seedlings actually overaccumulate nonphotoactive Pchl(ide)-F632 (8, 13), which cannot be reduced to Chlide upon illumination and therefore acts as a potent sensitizer for photooxidative damage to plastids (9).

Overexpression of either PORA or PORB in cFR-grown seedlings confers subsequent photooxidative protection in WL, as evidenced by significantly increased thylakoid membrane formation, Chl accumulation and seedling viability (9). POR overexpression apparently functions by decreasing the steady-state amount of total Pchl(ide) present in seedlings prior to the shift to WL. In the dark-grown *cop1* mutant overexpression of either PORA or PORB suffices for etioplast differentiation, as measured by PLB membrane assembly and the formation of Pchlide-F655 (10).

Here we examine the developmental accumulation of total Pchl(ide) during the growth of wildtype *Arabidopsis* seedlings in cFR, and expand our previous analyses (8, 9) of the effects of PORA- or PORB-overexpression on cFR seedling phenotypes before and after a shift to WL. In particular, we have characterized the Pchl(ide) pigment and pigment-protein species present in POR-overexpressing seedlings by in situ low temperature (LT) fluorescence to obtain a better understanding of the mechanisms by which PORA and PORB can protect cFR-grown plants against photooxidative damage.

2. Procedure

Experiments were performed with wildtype or transgenic *Arabidopsis thaliana* (L.) Heynh. plants of the ecotypes Landsberg *erecta* (Fig. 1) and the Columbia derivative C24 (Figs. 2, 3). The PORA-overexpressing line PAO-3, and the PORB-overexpressing lines PBO-1 and PBO-10 have been previously described (9, 10). Seeds were surface-sterilized, plated on Murashige-Skoog agar, placed in the dark at 4°C for 24 h, and finally exposed to 1 h of intense WL at 22°C (defined as 0 d) to help synchronize germination. The petri plates were thereafter maintained at 22°C for various lengths of time in darkness or in 16 W m^{-2} of nonphotooxidative cFR with a maximum at 740 nm (9, 15) (Fig. 1). Alternatively, plated seeds were transferred from the dark to cFR after 1 d and grown under this light quality for another 3 d (Figs. 2, 3). Germination typically occurred at about 36 h. Seeds (0 d, 1 d) and seedlings (2 d and older) were harvested under a green safety light and processed immediately for fluorescence measurements and electron microscopy.

The total Pchl(ide) content of acetone extracts of seeds or seedlings of various ages was determined with a luminescence spectrophotometer by monitoring the room temperature fluorescence emission at 628 nm upon excitation at 440 nm. To obtain relative Pchl(ide) concentrations the pigment amounts were normalized to the total protein content of each extract, as described elsewhere (3). LT fluorescence emission measurements of individual pairs of intact cotyledons from 4-d-old seedlings of various genotypes were collected with an excitation wavelength of 440 nm and processed as previously described (10). For analyses of plastid ultrastructure, cotyledons of 4-d-old seedlings were harvested, prepared and examined by transmission electron microscopy as described earlier (9, 13).

3. Results and Discussion

During skotomorphogenic (ie. etiolated) seedling growth the *Arabidopsis PORA* and *PORB* mRNAs are up- and down-regulated in parallel by endogenous factors (3, 11). Developmental regulation also occurs during growth in cFR, but is superimposed on a differential negative regulation of *POR* mRNA amounts by light, which leads to a POR-depleted state after 4 d (8, 11). As seen in the time courses presented in Fig. 1, total Pchl(ide) content increased steadily during the development of both etiolated and cFR-grown seedlings. However, 2- to 4-d-old cFR-grown seedlings accumulated several-fold more total Pchl(ide), despite their POR deficiency, than did etiolated seedlings of the same ages (9). Significant pigment accumulation also began at least 1 d earlier in cFR-grown than in etiolated seedlings, and was readily detectable during germination at 2 d. Therefore, growth of *Arabidopsis* seedlings in cFR uncouples POR content from the amount of total Pchl(ide).

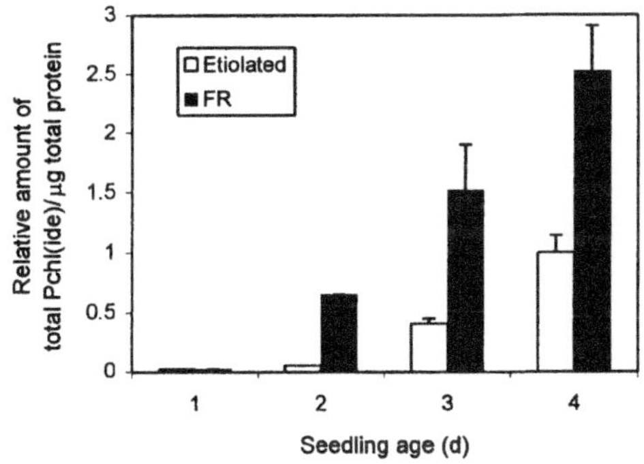

Figure 1. Developmental regulation of total Pchl(ide) accumulation during skotomorphogenic and cFR-induced photomorphogenic growth of wildtype seedlings. Pigment amounts determined from triplicate acetone extracts of seedlings are given relative to the 4-d-old etiolated sample. Error bars represent the standard deviations.

In situ LT fluorescence spectroscopy was performed to determine the qualitative effect of PORA or PORB overexpression on the relative amounts of photoactive Pchlide-F655 and nonphotoactive Pchl(ide)-F632 in 4-d-old cFR-grown seedlings. Fig. 2 illustrates that the overexpression of either POR enzyme significantly increased the steady-state ratio of Pchlide-F655 to Pchl(ide)-F632, an effect also observed in etiolated seedlings of the same transgenic lines (10). Furthermore, although cFR is used very inefficiently for the POR-mediated reduction of Pchlide-F655 to Chlide in the wild type (8, 9), constitutive POR overexpression led to the formation of readily detectable amounts of Chl-containing pigment-protein complexes (Fig. 2B-2D) that were not observed in the wild type (Fig. 2A). Thus, POR overexpression in cFR-grown seedlings decreases not only the steady-state amount of total Pchl(ide) (9) but also the steady-state ratio of nonphotoactive Pchl(ide)-F632, a sensitizer for photooxidative damage in WL, to

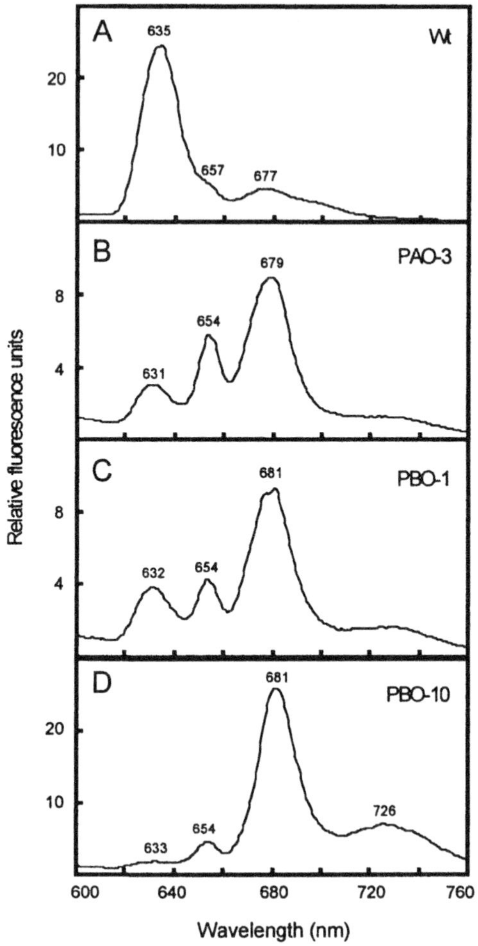

Figure 2. Pchl(ide) and Chl(ide) pigment forms present in intact cotyledons of cFR-grown *Arabidopsis* seedlings. (**A**) Wild type (Wt), (**B**) PORA-overexpressing line PAO-3, (**C**, **D**) PORB-overexpressing lines PBO-1 and PBO-10. LT fluorescence emission bands observed include nonphotoactive Pchl(ide)-F632, photoactive Pchlide-F655, and photoreduction products resulting from prolonged seedling growth in cFR, including Chl(ide)-F675 (8), and Chl-F681 and Chl-F726, corresponding to the light-harvesting antenna complexes of photosystems II and I, respectively.

photoactive Pchlide-F655, the substrate for light-dependent Chl biosynthesis. POR overexpression leads to a higher capacity for the utilization of cFR for Pchlide photoreduction, and hence formation of Chl-protein complexes.

Fig. 3 demonstrates the effects of PORB overexpression on plastid ultrastructure during seedling growth in cFR. As seen in other PORA- and PORB-overexpressing lines (9), the constitutive availability of PORB allowed the maintenance of large PLB membranes (Fig. 3A). Furthermore, when transgenic seedlings were transferred to strong WL, extensive unstacked thylakoid membranes formed (Fig. 3B). Plastids of the

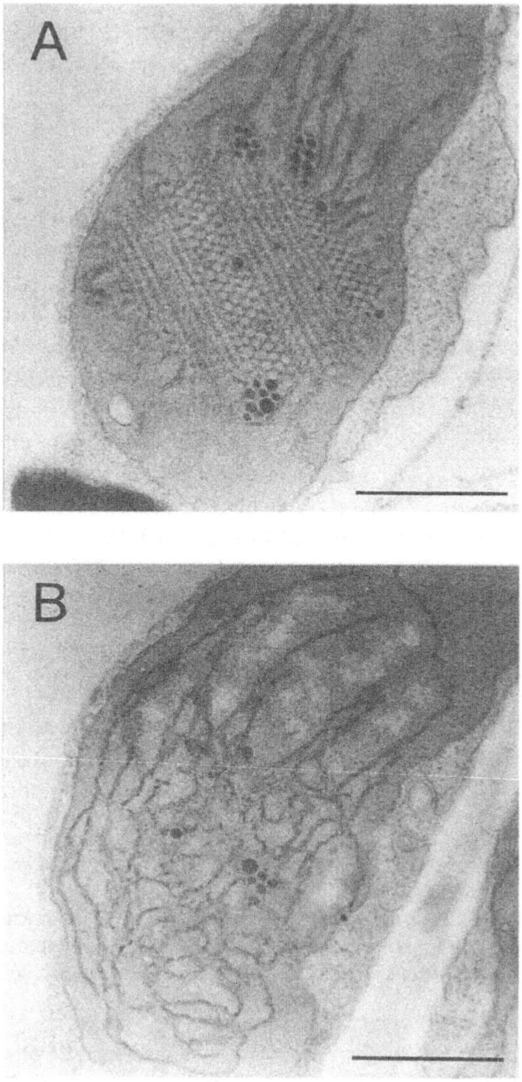

Figure 3. Influence of PORB overexpression on PLB and thylakoid membrane architecture in *Arabidopsis* cotyledon plastids. Plastid ultrastructure was examined in 4-d-old seedlings of line PBO-1 grown in cFR (**A**) and thereafter transferred to 4 h of WL with a fluence rate of 160 $\mu E\ m^{-2}\ s^{-1}$ (**B**). Magnifications were 20,000 x (bars = 1 μm).

cFR-grown wild type, in contrast, contain only a few small PLBs and produce no thylakoid membranes when shifted to WL (9, 12). We interpret this WL-phenotype to reflect severe photooxidative damage to cotyledon plastids.

According to our model, the cFR-mediated overaccumulation of total Pchl(ide) (Fig. 1) (9) combined with the negative light regulation of *POR* gene expression (8, 11,

12) lead to a dramatic increase in the ratio of Pchl(ide)-F632 to Pchlide-F655 in wildtype seedlings (Fig. 2) (8). Although cFR itself is nonphotooxidative (15), such seedlings are sensitive to photooxidative damage during a subsequent WL treatment. PORA or PORB overexpression can at least partially rescue this phenotype (Fig. 3) (9).

References

1. von Wettstein, D., Gough, S., and Kannangara, G.C. (1995) Chlorophyll biosynthesis, *Plant Cell* **7**, 1039-1057.
2. Armstrong, G. A. (1998) Greening in the dark: light-independent chlorophyll biosynthesis from anoxygenic photosynthetic bacteria to gymnosperms, *J. Photochem. Photobiol. B: Biol.* **43**, 87-100.
3. Armstrong, G. A., Runge, S., Frick, G., Sperling, U., and Apel, K. (1995) Identification of NADPH:protochlorophyllide oxidoreductases A and B: a branched pathway for light-dependent chlorophyll biosynthesis in *Arabidopsis thaliana*, *Plant Physiol.* **108**, 1505-1517.
4. Holtorf, H., Reinbothe, S., Reinbothe, C., Bereza, B., and Apel, K. (1995) Two routes of chlorophyllide synthesis that are differentially regulated by light in barley (*Hordeum vulgare* L.), *Proc. Natl. Acad. Sci. USA* **92**, 3254-3258.
5. Sundqvist, C. and Dahlin, C. (1997) With chlorophyll pigments from prolamellar bodies to light-harvesting complexes, *Physiol. Plant.* **100**, 748-759.
6. Armstrong, G. A. and Apel, K. (1998) Molecular and genetic analysis of light-dependent chlorophyll biosynthesis, *Meth. Enzymol.*, **297**, in press.
7. Frick, G., Apel, K., and Armstrong, G. A. (1995) Light-dependent protochlorophyllide oxidoreductase, phytochrome and greening in *Arabidopsis thaliana*, in P. Mathis (ed.), *Photosynthesis: from Light to Biosphere, Volume III*, Kluwer Academic Publishers, Dordrecht, The Netherlands, pp. 893-898.
8. Runge, S., Sperling, U., Frick, G., Apel, K., and Armstrong, G. A. (1996) Distinct roles for light-dependent NADPH:protochlorophyllide oxidoreductases (POR) A and B during greening in higher plants, *Plant J.* **9**, 513-523.
9. Sperling, U., van Cleve, B., Frick, G., Apel, K., and Armstrong, G. A. (1997) Overexpression of light-dependent PORA or PORB in plants depleted of endogenous POR by far-red light enhances seedling survival in white light and protects against photooxidative damage, *Plant J.* **12**, 649-658.
10. Sperling, U., Franck, F., van Cleve, B., Frick, G., Apel, K., and Armstrong, G. A. (1998) Etioplast differentiation in Arabidopsis: both PORA and PORB restore the prolamellar body and photoactive protochlorophyllide-F655 to the *cop1* photomorphogenic mutant, *Plant Cell* **10**, 283-296.
11. Frick, G., van Cleve, B., Apel, K., and Armstrong, G. A. (1998) Developmental regulation of the *PORA* and *PORB* genes in wildtype *Arabidopsis* seedlings grown in far-red light and in the dark-grown *cop1* mutant, in G. Garab and J. Pusztai (eds.), *Proceedings of the XIth International Congress on Photosynthesis*, Kluwer Academic Publishers, Dordrecht, The Netherlands, in press.
12. Barnes, S.A., Nishizawa, N.K., Quaggio, R.B., Whitelam, G.C., and Chua, N.-H. (1996) Far-red light blocks greening of Arabidopsis seedlings via a phytochrome A-mediated change in plastid development, *Plant Cell* **8**, 601-615.
13. Lebedev, N., van Cleve, B., Armstrong, G., and Apel, K. (1995) Chlorophyll synthesis in a deetiolated (det340) mutant of Arabidopsis without NADPH-protochlorophyllide (Pchlide) oxidoreductase (POR) A and photoactive Pchlide-F655, *Plant Cell* **7**, 2081-2090.
14. von Arnim, A. and Deng, X.-W. (1996) Light control of seedling development, *Annu. Rev. Plant Physiol. Plant Mol. Biol.* **47**, 215-243.
15. Mohr, H. (1966) Untersuchungen zur phytochrominduzierten Photomorphogenese des Senfkeimlings, *Z. Pflanzenphysiol.* **54**, 63-83.

SMALL CAB-LIKE PROTEINS: RELATIVES TO THE CHLOROPHYLL *A/B* BINDING PROTEINS IN CYANOBACTERIA

C. FUNK[1,2], V. LINDSTRÖM[1], W. VERMAAS[2]
[1]Arrhenius Laboratories, Dept of Biochemistry, Stockholm University, S - 106 91 Stockholm, Sweden; [2]Dept of Plant Biology / Photosynthesis Center, Arizona State University, AZ 85287-1601, USA

1. Introduction

In higher plants, the light harvesting complex (LHC) is an antenna complex encoded by a multi-gene family of at least 10 nuclear *cab* genes [1]. Most of these chlorophyll *a/b*-binding proteins (Cab proteins) have three membrane-spanning regions, the first helix being similar to the third one. However, the Cab family also includes the PS II-S protein predicted to have four membrane-spanning regions.

Cyanobacteria contain chlorophyll *a* but, in contrast to higher plants, lack chlorophyll *b* as well as Cab proteins. The major light-harvesting complex in cyanobacteria is the phycobilisome, located peripherally to the thylakoid membrane. However, recently a small protein of 8 kDa was identified in the cyanobacterium *Synechococcus* sp. PCC 7942. This protein was predicted to have a single membrane-spanning helix and showed significant sequence similarity to the first and third membrane-spanning region of Cab proteins [2]. The small cyanobacterial protein was designated a high-light inducible protein (HLIP) since its RNA level increased after transfer of cells to high light. Deletion of the corresponding gene (*hli*A) had no effect on the phenotype of the cyanobacterium. The similarity of this gene to members of the Cab gene family suggests that *hliA* codes for a single-helix Cab relative.

Here the existence of five genes and the presence of corresponding transcripts coding for small Cab-like proteins (SCPs) is shown in the cyanobacterium *Synechocystis* sp. PCC 6803. For several of these genes, transcription levels and presence of corresponding polypeptides was affected by the presence of photosystem I (PS I) and thereby by the chlorophyll level in the cell. In addition, for one member of this gene family, a correlation with the light-harvesting efficiency of phycobilisomes could be demonstrated. Therefore, we suggest that the one-helix Cab-like proteins in cyanobacteria are involved in pigment binding, and a role in transient chlorophyll binding upon increased chlorophyll availability is discussed.

J. H. Argyroudi-Akoyunoglou and H. Senger (eds.),
The Chloroplast: From Molecular Biology to Biotechnology, 103–106.

2. Material and Methods

Synechocystis sp. PCC 6803 cells were cultivated at 30°C in BG-11 medium. The ferrochelatase gene (*scp*A) was interrupted at the *Bsp*1286I site in the gene by insertion of a 1.6 kb kanamycin-resistance marker. This site in the *scp*A gene is in the region connecting the ferrochelatase and SCP domains. The *scp*D gene was interrupted at the *Sac* II site in the gene. Southern hybridization was performed with genomic DNA from the different transformants. Expression of the *scp* genes at the RNA level was tested by Northern blotting. *The scp* genes correspond to the following nucleotide numbers of the *Synechocystis* sp. PCC 6803 genomic sequence as they are accessible through CyanoBase (http://www.kazusa.or.jp/cyano/cyano.html): *scp*A: 2849096-2850259; *scp*B: 1142015-1141803; *scp*C: 701350-701138; *scp*D: 982968-983180 and *scp*E: 398188-398361. Peptide-directed antibodies against the N-terminal region of ScpC and ScpD were used to identify the polypeptides in thylakoids, which were separated via high-resolution SDS-PAGE [3]. Oxygen evolution measurements were carried out on intact cells.

3. Results and Discussion

A computer search for genes coding for Cab-like proteins in the genome of *Synechocystis* 6803 identified five genes of which the translated sequences are similar to each other, to HliA of *Synechococcus* sp. PCC 7942, and to the first and third helix of Cab proteins (Fig. 1). The predicted products of four of the five open-reading frames (*scp*B-E) are small (57-70 residues; 6-8 kDa) and have been named **s**mall **C**ab-like **p**roteins. The fifth SCP however, ScpA, is large (387 residues) with only the C-terminal 69 residues similar to Cab proteins; the N-terminal 80% of the sequence has strong similarity to ferrochelatase from other organisms. The same gene connection between an *scp* and ferrochelatase has been found in one of the two *Arabidopsis thaliana* ferrochelatase sequences in GenBank [4]. This apparent gene linkage between ferrochelatase and a possible pigment-binding component leads to the question whether this connection is of possible functional significance. After gene interruption of *scp*A leading to deletion of the Cab-like portion of the protein (not shown), significant

```
                                               *      *     ** *  **         *        *
ScpC      MTTRGFRLDQD NRLNNFAIEP EVYVDSSVQA .GWTKYAEKM NGRFAMIGFA SLLIMEVVTG HGVIGWLNS L..
ScpD      MTSRGFRLDQD NRLNNFAIEP PVYVDSSVQA .GWTEYAEKM NGRFAMIGFV SLLAMEVITG HGIVGWLNS L..
HliA      .MRSGRTVDEF GRQNNFAIEP QVVVAEADAS WGFHDRAEKL NGRLAMIGFV ALILTEVALG QGLLPFLAS GLS
ScpB      .MGAILCYIYL HRQPSQLVIT FLTMNNENSK FGFTAFAENW NGRLAMIGFS SALILELVSG QGVLHFFGI L..
ScpE      ........... ...MSEELQP NQTPVQEDPK FGFNNYAEKL NGRAAMVGFL LILVIEYFTN QGVLAWLGL R..
ScpA      .MDSLNDPPCT FETVPHPKKN MKMYPQERWE WGLTTAAEVW NGRLAMLGFI ALLV.ELISG QGPLHFVGL L..
PsbS(I)                                GG IGFTKENELF VGRVAMIGFA ASLLGEGITG KGI
PsbS(III)                           EGGPL FGFTKANELF VGRLAQLGFA FSLIGEIITG KG
```

Fig.1: Sequence alignment of the five small Cab-like proteins (SCPs) of *Synechocystis* sp. PCC 6803 and the *hliA* gene product of *Synechococcus* sp. PCC 7942. The sequence of the first (I) and third (III) presumed helix of PsbS from spinach has been included. Note that ScpA is a polypeptide expected to have both ferrochelatase and Cab-like characteristics, and only the C-terminal region of the polypeptide (starting at amino acid residue 319) has been shown.

phenotypic differences with the control strain are an increased light tolerance when in a PS I-less background and a decreased light-harvesting efficiency of phycobilisomes (Table 1), suggesting a role of ScpA in energy transfer. For this reason ScpA presumably binds pigments, most likely including chlorophyll. Most of the pigment binding residues of LHCII are conserved in the SCP sequence.

Table 1: Light intensity (μE m^{-2} s^{-1}) for half-saturation of oxygen evolution rates in PS I-less strains with and without *scp*A

strain	600 nm light	≥ 665 nm light
PS I - less	73	60
PS I - less / *scp*A$^-$	100	60

Since *hli*A found in *Synechococcus* sp. PCC 7942 was reported to be transcribed mainly under high light conditions [2], transcript and translation analyses were performed to monitor expression of the Scp genes and proteins as a function of light intensity. Wild-type as well as PS I-less, PS II-less, PS I-less/*chl*L$^-$ and PS I-less/PS II-less mutants were grown at different light intensities (such as high light (250 μE m^{-2} s^{-1}), normal light (50 μE m^{-2} s^{-1}), low light (5 μE m^{-2} s^{-1}) and light activated heterotrophic growth) and total RNA as well as thylakoids were prepared from these mutants. Peptide-directed antibodies against ScpC and ScpD were used for immunodecoration.

Probes for *scp*A, B, C, D and E hybridized with RNA sized at 1 kb for *scp*A, 0.25 kb for *scp*B, C, D and 0.3 kb for *scp*E, suggesting monocistronic transcription of these genes (Fig. 2). The *scp*A transcript level was low but similar in the different strains grown at different light conditions, suggesting a constitutive expression of *scp*A. Transcript levels of the other four *scp* genes were very low in the wild type and PSII-less strain at any light intensity, but were increased when PS I was absent. Significant transcript levels were found in the PS I-less and PS I-less/PS II-less strains, and to a lesser extent also in the PS I-less/*chl*L$^-$ strain. Transcript levels in cells grown under LAHG conditions were lower than those from cells grown at normal or low light. The level of transcript correlated widely with the presence of SCP protein [5].

In our hands, the only major factor modifying the expression of at least some SCPs seems to be the presence or absence of PS I. As in cyanobacteria most chlorophyll is associated with PS I, it is possible that *scp*s are expressed preferentially when less chlorophyll is bound by photosystems and thus availability of newly synthesized chlorophyll may be higher. This might also account for expression of *hli*A in *Synechococcus* sp. PCC 7942 at very high light intensity, as under these conditions photoinhibition (and thereby increasing availability of chlorophyll that is not linked to photosystems) may occur. PS I-less/PS II-less mutants still retain a significant amount of chlorophyll (about 3% of wild type and 20% of the amount in PS I-less strains), and increased levels of *scp* expression in such mutants suggest that SCPs may serve as pigment-binding proteins that are sinks for chlorophyll and can function as antenna protein as well.

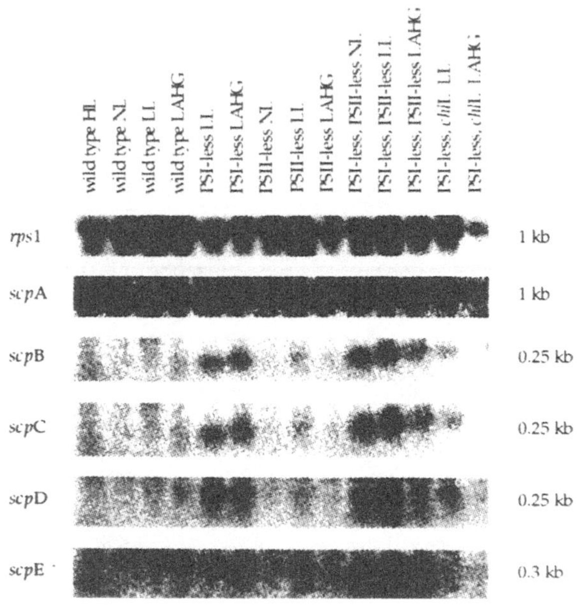

Fig. 2: Northern hybridization of *scp*A, B, C, D and E with RNA isolated from *Synechocystis* 6803 wild type, PS I-less, PS II-less, PS I-less/PS II-less and the PS I-less, *chl*L$^-$ strains grown at high (HL; 250 μE $m^{-2}s^{-1}$), normal (NL; 50 μE $m^{-2}s^{-1}$) and low light (LL; 5 μE m^{-2} s^{-1}) intensity and under light activated heterotrophic growth (LAHG) conditions. Northern hybridization with *rps1* was performed as a control.

Eliminating the existence of free chlorophyll is very important for the cell as chlorophyll without nearby triplet quenchers can catalyze the light-induced formation of the highly oxidizing singlet oxygen. However, this scenario would necessitate binding of carotenoids close to the chlorophyll, in order to have efficient triplet quenching. Carotenoid binding to SCPs is feasible as carotenoids in LHC crosslink two transmembrane helices, and as an oligomeric structure of SCPs is anticipated based on the charged residues in putative transmembrane domains. SDS-PAGE analysis under non-denaturing conditions as well as reconstitution studies are in progress to investigate the pigment binding of SCPs.

4. References

1 Green, B.R. and Durnford, D.G. (1996) Annu. Rev. Plant Physiol. Plant Mol. Biol. 47, 685-714.
2 Dolganov, N.A.M., Bhaya, D. and Grossman, A.R. (1995) Proc. Natl. Acad. Sci. USA 92, 636-640.
3 Schägger, H. and von Jagow, G. (1987) Anal. Biochem. 166, 368-379.
4 Chow, K.S., Singh, D.P., Roper, J.M. and Smith, A.G. (1997) J. Biol. Chem. 272, 27565-27571.
5 Funk, C. and Vermaas, W. (1998) in: Proceedings of the XIth International Congress on Photosynthesis, Garab, G. & Pusztai, J (eds), in press.

HIGH-LIGHT INDUCED PROTEINS HL#2 OF BARLEY ARE EXPORTED TO THE APOPLAST

S. K. MISHRA[1], A. R. MENHAJ, S. BEZHANI[2], K. KLOPPSTECH
Institut für Botanik der Universität, Herrenhäuser Straße 2, 30419 Hannover, Germany,
[1] School of Life Sciences, Jawaharlal Nehru University, New Delhi, India,
[2] Universiteti 'A. Xhuvani, Elbasan, Albania

1. Introduction

Light, when supplied in excess over the capacity of the electron transport chain, will be harmful to plants. This situation is tried to be prevented by the plants using different strategies. The first is to use uncoupling of the light harvesting antennae from the reaction centre on the short range and destruction of the light harvesting antennae during long-term adaptation to high light. During the process of adaptation also the composition in the Xanthophylls is changed by increasing the stationary concentrations of antheraxanthin (A), zeaxanthin (Z) and violaxanthin (V) as well as by simultaneously increasing the ratio of A+Z/V [6 and references therein]. But besides the mentioned and presumably some other short term regulations, long term changes in gene expression can be visualized which will include a great number of proteins. To these proteins belong obviously the ELIPs [1]. Starting from the assumption that these ELIPs would not be the only genes regulated by high light we used a differential cloning strategy by enriching for transcripts which are up-regulated by high light fluxes. Using this method clones were isolated which we termed "HL" for high light induced. Out of these, HL#2 represents a gene whose mRNA is up-regulated by high light fluxes. As described [8], sequence comparison showed similarity to proteins of rice. In the meantime cloning has considerably increased the number of related genes in the databank. We find lectins and jasmonate-induced proteins among the related proteins some of which are also known to be regulated by light.

When we studied the expression of HL#2 in green barley we observed an increase in the mRNA of HL#2. This occurred mainly in the basal section of the primary leaf when the leaves were illuminated floating on water exposed to high light fluxes. The pattern of RNA induction was somewhat similar to that of ELIPs as the highest mRNA levels were found in the basis of the leaf. We did not know, however, how the corresponding protein would be regulated. In order to study HL#2 abundance, antibodies were developed which recognize the protein. Some of the results obtained with this antibody are presented in this study.

J. H. Argyroudi-Akoyunoglou and H. Senger (eds.),
The Chloroplast: From Molecular Biology to Biotechnology, 107–112.

2. Methods

Plant growth and high light treatment

Barley (*Hordeum vulgare* L. cv. Apex, v. Lochow-Petkus, Bergen, Germany) was grown for 6 days on Vermiculite at 25°C at a constant irradiance of 100 µmol m^{-2} s^{-1} in a 12h-light/12h-dark cycle. High light stress treatments were performed on isolated leaves of equal length at an irradiance of 1.800 µmol m^{-2} s^{-1} in white light floating on water of 25°C and under a water filter [7]. At the end of the experiment, the leaves were directly frozen in liquid nitrogen and stored at -70°C for further use.

Analysis of proteins

These procedures were carried out exactly as described by Pötter et al. [7]. For total protein extraction liquid nitrogen homogenized leaf material was extracted in 5 volumes of 3 x LDS medium, incubated for 20 min at 65°C and centrifuged at 12000x g for 5 min at room temperature and the supernatants were used as total protein.

3. Results

3.1. SEQUENCE COMPARISON OF HL#2 WITH OTHER GENES

The derived amino acid sequence of HL#2 was rerun against the Swiss protein database and gave, in addition to the observations published [8] and besides others less apparent ones, similarities with Ipomoelin [4], JIPs 32.6 and 32.7 [5] and a lectin of *Calistegia sepium* [9]. The similarities were found over the entire protein of *Calistegia* and Ipomoelin while the similarities with HL#2 were restricted to the carboxy-terminal ends of the JIPs as shown in Figure 1.

The important observation was the relationship which could be found with the lectins. This statement is based on the findings that lectins have to be considered proteins which are exported to the outside of the cells. Whether the same holds true for the two JIPs is not yet known. These facts as well as the evidence of our own biochemical work [S. Mishra et al., unpublished data] provide evidence that at least the protein HL#2* which we consider at the moment, on the basis of immune reactivity and apparent molecular mass, as the processed form of HL#2 should reside outside the cells that is within the apoplast [2]. This statement, however, needs more experimental support e. g. by immuno-fluorescence or immuno gold-electron microscopy. Indeed, using the extracellular fluid of control and light stressed barley leaves we found the form HL#2* outside the cells. This is to our knowledge the first evidence of an interference of oxidative stress caused by high light with components outside the cells and within the extra-cellular fluid. The finding might eventually be related to the well known oxidative defence directed against phytopathogens during the hypersensitivity defence reaction. Such relationship remains to be established.

3.2. ACCUMULATION OF HL#2 IN GREEN PLANTS

When we studied the effect of high light on the amount of HL#2 protein that accumulated within the primary barley leaves we obtained a rather surprising result which was in clear contrast to the accumulation of HL#2-mRNA induced by high light as described earlier [8]. Despite several attempts we could not observe any effect of light-stress on the accumulation of the HL#2 protein within the primary leaves of green

```
HL#2              ...MSSSPVV KLGPWGGDDG .VAHDITVAP QRLESITIRW G.KVLHSVAF   45
Ipomoelin         ALQLAAHSDA RSGPVGSNGG QFWSFRPVRP LNKIVLSFSG SPDQTLNLIS   51
Jip 32.6          PVLKSESLLT KLGPWGGNGG GDKDILEAVP RRLESITVSS G.SIIDSIKF  199
Lectin (C.s.)     ...MAVPMDT ISGPWGNNGG NFWSFRPVNK INQIVISYGG GGNNPIALTF   47
Consensus                      GPwGg  g        v p    esit    g     s  f

HL#2              TYRDKDNQLH TAGPWGGAGG EKEDPDTITL GPSEYITQVD WSVGPFKLKE   95
Ipomoelin         ITFSSN.PTD IITVGGVGPE PLTYTETVNI ..DGDIIEIS GMIANYK...   95
Jip 32.6          SYVDQTGQKH NAGPWGGSGG NQ...NTFVL GASEFVKEVS GTFGIYD.KD  245
lectin(C.s.)      SSTKADGSKD TITVGGGGPD SITGTEMVNI GTDEYLTGIS GTFGIYL...   94
consensus          y d   q h  agpwGg gg       t  l g seyit v     g  k k

HL#2              IEHCITSLKF VTNQASYGPF GYAVDSTHYS LPVLNNGSVV GMFGRAGDYL  145
Ipomoelin         GYNVIRSIKF TTNKKEYGPY GANAG.TPFN IKIPDGNKIV GFFGNSGWYV  144
Jip 32.6          RHNIITSLKF ITNVKTYGPF GEAKG.TPFT IAVQKNSSIV GFFARSGIYL  294
lectin(C.s.)      DNNVLRSITF TTNLKAHGPY GQKVG.TPFS SANVVGNEIV GFLGRSGYYV  143
consensus             itSlkF  TN   yGPf G av  T  s   v  n s V G fgr G Yl

HL#2              HAIGFYVLPE TSLAS  160
Ipomoelin         DAIGAYYTAK .....  154
Jip 32.6          DALGVYVRPL .....  303
lectin(C.s.)      DAIGTYNHRK .....  153
consensus          AiG Yv p
```

Figure 1. Sequence comparison of the derived amino acid of HL#2 with other sequences from the Swiss data bank.

The derived amino acid sequence was run against the NCBI data bank (edition Blast 2.0). Similarities which we found are displayed bold and underlayed with grey fields. References are: Ipomoelin (accession number 1752827), lectin from *Calystegia sepium* (accession number 1730286), and JIP 32.6 (accession number 1167953).

7 day-old plants as shown in Figure 2. The linearity of the western blots has been tested by applying ½ and ¼ of the regular amounts applied to the gels; lanes 8 x 0.5 and 8 x 0.25 in Figure 2.

3.3. ACCUMULATION OF HL#2 IN ETIOLATED PLANTS

The possible explanation for this failure eventually stems from an experiment in which etiolated plants after 5 days of culture were exposed to light of 100 µmoles m^{-2} s^{-1}. In contrast to the light induction of ELIPs which in etiolated plants had been found to saturate at rather low light fluxes and which have a short half life time [3] the can be seen in Figure 3. Nothing is known about the half life of the protein. According to our experience the HL#2 proteins increase to even higher levels under light fluxes of 1500 µmoles m^{-2} s^{-1} during a period for up to 48 hours (data not shown). This is the longest possible period of light stress that can be tolerated by etiolated plants. After this period the primary leaves of the plants tend to dry out at their tips and eventually the secondary leaves will develop and appear almost normal. These etiolated, stressed

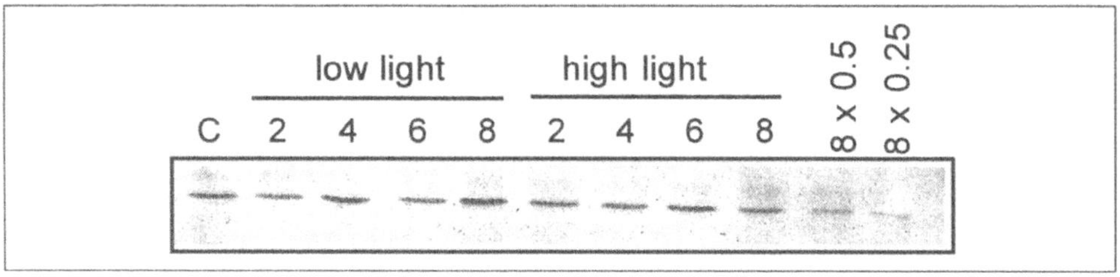

Figure 2. The effect of high light on the accumulation of the high light induced protein HL#2 in green plants.
Total proteins were isolated in LDS mixture by homogenization and heating up to 65°C for 15 min. Proteins were determined and equal amounts of protein applied to a 14% PAGE. A Western blot decorated with the antibody from goat against HL#2 is shown.

plants are also unable to accumulate chlorophylls under such extreme conditions of light. Control plants which received 100 µmoles m^{-2} s^{-1} accumulated normal amounts of chlorophyll during the 48 h period of the experiment (data not shown). After an initial increase the accumulation of HL#2 remains almost the same as observed at the beginning of the experiment (Figure 3).

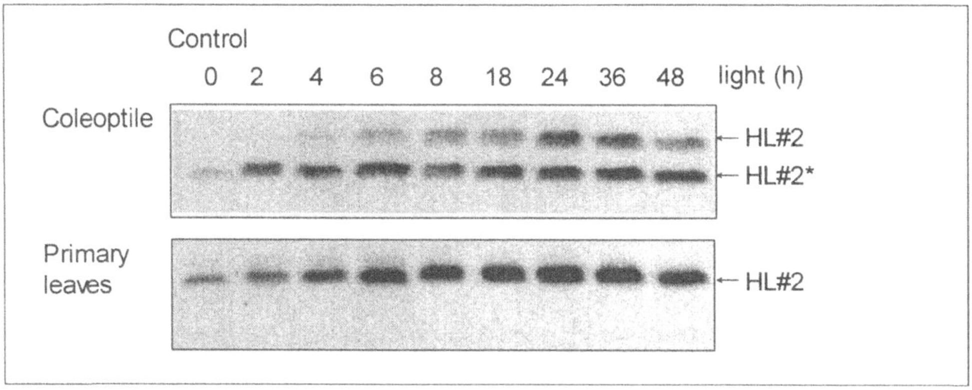

Figure 3. Influence of light (100 µmoles m^{-2} s^{-1}) on the induction of HL#2 proteins in etiolated barley seedlings
Plants were grown for 5 days in the dark and then transferred to light of 100 µmoles m^{-2} s^{-1}. Thereafter the plants were illuminated for the indicated times and the isolated protein quantified by Western blotting.

3.4. HL#2* IS FOUND IN THE APOPLAST

For this experiment leaves were exposed to light stress. The leaves were then soaked in 0.1 M phosphate buffer (pH 7.6), the coleoptiles removed and the primary leaves fractionated into three parts. The liquid was collected from the apoplasts of all segments by centrifugation at 200 x g. The concentrated and fractionated proteins were blotted and detected with the antibody against HL#2. As can be seen from the data of Figure 4 the smaller form of HL#2 is present in the washing fluid obtained from the coleoptile and the basal part of the primary leaf. The concentration of HL#2* in the washing fluid (IWF) is somewhat over-represented as the proteins in the apoplast are much fewer in number than in the cells; thus application of the same amount of protein will favour the appearance of individual proteins. But also the higher molecular mass form of HL#2 appears to reside in the interstitial fluid of all parts of the leaf. Control experiments carried out with the cytosolic form of malate dehydrogenase showed that proteins from within the cell did not contribute to the protein composition of the IWF.

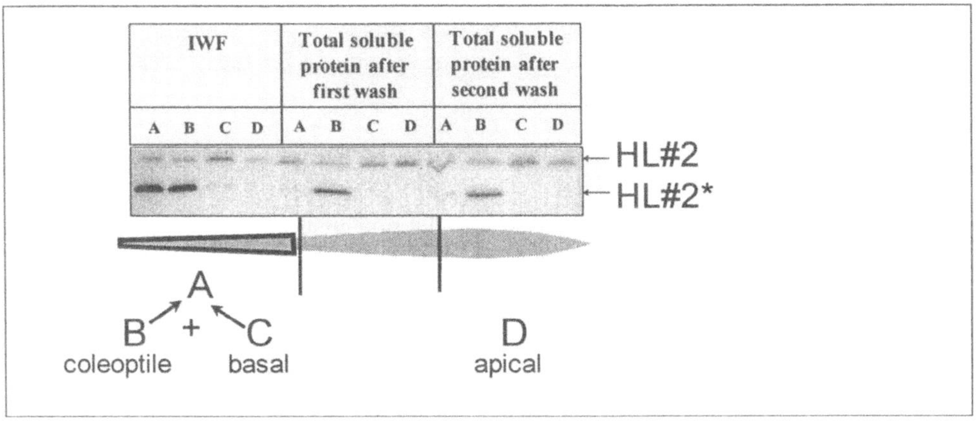

Figure 4. Distribution of HL#2 in the leaves, the coleoptiles, and the apoplast of 5 day-old dark/light grown barley seedlings.

4. Discussion

The data of this publication provide evidence for the existence of two related proteins of 15.5 and 18.0 kDa in barley whose mRNA is induced by high light stress but also by jasmonic acid (not shown). The relationship between the two immuno-reactive proteins is not quite clear, amino-terminal sequencing will clarify this point. But so far there can be hardly any doubt about the relationship. The blots are devoid of other protein bands and also the mRNAs from leaves and coleoptiles react with the same labelled cDNA fragment and show the same molecular mass. As in addition also the southern analysis provided evidence for the existence of only one gene this point should be rather established. Nevertheless, the relationship is not entirely clear. The small form of

HL#2* has been found outside the cell in the interstitial washing fluid, the use of the cytosolic enzyme MDH as an indicator excludes that the protein had been released due to cellular damage caused by the method of centrifugation that was used during the preparation of the washing fluid. Less clear is the relation between the two proteins. Figure 4 presents evidence for the occurrence of the high molecular mass protein outside the cell. This would most likely exclude a precursor product relationship. Indeed, also the kinetic of occurrence of both proteins and their localization are not in accordance with this idea. Since there is only one gene we have to assume that both proteins derive from the same gene. This could either be achieved by a processing step which should occur during or prior to transport. Alternatively, one could envisage also the possibility that different codon usage might explain the data. So far there is no indication for an alternative translation start. Again sequencing will clarify this point.

There exist interesting modes for the expression of HL#2 mRNA and protein. In the primary leaf of the green plant we find the mRNA for the 18 kDa protein under light control while the protein does not accumulate at the same time eventually due to lack of light stress. The HL#2 proteins accumulate in etiolated plants and in this respect they behave differently from the well investigated ELIPs. While the expression of ELIP-mRNA is maximal already at rather low light fluxes of approximately 5% of full sunlight, the expression of HL#2 protein continues up to light fluxes of at least 1500 µmoles m^{-2} s^{-1}.

5. Acknowledgements

The authors are grateful to Deutsche Forschungsgemeinschaft (DFG) and to the Deutscher Akademischer Austauschdienst (DAAD) for the financial support.

6. References

1. Adamska, I., Kloppstech, K. and Ohad, I. (1992) UV light stress induces the synthesis of the early light-inducible protein and prevents its degradation. J. Biol. Chem. **267**, 24732-24737.
2. Boller, T. and Métraux, J.P. (1988) Extracellular localization of chitinase in cucumber. Physiol. Mol. Plant Pathol. **33**, 11-16.
3. Grimm, B., Kruse, E. and Kloppstech, K. (1989) Transiently expressed early light-inducible thylakoid proteins share transmembrane domains with light harvesting chlorophyll binding proteins. Plant Mol. Biol. **13**, 583-593.
4. Imanishi, S., Kito-Nakamura, K., Matsuoka, K., Morikami, A. and Nakamura, K. (1997) A major jamonate-inducible protein of sweet potato, Ipomoelin, is an ABA-independent wound inducible protein. Plant Cell Physiol. **38**, 643-652.
5. Lee, J.E., Parthier, B., Loebler, M. (1995) Accession number U43496 in NCBI-blastserver databank.
6. Montané, M.H., Dreyer, S., Triantaphyllides, C. and Kloppstech, K. (1997) Early light-inducible proteins during long-term acclimation of barley to photooxidative stress caused by light and cold: High level of accumulation by posttranscriptional regulation. Planta **202**, 293-302.
7. Pötter, E. and Kloppstech, K. (1993) Effects of light stress on the expression of early light-inducible proteins in barley. Eur. J. Biochem. **214**, 779-786.
8. Pötter, E., Beator, J. and Kloppstech, K. (1996) The expression of mRNAs for light-stress proteins in barley: Inverse relationship of mRNA levels of individual genes within the leaf gradient. Planta **199**, 314-320.
9. Van Damme, E.J.M., Barre, A., Verhaert, P., Rouge, P. and Peumans, W.J. (1996) Molecular cloning of the mitogenic mannose/maltose-specific rhizome lectin of *Calystegia sepium*. FEBS Lett **397**: 352-356.

EXPRESSION OF *rbcS*, *rbcL* AND *rca* GENES DURING SENESCENCE AND REGREENING OF *ZANTEDESCHIA AETHIOPICA* FLORAL SPATHE

M.C. PIQUES[1], T. LINO-NETO[1], K. PALME[2], M.S. PAIS[3] AND R.M. TAVARES[1]
[1]*Plant Molecular Biology Laboratory, Department of Biology, University of Minho, Braga, PORTUGAL;* [2]*Max-Delbrük-Laboratorium in der Max-Planck-Gesellschaft, Köln, GERMANY;* [3]*Department of Plant Biology, Science Faculty of Lisbon, Lisboa, PORTUGAL*

1. Introduction

The spathe of *Zantedeschia aethiopica* (L.) Spreng. (Araceae) undergoes a senescence process soon after its formation, leading to the death if fructification doesn't occur [1]. During fructification, spathe regreening occurs being a consequence of a senescence-inhibition process. It was suggested that spathe senescence inhibition and regreening at time of fructification is due to the action of cytokinins produced by fruits and translocated to the spathe [2-4].

The spathe undergoes markedly changes in its photosynthetic capacity, due to variations in photosystem II and I and activities as well as in photophosphorylation rate [5]. Regreened spathes and leaves show chloroplasts with similar ultrastructure, but exhibit significative changes in its lipid composition, namely the absence of Δ^3-*trans*-hexadecenoic acid, which apparently does not affect *grana* stacking [6]. Ribulose-1,5-bisphosphate carboxylase/oxygenase (RuBisCO) activity was detected in the early stage of spathe development decreasing to undetectable values during senescence. During regreening, the increase in RuBisCO activity and content suggest *de novo* synthesis of this enzyme (unpublished results). With the aim to understand the variations in photosynthetic activity during spathe development, RuBisCO was further studied concerning expression of both subunits. In addition, due to importance of RuBisCO activase in the activation of RuBisCO [7], the expression of *rca* genes, which code for

J. H. Argyroudi-Akoyunoglou and H. Senger (eds.),
The Chloroplast: From Molecular Biology to Biotechnology, 113–116.

this protein was also studied during the time course of *Z. aethiopica* spathe development.

2. Identification of the *rbcL* gene and *rbcS* cDNA

For the identification of the *rbcL* gene, a *Z. aethiopica* cpDNA subgenomic library was constructed and screened, allowing the isolation of seven positive clones. After sequencing of the corresponding inserts it was possible to obtain the full coding sequence of *rbcL* gene.

rbcS cDNA was identified after random sequenced of several clones of a *Z. aethiopica* young leaf cDNA library.

3. Identification of cDNAs encoding RuBisCO activase

The identification of cDNAs encoding RuBisCO activase was performed by screening a *Z. aethiopica* young leaf cDNA library, using an homologous probe. This probe was obtained by PCR, using degenerated primers for the most conserved regions of RuBisCO activase coding region from different plants. Ten positive cDNA clones were isolated and sequenced, allowing the identification of four different incomplete nucleotide sequences homologous to *rca* genes.

- *rca1* and *rca3* sequences are similar but have five silent nucleotidic substitutions. The stop codon is located at positions 1455-1457.
- *rca2* and *rca4* sequences, are similar and have the same five silent nucleotide substitutions referred for the other ones. In addition, they have a 38pb insertion located at position 1305. This insertion has a stop codon (TAA) in position 1306-1308, which is responsible for the anticipation of protein synthesis termination (when compared to *rca1* and *rca3* sequence.

From these four sequences, two different polypeptides could be deduced, one having a longer 37 aminoacidic C-terminal than the other. In other species it was also reported two different polypeptides for RuBisCO activase [8]. In order to explain the presence of two different polypeptides in plants where only one gene was described, it was proposed a conserved splicing mechanism [9].

Our results suggest that in *Z. aethiopica*, RuBisCO activase is encoded by two distinct nuclear genes, each one coding for two types of transcripts, which could result from an alternative splicing mechanism of the initial transcript.

4. Expression analysis of *rbcL*, *rbcS* and *rca* genes during *Z. aethiopica* spathe development

The expression levels of *rbcS*, *rbcL* e *rca* genes during the time course of *Z. aethiopica* spathe development were quantified by Northern blotting, followed by hybridisation with the corresponding homologous probes (Fig. 1). The results indicate that during spathe senescence (whitening), no transcripts from all these genes were observed, suggesting that this developmental process is an example of

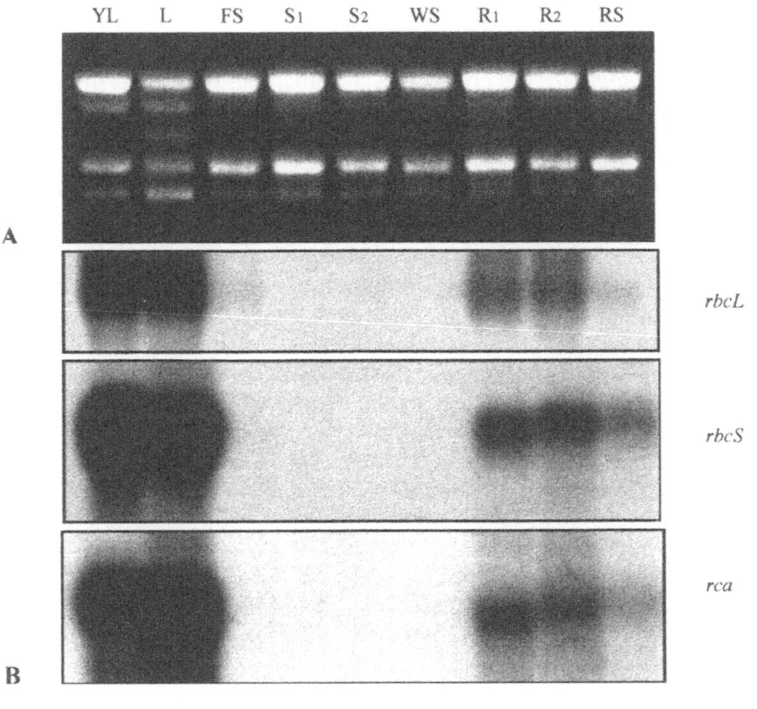

Figure 1. Expression of *rbcL*, *rbcS* and *rca* genes during the time course of *Z. aethiopica* spathe and leaf development. Total RNA (10 µg) from each stage was fractionated on a agarose-formaldehyde gel (A), transferred to a nylon membrane and hybridised with ^{32}P-labelled homologous probes (B). YL - young leaf; L - leaf; FB - floral bud spathe; S1 and S2 - intermediate senescence stages; WS - white spathe; R1 and R2 - intermediate regreening stages; RS - regreened spathe.

a programmed cell death. During spathe regreening, the coordenate increase of expression levels of RuBisCO activase and RuBisCO genes, could be a consequence of their activation due to the action of endogenous cytokinins.

5. Acknowledgements

We thank S. Rodermel and. A. Barkan for pJQ11 and rbcLA probes, respectively. M.C. Piques thanks JNICT for her Praxis XXI Master fellowship.

6. References

1. Pais, M.S., Novais, M.C., and Abreu, I. (1976) Sur le reverdissement de la spathe de *Zantedeschia aethiopica* au cours de la fruitification. *Port. Acta Biol., (A)*, **15**, 1-22.
2. Chaves Neves, H.J. and Pais, M.S.S. (1980) A new cytokinin from the fruits of *Zantedeschia aethiopica. Tetrahed. Lett.* **21**, 4387-4390.
3. Chaves Neves, H.J. and Pais, M.S.S. (1980) Identification of spathe regreening factor in *Zantedeschia aethiopica. Biochim. Biophys. Res. Commun.* **95**, 1387-1392.
4. Pais, M.S.S. and Chaves Neves, H. (1982/83) Regreening of *Zantedeschia aethiopica* Spreng. spathe induced by reapplied cytokinins. *Plant Growth Reg.* **1**, 233-242.
5. Tavares, R.M., Morais, F., Melo, N. and Pais, M.S.S. (1998) Thylakoid membrane reorganization during *Zantedeschia aethiopica* spathe regreening: consequence of the absence of Δ3-*trans*-hexadecenoic acid in photochemical activity. *Phytochemistry* **47**, 979-984.
6. Melo, N., Tavares, R.M., Morais, F., Barroso, J.G. and Pais, M.S.S. (1995) Lipid composition of thylakoid membranes from leaves and regreened spathes of *Zantedeschia aethiopica. Phytochemistry* **40**, 1367-1371.
7. Portis, A.R., Salvucci, M.E. and Ogren, W.L. (1986) Activation of ribulose bisphosphate carboxylase/oxygenase at physiological CO_2 and ribulose bisphosphate concentrations by RuBisCO activase. *Plant Physiol.* **82**, 967-971.
8. Salvucci, M.E., Werneke, J.M., Ogren, W.L and Portis, A.R. Jr. (1987) Purification and species distribution of rubisco activase. *Plant Physiol.* **84**, 930-936.
9. Werneke, J.M., Chatfield, J.M. and Ogren, W.L. (1989) Alternative mRNA splicing generates the two ribulosebiphosphate carboxylase/oxygenase activase polypeptides in spinach and *Arabidopsis. Plant Cell* **1**, 815-825.

EXPRESSION OF GLUTATHIONE PEROXIDASE DURING *ZANTEDESCHIA AETHIOPICA* SPATHE SENESCENCE AND REGREENING

T. LINO-NETO[1], R.M. TAVARES[1], K. PALME[2], AND M.S. PAIS[3]
[1]*Plant Molecular Biology Laboratory, Department of Biology, University of Minho, Braga, Portugal;* [2]*Max-Delbrück-Laboratorium in der Max-Planck-Gesellschaft, Köln, Germany;* [3]*Department of Plant Biology, Science Faculty of Lisbon, Lisboa, Portugal*

1. Introduction

Senescence is the terminal phase in the life of a plant or a plant organ that leads ultimately to death. From many physiological, biochemical, and molecular studies of leaf senescence, it becomes clear that leaf cells undergo highly coordinated changes in cell structure, metabolism, and gene expression during senescence [1].

The higher levels of active oxygen species (AOS) appears to have an important role during senescence [2]. These species are produced during the normal plant cell metabolism, as a result of several electron transfer reactions. To counteract the toxicity of AOS, a highly efficient antioxidative defence system is present in all plant cells, which includes several oxidative stress enzymes (such as superoxide dismutases, catalase, glutathione reductase and peroxidases) [3].

In senescence, the breakdown of the photosynthetic apparatus could lead to the production of active oxygen species (AOS). In addition, there are major alterations in the molecular organisation of the lipid bilayer, which also leads to production of more AOS. Using the enzyme lipoxygenase, fatty acids liberated after lipid degradation are converted to the corresponding hydroperoxides and thus may become the substrate for chain reactions generating more free radicals [4].

The higher levels of AOS detected during senescence could reflect a decline in the efficacy of the various defence mechanisms that normally afford protection against oxidative damage [2]. In fact, the capacity of plant tissue to scavenge oxygen active species appears to decrease as senescence progresses [5, 6]. The increase of AOS level could be also a result from the fatty acid hydroperoxides produced after lipid peroxidation by lipoxygenase.

Although glutathione peroxidase (GPX, EC 1.11.1.9) is considered as one of the key enzyme involved in scavenging oxyradicals in animals, it was not known in

J. H. Argyroudi-Akoyunoglou and H. Senger (eds.),
The Chloroplast: From Molecular Biology to Biotechnology, 117–120.

plants until recently [7]. It has been speculated that this enzyme may have an important role as a scavenger of lipid hydroperoxydes, as well as H_2O_2, produced by lipoxygenase during membrane desintegration.

Leaf senescence has been mainly studied using senescence-induction artificial systems. However, experiments using different induction methods has often given conflicting results that are sometimes difficult to interpret [8, 9]. In this way, by exhibiting a natural foliar senescence and self-inhibition senescence processes, *Z. aethiopica* is an excellent experimental model for the study of the mechanisms associated to senescence and regreening.

Z. aethiopica (L.) Spreng. is an ornamental plant from Araceae family. This plant exhibits a natural senescence process followed by a senescence self-inhibition of a leaf-like organ (spathe). Soon after its formation, the chlorophyllous structure floral bud spathe (FB) evolves to a non-chlorophyllous structure, the white spathe (WS). When fructification occurs, white spathe enveloping fructified spadix, undergoes regreening, reaching the final stage of spathe development, the regreened spathe (RS). Spathe regreening occurs only when fructification takes place; otherwise, the spathe dies [10].

2. Sequence analysis a *gpx Z. aethiopica* cDNA

The screening of a young leaf cDNA library from *Z. aethiopica* with a gpx probe from *Nicotiana sylvestris* (Accession number X60219, GenBank) allowed us to isolate two identical cDNA clones. The nucleotide sequence (Accession number AF053311, GenBank) of these two full length cDNA clones and the deduced aminoacid sequence are very similar to the others plant GPx proteins. Furthermore, similarity with other chloroplastidial targeted GPx protein (*Arabidopsis thaliana*; Accession number AJ000469, GenBank) suggests that the identified cDNA codes for a targeted chloroplastidial glutathione peroxidase protein. In addition, from the deduced aminoacidic sequence it is possible to perceive a peptide signal for chloroplast translocation [11].

This protein is very well studied in animals, being one of a few proteins known to contain the rare aminoacid Sec which is encoded by TGA - normally a stop codon. In contrast, in all homologous plant gene (and also in *Z. aethiopica*) the Sec residue is replaced by Cys encoded by TGT [7].

3. GPx activity during *Z. aethiopica* spathe regreening

It has been speculated that reactive oxygen is causally involved in the initiation of leaf senescence. In plants, chloroplasts are potentially the most powerful source of AOS as they are produced as a consequence of both inefficient transfer of electrons through the photosystems and breakdown of thylakoid components due to lipoxygenation. Oxidative damage in the chloroplasts is minimised by the combined activities of oxidative stress enzymes. It was expected that GPx contributes to this defence mechanism by reducing H_2O_2, organic hydroperoxides and lipid hydroperoxydes via reduced glutathione.

GPx activity determination during *Z. aethiopica* spathe development (Fig. 1), revealed that there is no significant variation of GPx activity during spathe senescence. In spite of some authors suggest that senescence is partially promoted by the decline of detoxifying systems, in our system, it seems there is no contribution of GPx activity for senescence promotion.

Our results showed that during regreening an evident increase of GPx activity occurs. From white stage (WS) to the first stage of regreening (R1), GPx activity increased 1.4-fold. After the first stage of regreening until regreened spathe (RS), GPx activity decreases gradually. Probably the induction of this enzyme is important to counteract the toxicity of reactive oxygen species produced during spathe senescence, protecting thylakoid membranes reorganisation.

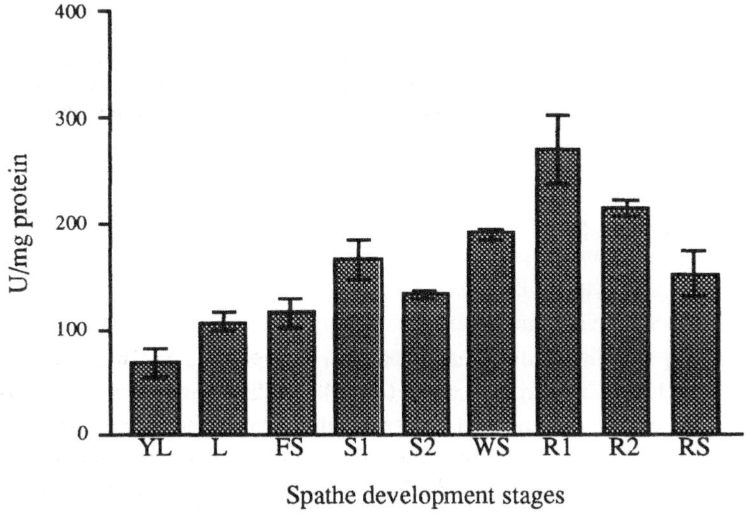

Figure 1. Glutathione peroxidase (GPx) activitiy during *Zantedeschia aethiopica* leaf and spathe development. Three pools of independent leaves and spathes in different development stages were used for protein extraction and activity determination. FB - floral bud spathe; S1 and S2 - intermediate senescence stages; WS - white spathe; R1 and R2 - intermediate regreening stages; RS - regreened spathe.

4. Conclusions

Our results suggest that glutathione peroxidase is present in *Z. aethiopica* chloroplasts, being detected a full-length cDNA comprising a signal peptide.

It seems that this enzyme is not involved in spathe senescence promotion, but could contribute significantly to the spathe regreening. Glutathione peroxidase could be important in scavenging oxyradicals, like H_2O_2, organic hydroperoxides and lipid hydroperoxydes, produced during senescence. In this way, this enzyme will contribute for the restoration of AOS environment of chloroplast, that will permit the thylakoid reorganisation.

5. Acknowledgements

We would like to thank Dr Dirk Inzé for kindly supplied *Nicotiana sylvestris* gpx probe (Accession number X60219, GenBank).

6. References

1. Gan, S. and Amasino, R.M. (1997) Making sense of senescence. Molecular genetic regulation and manipulation of leaf senescence. *Plant Physiol.*, **113**, 313-319.
2. Brown, J.H., Paliyath, G., and Thompson, J.E., (1991) Physiological mechanisms of plant senescence, in R.G.S. Bidwell (ed.), *Plant Physiology - A Treatise*, Academic Press Inc, New York, pp. 227-275.
3. Foyer, C.H., Lelandais, M., and Kurnet, K.J. (1994) Photooxidative stress in plants. *Physiol. Plant.*, **92**, 696-717.
4. Lynch, D.V. and Thompson, J.E. (1984) Lipoxygenase-mediated production of superoxide anion in senescing plant tissue. *FEBS Lett.*, **173**, 251-254.
5. Dhindsa, R.J., Dhindsa, P.P., and Thorpe, T.A. (1981) Leaf senescence correlated with increased levels of membrane permeability and lipid peroxidation and decreased levels of superoxide dismutase and catalase. *J. Exp. Bot.*, **32**, 93-101.
6. Pauls, K.P. and Thompson, J.E. (1984) Evidence for the accumulation of peroxidized lipids in membranes of senescing cotyledons. *Plant Physiol.*, **78**, 690-695.
7. Eshdat, Y., Holland, D., Faltin, Z., and Ben-Hayyim, G. (1997) Plant glutathione peroxidases. *Physiol. Plant.*, **100**, 234-240.
8. Becker, W. and Apel, K., (1993) Differences in gene expression between natural and artificially induced leaf senescence. *Planta*, **189**, 74-79.
9. Oh, S.A., Lee, S.Y., Chung, I.K., Lee, C.-H., and Nam, H.G. (1996) A senescence-associated gene of Arabidopsis thaliana is distinctively regulated during natural and artificially induced leaf senescence. *Plant Mol. Biol.*, **30**, 739-754.
10. Pais, M.S., Novais, M.C., and Abreu, I. (1976) Sur le reverdissement de la spathe de *Zantedeschia aethiopica* au cours de la fruitification. *Port. Acta Biol., (A)*, **15**, 1-22.
11. Keegstra, K., Bruce, B., Hurley, M., Li, H., and Perry, S. (1995) Targeting of proteins into chloroplasts. *Physiol. Plant.*, **93**, 157-162.

MOLECULAR EFFECTS OF SOME STRESS FACTORS ON THE CHLOROPLAST GENETIC APPARATUS OF THE FLAGELLATE *Euglena gracilis*

J. KRAJČOVIČ, R. VACULA, J.M. STEINER*, W. LÖFFELHARDT*, A. BELICOVÁ, S. SLÁVIKOVÁ, L. EBRINGER, E. STUTZ **
*Institute of Molecular and Subcellular Biology, Comenius University, Bratislava, Slovakia. *Institute of Biochemistry & Molecular Cell Biology and Ludwig Boltzmann Forschungstelle für Biochemie, Vienna, Austria. ** Laboratoire de Biochimie, Université de Neuchâtel, Switzerland.*

Abstract

The chloroplast system of the flagellate *Euglena gracilis* is particularly sensitive to various chemical and physical factors. Ofloxacin, a quinolone antibacterial chemotherapeutical, the mutagen nitrosoguanidine, as well as the hyperthermic cultivation at 33°C for 7 days, induce complete bleaching of *Euglena* cultures. Treatment with all those factors results in a mass degradation of *E. gracilis* chloroplast DNA. Significant decrease in the hybridisation signal intensity is connected with prolongation of cell exposure to the stress factors. No evidence for chloroplast genome rearrangement was obtained for chloroplasts of chemically stressed cells. On the contrary, heat induced loss of chloroplast functions is associated with partial restructurations of the chloroplast genome. Electrophoretograms of *E. gracilis* total RNA show distinct difference between wild type cells and bleached mutants. Bands for 23S and 16S rRNA are absent in all mutants. Subsequent hybridisation with DNA probes for nuclear encoded chloroplast proteins as Rubisco SSU and LHCPII provided identical signals for wild type cells and all mutants. However, in the case of the *petJ* gene for cytochrome c_6 variable hybridisation signals were obtained with northern blots from an ofloxacin-bleached mutant depending on the light régime.

1. Introduction

The chloroplast system of the unicellular photosynthetic flagellate *Euglena gracilis* is particularly sensitive to various chemical and physical factors which can induce irreversible mass bleaching of the cultures and cause depletion of chloroplast DNA (ctDNA) (for a review see [2]). The complete sequence of *E. gracilis* chloroplast genome (143 170 bp) has recently been published [6]. Antichloroplastic activity of various agents is macroscopically manifested by permanent loss of capability of cells to form green colonies (bleaching phenomenon). The ultrastructural studies showed chloroplast degradation and their substitution by membranous structures [3]. Only partial knowledge exists on bleaching process on the level of chloroplast genome changes. It was shown that UV light [15], streptomycin [9], nalidixic acid [7] as well as permanent hyperthermia [1] result in decreasing the total amount of ctDNA, in its rearrangement, degradation or complete loss.

J. H. Argyroudi-Akoyunoglou and H. Senger (eds.),
The Chloroplast: From Molecular Biology to Biotechnology, 121–128.

We have observed that bleaching agents with different mode of action (quinolones - antibacterial chemotherapeuticals based on DNA-gyrase inhibitors *versus* nitrosoguanidine - classical mutagen) respond to the changes of experimental conditions during the bleaching process (heat shock, hyperthermia, growth *versus* resting conditions) by different, even opposed mechanisms [13, 14]. These findings initiated our study of molecular effects of some selected xenobiotics on the chloroplast system of *E. gracilis*. Our goals were: (i) to test whether typical structural modifications of the chloroplast genome occur when xenobiotics with different modes of action are applied; (ii) to analyze total RNA profiles from wild type cells *versus* various mutants with damaged plastids.

2. Material and Methods

Chloroplast DNA (ctDNA) was isolated from mixotrophically grown *E. gracilis*, strain Z, in large scale via purification of chloroplasts by flotation on sucrose gradient. Cells were treated with ofloxacin (100 μg/ml) and harvested after various length of growth in the presence of the drug (1, 2, 5, 6 and 7 days) and in the case of nitrosoguanidine (12.5, 15 and 22 μg/ml) after 8 and 24 hours. ctDNA was isolated according to the same procedure also from wild type cells (as a negative control) and from *E. gracilis* growing at elevated temperature (33° C) for 7 days (as a positive control).

ctDNA was digested with selected restriction endonucleases and separated by electrophoresis on 0.6% agarose gels. Southern blots onto Hybond-N+ membrane (Amersham) were probed with labeled (i) total *E. gracilis* ctDNA digest, (ii) specific cloned fragments of ctDNA. For hybridisation and detection the Enhanced Chemoluminiscent direct nucleic acid Labeling method (ECL) of Amersham was used.

Total RNA was prepared from wild type *E. gracilis* and its various stable mutants (originally irreversibly bleached wt cells by: ofloxacin - WZOflL, acridine orange - WZAoL, nitrosoguanidine - WZNgL, hyperthermia - 33°C - WZHL, and the green mutant GZSmL, resistant to streptomycin) by a single step method of acetate-guanidine/phenol-chloroform extraction. Intact total RNA was electrophoresed on 1.1% agarose (2.2M formaldehyde) and transferred to uncharged nylon membrane (Qiabrane, Qiagene). Northern blots were: (A) visualized by methylene blue, (B) hybridized with four different DNA probes for nuclear encoded chloroplast proteins: two different parts of small subunit of Rubisco (SSU - 4.3 kbp fragment, and mSSU - 420 bp fragment, chlorophyll *a/b*-binding protein (LHCPII) - 600 bp fragment, and 248 bp fragment of the *petJ* gene for cytochrome c_6. The *petJ* gene fragment was originally prepared by a PCR strategy with degenerate primers derived from conserved regions using our *E. gracilis* cDNA bank (λZAP II - cDNA kit, Stratagene) as a template. The β-subunit of the cytoskeletal protein tubulin was used as a control. Signals were detected by the DIG system (Boehringer, Mannheim).

3. Results and Discussion

In contrast to the quinolone ofloxacin which produce 100% white cells after 24 hours treatment in concentrations not affecting cell viability, the mutagen nitrosoguanidine acts more rapidly (6-8 hours), but only in near toxic concentrations (100% lethality after 48 hours). Complete bleaching of *Euglena* cultures was also obtained through hyperthermic cultivation at 33 °C for 7 days.

3.1. CHLOROPLAST DNA IN CHEMICALLY STRESSED CELLS

Treatment with both substances, quinolone ofloxacin and mutagen nitrosoguanidine results in a mass degradation of *E. gracilis* ctDNA. At the level of detection by ethidium bromide staining of the agarose gels under UV, restriction endonuclease patterns become unreadable, mostly only smears are observed (Figure 1A, 2A). However subsequent Southern blotting of these gels, their hybridisation and detection by chemoluminiscence method however demonstrate a presence of ctDNA in all the samples analyzed (Figure 1B, 1C, 2B). The restriction endonuclease patterns of ctDNA from treated cells are identical with those of wild type *E. gracilis*. Significant losses in the hybridisation signal intensities are evident in all ctDNA samples from treated cells and correlate to the prolongation of cell exposure to the chemicals. No indications evidence for chloroplast genome rearrangements could be found in chloroplasts of chemically stressed *E. gracilis*.

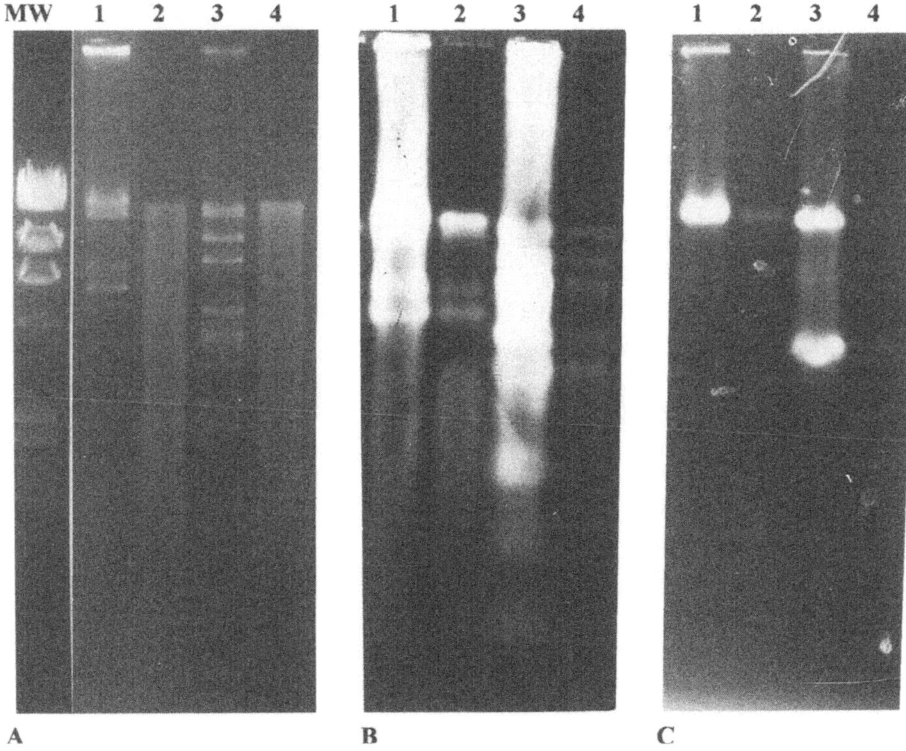

Figure 1. Restriction endonuclease analysis of *E. gracilis* ctDNA wild type (wt) *versus* treated cells with nitrosoguanidine (NG - 12.5 μg/ml) for 8 hours. **1** - ctDNA-wt/*Bam*HI, **2** - ctDNA-NG/*Bam*HI, **3** - ctDNA-wt/*Bg*lII, **4** - ctDNA-NG/*Bgl*II. **MW** - λDNA/*Hind*III. **A** - detection by ethidium bromide staining under UV. **B** and **C** - lumigraphs of the gel A. (Detection by chemiluminiscence - ECL Amersham). **B** - probing with labelled total ctDNA from wt *E. gracilis* digested with *Eco*RI. **C** - probing with labelled *Eco*RI fragment "E" containing genes *psbC* and *psbD*.

The detection of reduced amounts of ctDNA in chemically treated *E. gracilis* is in agreement with the reported presence of the chloroplast genome in bleached *Euglena* mutants [8, 11]. A colourless heterotrophic relative, *Astasia longa,* contains a chloroplast genome that has lost half of its genome complexity compared with its photosynthetic counterpart *E. gracilis* [5]. Grossly reduced plastid genomes are present in nonphotosynthetic parasitic higher plants [16]. Residual plastid genomes have recently been detected also in apicomplexan parasites [12]. These apicoplast genomes are considered as welcome new parasite-specific targets for therapeutic agents inclusive quinolone drugs [4].

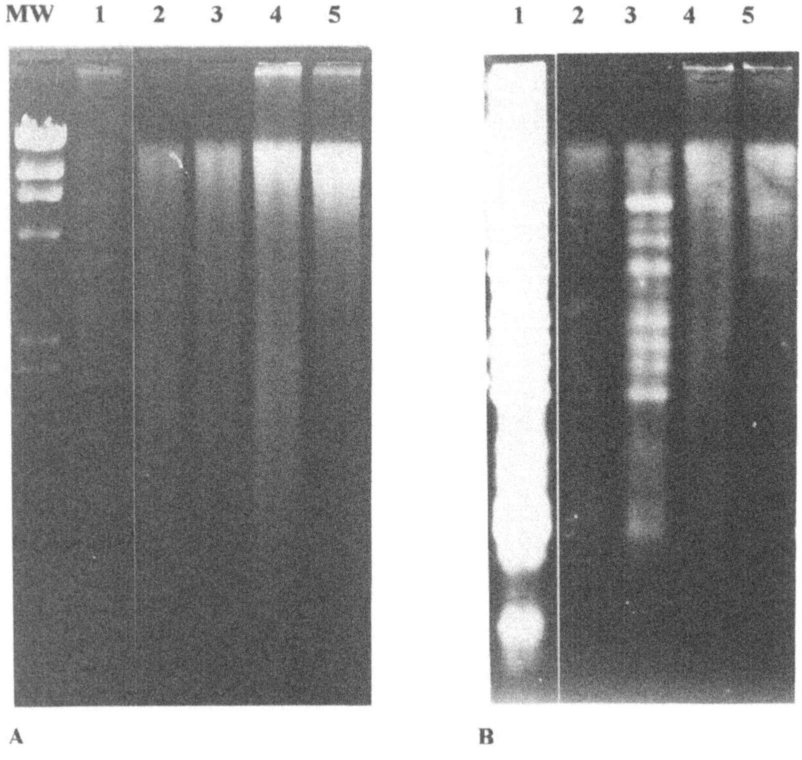

Figure 2. Agarose gel electrophoresis of ctDNA from *E. gracilis* treated with nitrosoguanidine (NG) and ofloxacin (OFL). Digestion with *Hind*III. **1** - ctDNA-wt , **2** - ctDNA-NG - 15μg/ml for 24 hours, **3** - ctDNA-NG - 12.5μg/ml for 8 hours, **4** - ctDNA-OFL - 100μg/ml for 7 days, **5** - ctDNA-OFL - 100μg/ml for 5 days. **MW** - λDNA/*Hind*III. **A** - detection by ethidium bromide staining under UV. **B** - lumigraph of the gel A. Southern blot was probed with labelled total ctDNA from wild type *E. gracilis* digested with *Eco*RI.

3.1.1. *Chloroplast DNA during heat bleaching*

Results from heat bleaching experiments are partly different from those with chemically stressed cells. We have observed that heat induced loss of chloroplast functions is associated with partial chloroplast genome rearrangements; i.e. novel hybridizing fragments (smaller than the expected size for wt ctDNA fragments) appeared (Figure 3A). However, after hybridisation also typical wild type ctDNA restriction fragment patterns were detected (Figure 3B). In contrast, Conkling *et al.* [1] observed delayed but complete loss of chloroplast DNA in heat-bleached cultures of *E. gracilis* (ctDNA was undetectable at 100 h). This discrepancy might be due to the different DNA isolation method used, i.e. our large scale preparation of ctDNA from purified chloroplasts *versus* their minipreparation of total DNA from 10^5 cells.

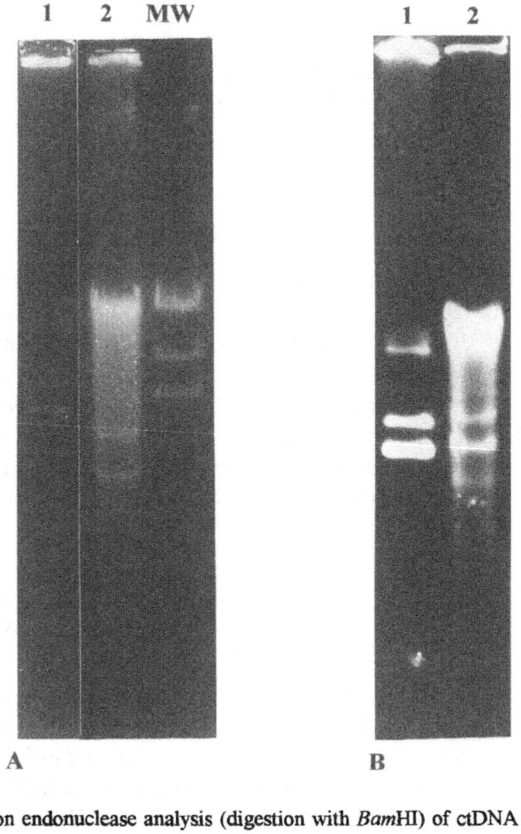

Figure 3. Restriction endonuclease analysis (digestion with *Bam*HI) of ctDNA from wt *E. gracilis* and cells cultivated under hyperthermic conditions (33°C) for 7 days. **1** - ctDNA-wt, **2** - ctDNA-33°C, **MW** - λDNA/*Hind*III. **A.** Detection by ethidium bromide staining under UV. **B** - lumigraph of the gel A. Southern blot was probed with labelled total ctDNA from wt *E. gracilis* digested with *Hae*III.

3.2. RNA IN BLEACHED MUTANTS

Electrophoretograms of *E. gracilis* total RNA show that there is an obvious difference between wild type cells and bleached mutants. Sharp bands of the 28S, 23S, 18S+16S ribosomal RNA from cytoplasmic and plastidial ribosomes are present in wt cells with photosynthetically active chloroplasts and also in the green mutant GZSmL resistant to streptomycin. Bands for 23S and 16S rRNA are lost in all bleached mutants (Figure 4A).

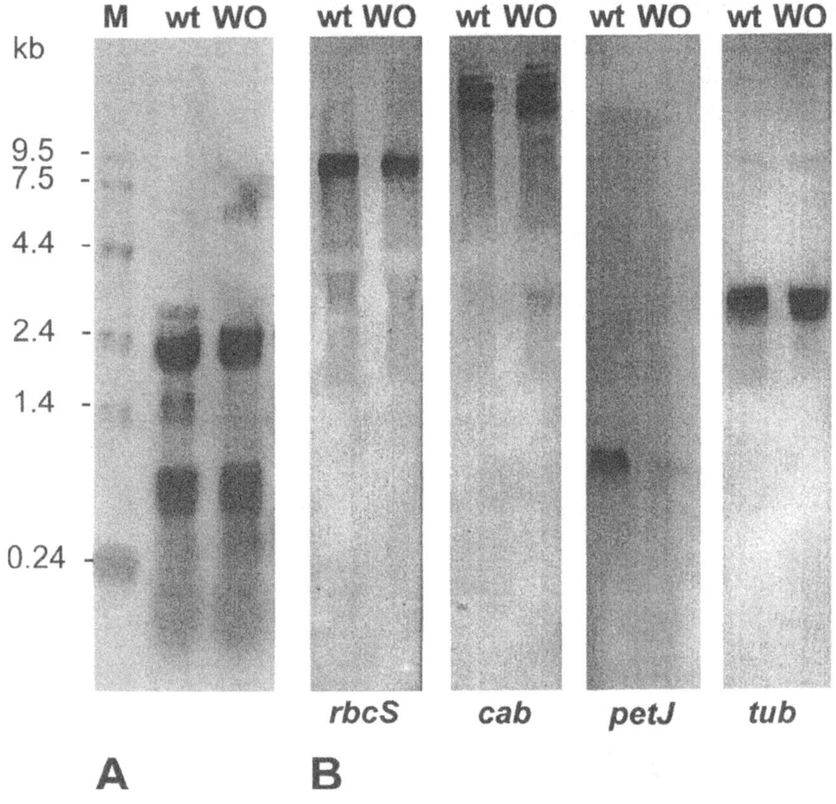

Figure 4. Northern blot analysis of transcripts for nuclear-encoded chloroplast proteins. Total RNA isolated from light-grown *E. gracilis* wild type (wt) and from the white mutant WZOflL (WO) induced by ofloxacin as a representative example of our collection of bleached mutants in following respect: **A**. Visualisation by methylene blue (M - RNA ladder, Gibco BRL). **B**. Hybridisation and detection (DIG system, Boehringer Mannheim) with homologous DNA probes: Rubisco SSU (*rbcS*), chlorophyll a/b-binding protein - LHCPII (*cab*) and cytochrome c_6 (*petJ*). Probe for cytosceletal protein β-tubulin (*tub*) is added as a positive control.

Subsequent hybridisation with three DNA probes for nuclear encoded chloroplast proteins: two different regions of the gene for the small subunit of Rubisco (SSU and mSSU) and chlorophyll a/b-binding protein (LHCPII), as well as with the β-subunit of tubulin provided identical signals in the case of wt cells, *A. longa* and all mutants studied (Figure 4B). These findings suggest that in euglenoids there is no feedback between the stage of chloroplasts and transcription of these nuclear-encoded genes for chloroplast proteins. If this is the case, then there is a principal difference in regulation of expression of those photosynthetic genes between *E. gracilis* and higher plants [10].

3.2.1. *Cytochrome c_6 gene expression*
When cytochrome c_6 (*petJ*) was used as the fourth probe variable hybridisation signals were obtained with northern blots from an ofloxacin-bleached mutant depending on the light régime. The weak signal under light conditions (Figure 4B) indicates that in this case there is a feedback between the state of plastids and the expression of nuclear-encoded photosynthetic genes which resembles the situation in higher plants [10]. Contrary, positive signal in the darkness suggests the existence of more complicated pathways in biosynthesis of chloroplast system in *E. gracilis*.

4. Acknowledgements

Gene probes for Rubisco SSU - 4.3 kbp fragment and β-tubulin were kindly provided by P. Imbault (Strasbourg). We thank S. Schwartzbach (Lincoln) for donating gene probes for LHCPII - 600 bp fragment and mSSU - 420 bp fragment. Part of the work reported was supported by a grant from Ministry of Education of Slovak Republic No. 1/4234/97.

5. References

1. Conkling, A.B., Thomas, E.J., and Ortiz, W. (1993) Delayed but complete loss of chloroplast DNA in heat-bleaching cultures of *Euglena gracilis*. *J. Plant Physiol.* **142,** 307-311.
2. Ebringer, L. and Krajčovič, J. (1994) *Cell Origin and Evolution*, Publishing House VEDA, Bratislava.
3. Ebringer, L., Polónyi, J., and Krajčovič, J. (1993) Influence of ofloxacin on chloroplasts and mitochondria in *Euglena gracilis*. *Arzneimittel-Forschung/Drug Research* **43,** 777-781.
4. Fichera, M.E. and Roos, D.S. (1997) A plastid organelle as a drug target in apicomplexan parasites. *Nature* **390,** 407-409.
5. Gockel, G., Hachtel, W., Baier, S., Fliss, C., and Henke, M. (1994) Genes for components of the chloroplast translation apparatus are conserved in the 73 kb plastid DNA of the nonphotosynthetic euglenoid flagellate *Astasia longa*. *Curr. Genet.* **26,** 256-262.
6. Hallick, R.B., Hong, L., Drager, R.G., Favreau, M.R., Monfort, A., Orsat, B., Spielman, A., and Stutz, E. (1993) Complete sequence of *Euglena gracilis* chloroplast DNA. *Nucleic Acid Res.* **21,** 3537-3544.
7. Hashimoto, H. and Murakami, S. (1982) Chloroplast replication and loss of chloroplast DNA induced by nalidixic acid in *Euglena gracilis*. *Cell Struct. Func.* **7,** 111-120.
8. Heizmann, P., Doly, J., Hussein, Y., Nicolas, P., Nigon, V., and Bernardi, G. (1981) The chloroplast genome of bleached mutants of *Euglena gracilis*. *Biochim. Biophys. Acta* **653,** 412-415.
9. Heizmann, P., Hussein, Y., Nicolas, P., and Nigon, V. (1982) Modifications of chloroplast DNA during streptomycin induced mutagenesis in *Euglena gracilis*. *Curr. Genet.* **5,** 9-15.
10. Hess, W.R., Linke, B., and Börner, T. (1997) Impact of plastid differentiation on transcription of nuclear and mitochondrial genes, in H.E.A. Schenk, R.G. Herrmann, K.W. Jeon, N.E. Müller, and

W. Schwemmler (eds.), *Eukaryotism and Symbiosis,* Springer-Verlag, Berlin - Heidelberg - New York, pp. 233-242.
11. Hussein, Y., Heizmann, P., Nicolas, P., and Nigon, V. (1982) Quantitative estimations of chloroplast DNA in bleached mutants of *Euglena gracilis. Curr. Genet.* **6,** 111-117.
12. Köhler, S., Delwiche, C.F., Denny, P.W., Tilney, L.G., Webster, P., Wilson, R.J.M., Palmer, J.D., and Roos, D.S. (1997) A plastid of probable green algal origin in apicomlexan parasites. *Science* **275,** 1485-1489.
13. Krajčovič, J., Ebringer, L., and Polónyi, J. (1989) Quinolones and coumarins eliminate chloroplasts from *Euglena gracilis. Antimicrob. Agents Chemother.* **33,** 1883-1889.
14. Krajčovič, J., Chreňo, O., and Ebringer, L. (1990) Different effect of hyperthermia and heat shock on the action of quinolone drugs versus some mutagens against chloroplasts of *Euglena gracilis. Mutation Res.* **244,** 21-26,.
15. Nicolas, P., Hussein, Y., Heizmann, P., and Nigon, V. (1980) Comparative studies of chloroplastic and nuclear DNA repair abilities after ultraviolet irradiation of *Euglena gracilis. Mol. Gen. Genet.* **178,** 567-572.
16. Wolfe, K.H., Morden, C.W., and Palmer, J.D. (1992) Function and evolution of a minimal plastid genome from a nonphotosynthetic parasitic plant. *Proc. Natl. Acad. Sci. USA* **89,** 10648-10652.

CHLOROPLAST ENVELOPE MEMBRANE

PROTEIN IMPORT ACROSS CHLOROPLAST ENVELOPES

K. SOHRT, and J. SOLL
Botanisches Institut, Christian Albrechts-University
Am Botanischen Garten 1-9, 24118 Kiel, Germany

1. Introduction

The vast majority of chloroplast proteins is nuclear encoded and cytoplasmically expressed as precursor proteins. As a consequence preproteins destined for the organelle interior need to be posttranslationally imported into the chloroplast across both envelope membranes. The translocation involves two distinct import complexes at the outer and inner envelope membrane in cooperation with soluble factors in the cytosol and the stroma. Within the last years several proteins have been identified as members of the chloroplast import machinery, and understanding of the import process is rising.

2. Chloroplast targeting by transit peptides

Nuclear encoded chloroplast precursor proteins destined for translocation across the envelope membranes contain N-terminal extensions referred to as transit peptides. Even though these chloroplast specific targeting signals have no homologous sequences and vary enormously in length (20-120 amino acids), they share distinct common features. They exhibit an N-terminal moiety without glycine, proline or positively charged amino acids, a central part poor in acidic residues and with high amounts of serine and threonine, and a C-terminus with a predicted amphiphilic ß-strand and a cleavage site [1]. Furthermore, transit peptides contain a consensus motif, which is recognized by a cytosolic precursor protein protein kinase that phosphorylates a Ser/Thr-residue within this motif in an ATP-dependent manner [2]. In contrast to the nonphosphorylated mitochondrial or peroxisomal preproteins the phosphorylation might function as an additional regulator signal for chloroplast precursor protein import.

The import of some, but not all chloroplast precursor proteins is stimulated by other cytosolic factors such as the chaperone Hsp70 [3]. It is thought to keep precursor proteins in an import competent conformation and prevent protein aggregation. This function is maintained throughout the whole import process by several chaperones interacting with the precursor protein during different translocation steps (see below). Further cytosolic components are imaginable to add to a specific sorting and chloroplast recognition process, for the maintenance of the import competent state, or for regulation of precursor protein import.

J. H. Argyroudi-Akoyunoglou and H. Senger (eds.),
The Chloroplast: From Molecular Biology to Biotechnology, 131–136.

The information contained in transit peptides is both necessary and sufficient to target precursor proteins to the chloroplast. Transit peptides and chloroplast outer envelope membranes seem to have reciprocal effects in a protein-lipid interaction. The N- and C-proximal regions of transit peptides cause a transient reorganization of membrane lipids that is attributed to MGDG, an abundant non-bilayer forming lipid in the outer envelope [4]. Simultaneously, the random coil conformation of the N- and C-proximal portions of transit peptides in aqueous solution undergo an increase in secondary structure when brought into membrane lipid mimetic environments [5, 6]. These conformational changes might facilitate protein import initiation by inducing the necessary secondary structure motifs to bind to the receptor of the outer envelope translocation complex, called Toc complex.
Precursor protein translocation occurs in at least three steps. The first binding to the Toc complex takes place without any energy requirements and seems to be reversible. In the second stage low concentrations of ATP, 5-50 μM, change the weak interactions into strong binding of precursor proteins [7]. An increase to concentrations above 100 μM ATP finally drives precursor proteins into the stroma [8, 9, 10].

3. The Toc complex

Five outer envelope membrane or membrane associated proteins, namely Toc 34, Toc75, Toc86, Com44, and Hsp70 homologs, have been identified as components of the import machinery by direct interaction with chloroplast precursor proteins. On the cytosolic side of the outer envelope COM70 or SCE70, an Hsp70 homolog, is associated with the membrane and can be crosslinked to chloroplast precursor proteins [3, 11]. It might cause the strong unfolding activity of precursor proteins observed on the cytosolic surface of plastids [12, 13].
Toc86 is a highly protease sensitive protein [14] that becomes strongly crosslinked to bound precursor proteins in label transfer studies [15]. Fab fragments generated from antibodies raised against Toc86 inhibit binding and import of chloroplast precursor proteins [16]. Therefore, Toc86 is considered to be an import receptor. It exhibits a phosphorylation as well as distinct ATP and GTP binding sites indicating an extensive import regulation via Toc86. The only sequence found to show homologies to Toc86 belongs to Toc34, which exact function has not been established yet. Toc34 is a highly protease sensitive protein representing a new type of GTP-binding protein [17, 18]. Its C-terminus is inserted into the outer envelope, and the larger N-terminal moiety including the GTP-binding motif extends into the cytosol. The GTPase activity suggests a regulatory function that takes place in an early stage of protein import, because in label transfer studies interactions of Toc34 and precursor proteins could only be demonstrated in absence of NTP's [15], a condition under which precursor binding is still reversible (see above). Under nonreducing conditions Toc34 was observed to form a disulfide bridge with Toc75 [18], the major protein of the outer envelope [19]. Secondary structure prediction of Toc75 proposed 16 hydrophobic ß-sheets penetrating the outer membrane [20, 21] explaining the highly protease resistant behavior of Toc75 [14]. Under irreversible precursor binding conditions Toc75 was the most prominent protein crosslinked to chloroplast precursor proteins [22, 23]. These results led Toc75 to be discerned as the import

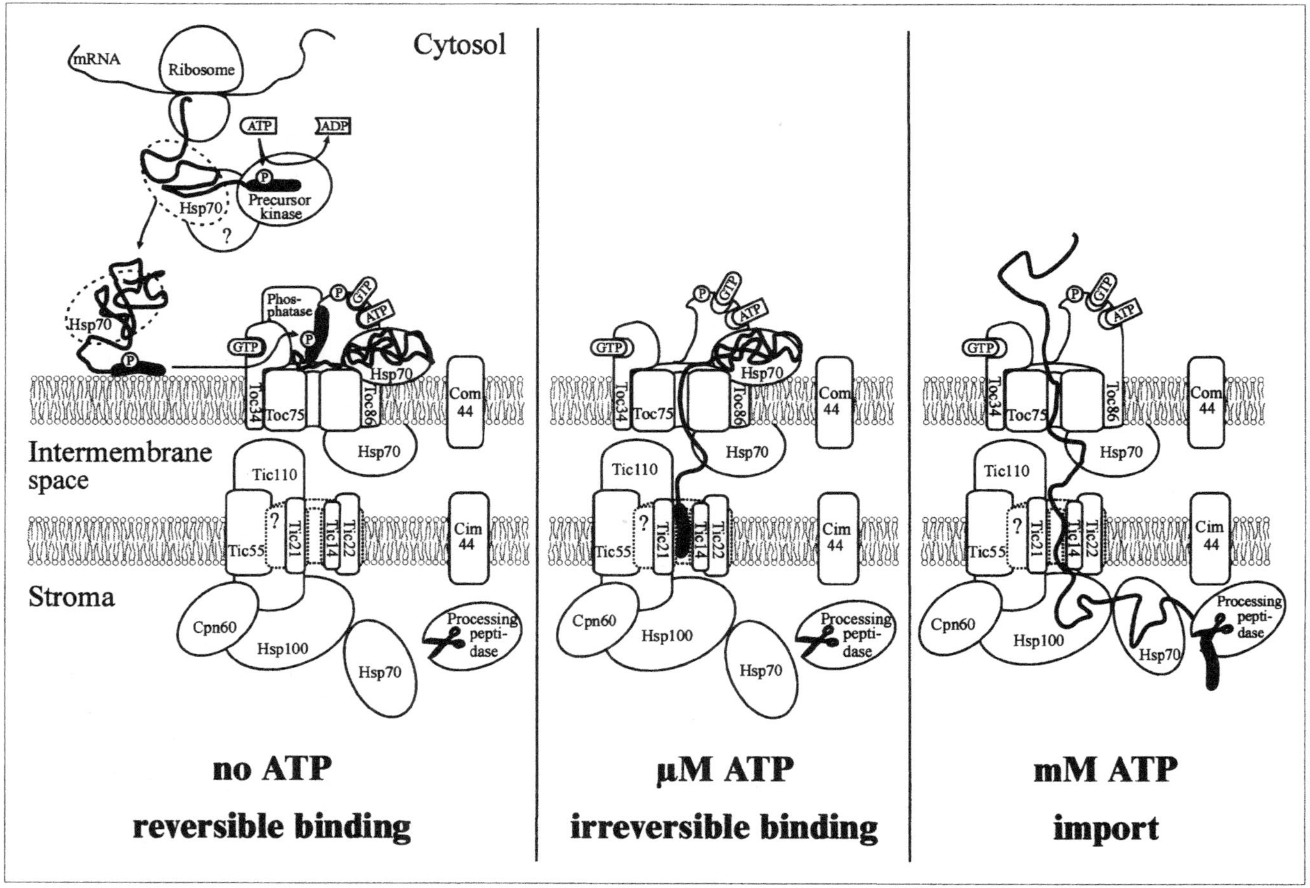

Figure 1. Model for chloroplast protein import

pore of the Toc complex. The function as a translocation channel was verified by conductivity measurements with heterologously expressed Toc75 reconstituted in liposomes. The electrophysiological properties indicated a rather narrow pore with a diameter of 8-9 Å [21]. However, the import of a tightly folded protein with a with of 18 Å [24] suggests a more flexible and regulated translocation pore *in vivo*. Toc75 is discussed to contain a specific precursor recognition site, because a low level of precursor import takes place even without Toc86 and the ion conductivity by reconstituted Toc75 in liposomes is reduced by the precursor of the small subunit of ribulose bisphosphate carboxylase/oxygenase but not by its mature form [21].
Com44 is a member of the Cim/Com44 family, which localization could not be assigned to just one, the outer or inner envelope membrane [25]. A crosslink product of Com44 with a chimeric precursor suggests a close proximity with the partially translocated protein during the import process [26]. The definite function of Com44 remains to be elucidated.
Furthermore, a phosphatase is proposed to be at least temporarily associated with the outer envelope translocon being a necessity for dephosphorylation of phosphorylated transit peptides to enable import of precursor proteins [2]. Since unphosphorylated precursors follow the import pathway without hindrance, phosphorylation is neither a prerequisite nor a direct part of the import mechanism.
An additional Hsp70 homolog was found at the inner leaflet of the outer membrane, where it behaved like an integral membrane protein within the Toc complex [27]. It might maintain the import competent state of translocating precursor proteins entering the intermembrane space and the import complex of the inner envelope, named Tic complex.

4. The Tic complex

During the import process the inner and outer envelopes form sites where the membranes stay near to each other with the Toc and Tic complexes in close vicinity [28]. Little is known about the organization and function of the Tic complex subunits. So far, six proteins, Tic14, Tic21, Tic22, Tic55, Tic110, and Cim44 were proposed to belong to the inner envelope translocon (see below).
Neither Tic55 nor Tic110 could be proven to directly interact with translocated precursor proteins, but both occurred in complexes with translocating proteins and components of the Toc complex [29, 30, 31]. Tic110 contains at least one hydrophobic membrane anchor located near its N-terminus. Protease treatment studies gave contradictory results whether the large hydrophilic part protrudes into the intermembrane space or the stroma [31, 32]. Extending into the intermembrane space Tic110 could mediate the formation of joint translocation sites between the outer and inner envelope translocons which would correspond with the copurification of Tic110 with precursor proteins and Toc components in a detergent-solubilized complex [31, 20]. The stroma regions of Tic110 seem to function as a binding site for chaperones and chaperonins which is supported by the coimmunoprecipitation with Tic110 [30, 33]. Since the Tic110 topology is under dispute further experimental evidence is needed to elucidate its function.

Tic55 was identified in association with Tic110, Hsp100, Toc34, Toc75, and Toc86 [29]. According to the primary structure Tic55 possesses a Rieske-type iron-sulfur-cluster and a mononuclear iron binding site. A prevention of iron binding by diethylpyrocarbonate resulted in a precursor translocation stagnation at the level of the Tic complex [29], which implies that Tic55 might represent a regulatory translocon subunit working as a biosensor of oxidants and iron.
Crosslinking to precursor proteins established Tic14, Tic21, Tic22 [15, 22] and Cim44, the counterpart of Com44 [26], as proteins involved in chloroplast precursor import. But nothing is known about their molecular structure, association with the Tic complex, or function.

On the stromal side of the inner envelope chaperonin cpn60 and chaperones Hsp70 and ClpC, an Hsp100, are associated with translocating precursor proteins and/or the Tic complex [30, 33]. The chaperones might exert a strong unfolding and inward moving force on precursor proteins and thereby pull them into the stroma in an ATP-dependent manner.
Reaching the stroma, precursor proteins undergo an endoproteolytical processing by a chloroplast processing enzyme, a stromal metalloendoprotease cleaving off the transit peptides [34, 35]. Released from the Tic complex the processed proteins continue to their destination at the thylakoids, inner envelope [1, 31], or fold into their functional conformation within the stroma.

5. References

[1] v.Heijne, G., Stepphuhn, J., Herrmann, R.G. (1989) Domain structure of mitochondrial and chloroplast targeting peptides, *Eur. J. Biochem.* **180**, 535-545.
[2] Waegemann, K. and Soll, J. (1996) Phosphorylation of the transit sequence of chloroplast precursor proteins, *J. Biol. Chem.* **271**, 6545-6554.
[3] Waegemann, K., Paulsen, H., and Soll, J. (1990) Translocation of proteins into chloroplasts requires cytosolic factors to obtain import competence, *FEBS-Letters* **261**, 89-92.
[4] Pinnaduwage, P. and Bruce, B.D. (1996) *In vitro* interaction between a chloroplast transit peptide and chloroplast outer envelope lipids is sequence-specific and lipid class-dependent, *J. Biol. Chem.* **271**, 32907-32915.
[5] Endo, T., Kawamura, K., and Nakai, M. (1992) The chloroplast targeting domain of plastocyanin transit peptide can form a helical structure but does not have a high affinity for lipid bilayers, *Eur. J. Biochem.* **207**, 671-675
[6] Horniak, L., Pilon, M., van't Hof, R., and de Kruijff, B. (1993) The secondary structure of the ferredoxin transit sequence is modulated by its interaction with negatively charged lipids, *FEBS Lett.* **334**, 241-246.
[7] Olsen, O.J. and Keegstra, (1992) The binding of precursor proteins to chloroplasts requires nucleoside triphosphates in the intermembrane space, *J. Biol. Chem.* **267**, 433-439.
[8] Flügge, U.I., and Hinz, G. (1986) Energy dependence of protein translocation into chloroplasts, *Eur. J. Biochem.* **160**, 563-570.
[9] Schindler, C., Hracky, R., and Soll, J. (1987) Protein transport in chloroplasts: ATP is prerequisit, *Z. Naturforsch.* **42c**, 103-108.
[10] Theg, S.M., Bauerle, C. Olsen, L.J. Selman, B.R. and Keegstra, K. (1989) Internal ATP is the only energy requirement for the translocation of precursor proteins across chloroplastic membranes, *J. Biol. Chem.* **264**, 6730-6736.
[11] Ko, K., Bornemisza, O., Ko, Z.W., Plaxton, W.C., and Cashmore, A.R. (1992) Isolation and characterization of a cDNA clone encoding a cognate 70-kDa heat shock protein of the chloroplast envelope, *J. Biol. Chem.* **267**, 2986-2993.

[12] America, T., Hageman, J., Guera, A., Rook, F., Archer, K., Keegstra, K., and Weisbeek, P. (1994) Methotreate does not block import of a DHFR fusion protein into chloroplasts, *Plant Mol. Biol.* **24**, 283-294

[13] Guera, A., America, T., van Waas, M., and Weisbeek, P.J. (1993) A strong protein unfolding activity is associated with the binding of precursor chloroplast proteins to chloroplast envelopes, *Plant Mol. Biol.* **23**, 309-324.

[14] Cline, K., Werner-Washburne, M., Andrews, and J. Keegstra, K. (1984) Thermolysin is a suitable protease for probing the surface of intact pea chloroplasts, *Plant Physiol.* **75**, 675-678.

[15] Kouranov, A. and Schnell, D.J. (1997) Analysis of the interactions of preproteins with the import machinery over the course of protein import into chloroplasts, *J. Cell. Biol.* **139**, 1677-1685.

[16] Hirsch, S., Muckel, E., Heemeyer, F., v. Heijne, G., and Soll, J. (1994) A receptor component of the chloroplast protein translocation machinery, *Science* **266**, 1989-1992.

[17] Kessler, F., Blobel, G., Patel, H.A., and Schnell, D.J. (1994) Identification of two GTP-binding proteins in the chloroplast protein import machinery, *Science* **266**, 1035-1039.

[18] Seedorf, M., Waegemann, K., and Soll, J. (1995) A constituent of the chloroplast import complex represents a new type of GTP-binding protein, *Plant J.* **7**, 401-411.

[19] Joyard, J., Billecocq, A., Bartlett, S.G., Block, M.A. and Chua, N.H. (1983) Localization of polypeptides to the cytosolic side of outer envelope membrane of spinach chloroplasts, *J. Biol. Chem.* **258**, 10000-10006.

[20] Schnell, D.J., Kessler, F., and Blobel, G. (1994) Isolation of components of the chloroplast protein import machinery, *Science* **266**, 1007-1012.

[21] Hinnah, S.C., Hill, K., Wagner, R., Schlicher, T., and Soll, J. (1997) Reconstitution of a chloroplast protein import channel, *EMBO J.* **16**, 7351-7360.

[22] Ma, Y., Kouranov, A., LaSala S.E., and Schnell, D.J. (1996) Two components of the chloroplast import apparatus, IAP86 and IAP75, interact with the transit sequence during the recognition and translocation of precursor proteins at the outer envelope, *J. Cell. Biol.* **134** (2), 315-327.

[23] Perry, S.E. and Keegsrta, K. (1994) Envelope membrane proteins that interact with chloroplastic precursor proteins, *Plant Cell.* **6**, 93-105.

[24] Clark, S.A. and Theg, S.M. (1997) A folded protein can be transported across the chloroplast envelope and the thylakoid membranes, *Mol. Biol. Cell* **8**, 923-934.

[25] Ko, K., Budd, D., Seibert, F., Kourtz, L., Ko, Z.W. (1995) Isolation and characterization of a cDNA clone encoding a member of the Com44/Cim44 envelope components of the chloroplast protein import apparatus. *J. Biol. Chem.* **270**, 28601-28608.

[26] Wu, C., Seibert, F.S., and Ko, K. (1994) Identification of chloroplast envelope proteins in close physical proximity to a partially translocated chimeric precursor protein, *J. Biol. Chem.* **269**, 32264-32271.

[27] Waegemann, K. and Soll, J. (1991) Characterization of the import apparatus in isolated outer envelopes of chloroplasts, *Plant J.* **1**, 149-158.

[28] Schnell, D., and Blobel, G. (1993) Identification of intermediates in the pathway of protein import into chloroplasts and their localization to envelope contact sites, *J. Cell.Biol.* **120**, 103-115.

[29] Caliebe, A. Grimm, R., Kaiser, G., Lübeck, J., Soll, J., and Heins, L. (1997) The chloroplastic protein import machinery contains a Rieske-type iron-sulfur cluster and a mononuclear iron-binding protein, *EMBO* **16**, 7342-7350.

[30] Kessler, F. and Blobel, G. (1996) Interaction of the protein import and folding machineries of the chloroplast, *Proc. Natl. Sci. USA* **93**, 7684-7689.

[31] Lübeck, J., Soll, J., Akita, M., Nielsen, E., and Keegstra, K. (1996) Topology of IEP110, a component of the chloroplastic protein import machinery present in the inner envelope membrane, *EMBO J.* **15**, 4230-4238.

[32] Jackson, D.T., Froehlich, J.E., and Keegstra, K. (1998) The hydrophyllic domain of Tic110, an inner envelope membrane component of the chloroplast protein translocation apparatus, faces the stromal compartment, *J. Biol. Chem.* **273**, 16583-16588.

[33] Nielsen, E., Akita, M., Davila-Aponte, J., and Keegsrta. K. (1997) Stable association of chloroplastic precursors with protein translocation complexes that contain proteins from both envelope membranes and a stromal HSP100 molecular chaperone, *EMBO J.* **16**, 935-946.

[34] Richter, S. and Lamppa, G.K. (1998) A chloroplast processing enzyme functions as the general stromal processing peptidase, *Proc. Natl. Sci. USA* **95**, 7463-7468.

[35] Smith, S. and Ellis, R.J. (1976) Processing of small subunit precursor of ribulose bisphosphate carboxylase and its assembly into whole enzyme are stromal events, *Nature* **278**, 662-664.

CHARACTERIZATION OF THE *ARABIDOPSIS PPI1* MUTANT

P. JARVIS, C. PETO and J. CHORY
The Salk Institute for Biological Studies and HHMI
La Jolla, California 92028, USA
L.-J. CHEN and H.-M. LI
Academia Sinica, Nankang
Taipei 11529, Taiwan, ROC

1. Abstract

Genetic screens for loci involved in the positive regulation of LHCP gene expression have identified a series of *Arabidopsis* mutants displaying altered plastid development. One such mutant, *ppi1* (plastid protein import), has a pronounced pale or yellow-green phenotype throughout its life-cycle, although mature leaves frequently appear darker green than younger tissues. Molecular analyses of the *ppi1* locus revealed the presence of a T-DNA insertion within the second intron of a gene for a protein with similarity to GTP-binding proteins of the plastidic outer envelope membrane. Specifically, the *PPI1* gene encodes a protein of 33 kD which is highly homologous with Toc34 (translocon at the outer envelope membrane of chloroplasts, 34 kD), a protein involved in the translocation of proteins into the plastid. Import experiments using isolated chloroplasts demonstrated that the new protein, referred to as Toc33, inserts into the outer envelope membrane in a manner similar to Toc34. Given (1) that the majority of proteins entering the chloroplast do so through the general import apparatus (of which Toc34 is a component), (2) that the *ppi1* mutation is null, and (3) that the *ppi1* mutant phenotype is not severe, we believe that Toc33 is involved in the general import process as a redundant translocon component which acts interchangeably with Toc34. Consistently, complementation of the *ppi1* genetic defect by over-expression of Toc34 demonstrated the functional similarity of the two homologs. Detailed characterization of *ppi1* plants has provided clear evidence for the perturbation of plastid protein import *in vivo*. Our data also suggest that the expression pattern of Toc33 may be responsible, at least in part, for developmental changes in the import capabilities of plastids.

2. Biochemical Studies of Chloroplast Protein Import

Mitochondria and chloroplasts are believed to have been derived from the endosymbiotic uptake of ancient prokaryotes by an early eukaryotic progenitor. Although both organelles have retained their own genomes, the evolutionary transfer of genetic material to the nucleus has meant that the vast majority of proteins required for the formation of either organelle is now encoded in the nucleus and translated on cytoplasmic ribosomes. The displacement of these organellar genes, combined with the fact that each organelle is surrounded by a double membrane system, has given rise to the need for complex mechanisms for the translocation of proteins from the cytoplasm to their site of function in either the mitochondria or the plastids. The routing of nucleus-encoded proteins to the

J. H. Argyroudi-Akoyunoglou and H. Senger (eds.),
The Chloroplast: From Molecular Biology to Biotechnology, 137–142.

chloroplast is an especially important process for plants since these plastids are the site of the energy-generating process, photosynthesis, and because such organelles account for about 50% of all soluble protein in leaves.

The majority of nucleus-encoded proteins destined for the plastid are synthesized as preproteins carrying cleavable, amino-terminal extensions referred to as transit sequences [1]. Transit sequences provide information necessary for the targeting of proteins to the plastid and are removed following transfer across the envelope membrane. A common route across the plastidic envelope membrane is followed by all proteins destined for interior locations within the organelle; this pathway is referred to as the general protein import pathway. General protein import is believed to be mediated by a common set of proteins (the general protein import apparatus) and recently a great deal of progress has been made in the identification of proteins involved in the translocation process. *In vitro*, biochemical analyses using isolated pea chloroplasts have been predominantly employed in this respect [1,2].

Three major components of the outer envelope segment of the general protein import apparatus (Toc34, Toc75 and Toc86) were identified by virtue of their association with preproteins arrested at early stages of envelope translocation; Toc stands for translocon at the outer envelope membrane of chloroplasts, and the suffix in each case refers to the predicted molecular weight of the protein. Chemical cross-linking studies have also been employed, providing further evidence for the interaction of these and other proteins with translocation intermediates. Such data suggest that Toc86 interacts with preproteins at a very early stage and the protein is thus assumed to play an important role in recognition; Toc86 can thus be regarded as being part of the receptor for transit sequence binding. Recent studies have provided compelling evidence for the role of Toc75 as the channel forming molecule through which translocation occurs, although the protein is also believed to play a role in recognition. The role of Toc34 remains uncertain but its involvement in the translocation process is inferred from its stoichiometric association with Toc75 and Toc86 as part of the translocon complex, and from its demonstrable interaction with translocation intermediates (albeit to a lesser extent than either Toc75 or Toc86). Toc34 has been proposed to act in concert with Toc86 in regulating the presentation of preproteins to the protein-conducting apparatus, and also to play a role in the gating of the Toc75 channel, although there is no direct evidence to support either model. That Toc34 and Toc86 are homologous GTP-binding proteins is consistent with their proposed roles in the regulation in precursor recognition. While several components of the inner envelope translocon complex have been identified, our understanding of events at the inner envelope membrane is much less advanced. Other proteins associated with both the outer and inner envelope segments of the translocon include several molecular chaperones. These proteins are required to maintain preproteins in an unfolded, translocation-competent state during import, and are also thought to provide the driving force for the passage of proteins across the envelope.

3. Genetic Studies of Mitochondrial and Thylakoidal Protein Import

Biochemical studies of chloroplast protein import have greatly advanced our understanding of the mechanisms employed for the translocation of proteins across the plastidic envelope membrane, and have identified many of the protein components involved in mediating the import process. While the value of such studies cannot be doubted, a major caveat is that the conclusions drawn are based on *in vitro* findings. In order to demonstrate conclusively the role of identified envelope proteins in the plastid protein import process, ways must be found to analyse their roles *in vivo*. Genetic analyses have been particularly useful in the dissection of protein import into mitochondria [3] and the internal thylakoid membranes of chloroplasts.

A positive screening procedure, similar to that used for the genetic dissection of the yeast secretory pathway, was employed for the identification of yeast mutants with defects in mitochondrial protein import. This screen involved an introduced chimeric gene comprising the matrix-targeting signal of a mitochondrial gene and a positively selectable reporter gene, the functionality of which is dependent upon its localization in the cytosol. Mutations affecting the activity of the mitochondrial import apparatus were found to disrupt translocation of the fusion protein into the organelle and therefore conveyed a selectable phenotype on the affected cells. Multiple alleles of four loci were identified using this strategy: Tim17, Tim23, Tim44 and mtHsp70; Tim stands for translocon at the inner membrane of mitochondria, and the suffix in each case refers to the predicted molecular weight of the protein. Tim17 and Tim23 are integral membrane proteins which are thought to constitute part of the inner membrane translocation channel. Tim44 is a peripheral membrane protein which interacts with translocating preproteins and may co-operate with the molecular chaperone, mtHsp70, in driving translocation across the mitochondrial membrane. In addition to the identification of novel loci and the provision of *in vivo* data to support the involvement of previously identified translocon components in mitochondrial import, genetic analyses have been successfully utilized to assess possible interactions between different translocon components. Strains carrying mild mutations in any two of the Tim17, Tim23 and Tim44 translocon genes were found to express a synthetic lethal phenotype; such synthetic lethality is regarded as valid genetic evidence for a functional interaction between gene products. A different genetic approach was used to provide evidence for a functional interaction between Tim44 and mtHsp70. Over-expression of mtHsp70 was shown to partially suppress the import defects of mild Tim44 mutants, but not those of a null Tim44 mutant; conversely, the over-expression of Tim44 was found to alleviate the import defects of a mild mtHsp70 mutant.

Several genetic studies have led to the successful identification of loci involved in the transport of proteins across the thylakoid membranes of chloroplasts. Six *Chlamydomonas* loci were identified in a screen for nuclear suppressors which permit the translocation of cytochrome *f* proteins with mutant thylakoid signal sequences, thereby restoring the ability of affected cells to grow photoautotrophically [4]. These suppressor loci, referred to as *tip* for thylakoid insertion protein, have yet to be characterized at the molecular level but it is assumed that they will identify components of the translocation apparatus. The *tha1* (thylakoid assembly) and *hcf106* (high-chlorophyll fluorescent) mutants of maize were identified in screens for transposon-induced nuclear mutations which cause the loss of components of the photosynthetic apparatus [5,6]. Both mutations are recessive and result in pale green, non-photosynthetic seedlings which die shortly after germination. The mutations affect the localization of different subsets of proteins to the thylakoid lumen and thus provide evidence in support of conclusions drawn from *in vitro* studies [1] which have provided evidence for at least three different pathways for protein translocation across the thylakoid membrane.

Most proteins destined for the thylakoid membranes carry lumen-targeting domains at their amino-termini which resemble bacterial signal peptides and which are removed following transfer. A subset of thylakoid proteins (*e.g.* the 33 kD subunit of the oxygen evolving complex, OE33) engage a thylakoid translocation system which requires ATP and the stromal ATPase, cpSecA, and which bears a striking resemblance to the bacterial secretory (Sec) pathway for protein export across the cytoplasmic membrane. A second pathway (that followed by the protein OE17) does not require ATP but is strictly dependent upon a pH gradient (ΔpH) across the thylakoid membrane. The third mechanism for thylakoid translocation (utilized by the light-harvesting chlorophyll protein, LHCP) is homologous to the signal recognition particle (SRP) system for protein targeting to the endoplasmic reticulum and the bacterial cytoplasmic membrane. This pathway requires GTP and the stromal protein cpSRP54, a homolog of the cytoplasmic SRP54 protein. The *tha1* mutation was shown to affect the translocation of proteins such as OE33 specifically, and not others such as OE17 or LHCP, suggesting that the affected locus is involved in the cpSecA-dependent pathway. The *Tha1* gene was later cloned and shown to encode cpSecA itself [5]; bacterial SecA functions to initiate insertion of precursor proteins into a core translocon made up of SecY/E/G polypeptides, and it is thus assumed that cpSecA performs a similar role in chloroplasts. By contrast with the *tha1* mutant, *hcf106* plants exhibit defective thylakoid translocation of OE17 and other proteins utilizing the ΔpH pathway. The *Hcf106* gene encodes a receptor-like thylakoid membrane protein with homology to open reading frames from all completely-sequenced bacterial genomes [6]. This suggests that the ΔpH pathway, like the Sec and SRP pathways, has been conserved since the endosymbiotic origin of chloroplasts.

A reverse genetic approach has also been successfully employed in the analysis of thylakoid protein targeting [7]. Maize *Mutator* lines were screened specifically, using a PCR-based strategy, for transposable element insertions in the cpSecY gene. The gene was selected as one likely to be involved in Sec-dependent thylakoid translocation since the bacterial homolog forms part of the translocon for protein secretion across the plasma membrane. Null cpSecy, or *csy1*, mutants were found to have near albino phenotypes with severely reduced thylakoid membrane development. This is in contrast with the relatively mild phenotype of the cpSecA-deficient *tha1* mutant (and indeed of the *hcf106* mutant) suggesting that cpSecY may play a role in multiple thylakoid targeting pathways. The *csy1* phenotype is also more pleiotropic than that of a double mutant with defects in both the Sec- and ΔpH-dependent pathways which implies that its role extends even beyond these two pathways. Since Sec61p, the endoplasmic reticulum homolog of SecY, functions in the translocation of proteins which engage SRP54, it seems likely that cpSecY is also involved in the targeting of proteins such as LHCP which follow the SRP pathway.

4. Characterization of the *Arabidopsis ppi1* Mutant

Given the informative nature of the described analyses of mitochondrial and thylakoidal protein translocation, the application of genetic techniques to the study of protein transport across the plastid envelope would appear to be an obvious next step. A major barrier to the initiation of such studies has been the conviction that mutations in envelope translocon components would result in lethal phenotypes due to the apparently unique role performed by the general protein import machinery. We describe unpublished data on the identification and characterization of an *Arabidopsis* mutant lacking such an envelope translocon component which demonstrate the potential utility of genetics for the analysis of plastid protein import.

Genetic screens for loci affecting the expression of nucleus-encoded photosynthetic proteins have identified several mutants with defects in plastid biogenesis. Here we describe one such mutant which was identified as having a recessive, pale phenotype in a population of T-DNA-mutagenized plants. While the mutant appears uniformly pale during the first two weeks of its life-cycle, the oldest leaves of mature plants frequently have an appearance closer to wild type. The gene disrupted by the T-DNA insertion was found to encode a 33 kD protein having substantial homology with Toc34 from both *Arabidopsis* and pea. Toc34, as was mentioned earlier, is a GTP-binding protein of the plastid outer envelope membrane and is believed to be involved in protein translocation across the envelope [2]. The new mutant was thus named *ppi1* for plastid protein import, but the protein encoded by the *PPI1* gene is referred to as Toc33 in accordance with the established nomenclature. Given their strong sequence similarity, it seemed likely that Toc33 and Toc34 mediate similar processes *in vivo*. Consistent with this notion, Toc33 was found to insert into the outer envelope membrane of isolated chloroplasts in a manner similar to Toc34. Over-expression of Toc34 was furthermore shown to complement *ppi1* plants. The existence of two functionally similar Toc proteins suggests that at least two distinct translocon complexes exist in *Arabidopsis*. This observation is of particular interest since biochemical studies have revealed the presence of only one Toc complex in pea.

We have examined the expression of Toc33 and Toc34 throughout *Arabidopsis* development and characterized the *ppi1* phenotype with respect to chlorophyll accumulation, plastid ultrastructure and plastid protein import. Messenger RNA levels of both Toc33 and Toc34 were found to be high in very young plants but to decline rapidly as the age of the plant increased. Toc34 was expressed at a low level relative to Toc33 throughout development. Expression of both genes was also found to be substantially higher in the young, expanding tissues of mature plants than in mature, fully-expanded leaves, suggesting that the elevated expression of Toc33 in very young plants is due to them having more young, expanding tissues. Mutant plants were found to contain reduced levels of chlorophyll at all stages of developmental, although they were most extremely pale when very young. Mutant chlorophyll levels improved steadily throughout development implying an involvement of Toc33 in the chloroplast development which occurs concomitantly with the expansion of newly formed tissues. The most pronounced defects in plastid ultrastructure were observed in the tissues of young plants, again indicating a role for Toc33 during the early stages of plastid development. Etioplasts from five day-old *ppi1* plants contained prolamellar bodies of a markedly reduced size, and chloroplasts from plants of a similar age were very small and contained few and under-developed thylakoid membranes. In contrast, the chloroplasts of mature *ppi1* leaves were in many cases indistinguishable from wild type. Mutant plastids never appeared aberrant to a degree indicative of a complete block in their development; this observation can be accounted for by the functional similarity of Toc33 and Toc34.

To determine the precise nature of the defect in *ppi1* plants, we assayed the import of several proteins into mutant chloroplasts isolated from plants grown for either ten days or four weeks. We found that protein import into the plastids of mature *ppi1* plants occurs with wild-type efficiencies. In contrast, the translocation efficiencies of plastids from ten day-old mutant plants were substantially reduced. These observations are consistent with other aspects of the *ppi1* phenotype and are indicative of a role for Toc33 early in plastid and leaf development. To corroborate the findings of these *in organello* studies, immunoblot experiments designed to assess plastid protein import defects *in vivo* were conducted. Five day-old, etiolated mutant plants were found to contain reduced levels of the mature protochlorophyllide oxidoreductase protein (POR) and a pronounced accumulation of non-processed preprotein (pPOR). The stability of accumulated pPOR for many hours after transferal of plants to light suggests that the precursor is localized outside of the plastid since a chloroplast-specific protease is known to degrade both POR and pPOR. Our data suggest that previously described developmental changes in plastid protein import are mediated, at least in part, by changes in Toc33 expression. They thus imply that changing import capabilities are the result of variations in the number of translocon complexes at the envelope membrane. The reason for the presence of the two genes, Toc33 and Toc34, remains unclear, but one possibility is that they represent the simplest means of achieving the necessary pattern of expression. It is also possible that, under normal conditions in wild-type plants, subtle functional differences between the homologs exist. Most significantly, these data demonstrate the role of an envelope protein in the translocation process *in vivo*.

Our analyses of the *Arabidopsis ppi1* mutant clearly demonstrate the feasibility of applying genetic techniques to the study of the plastidic general protein import pathway. *Arabidopsis* genes encoding several envelope translocon components have recently been identified as a result of the genome sequencing project; reverse genetic techniques should enable the identification of mutants lacking these components in the near future. Positive screening strategies, similar to those used in dissecting mitochondrial import in yeast, provide a complementary approach and will allow the identification of weak mutant alleles. Once several mutants have been isolated, double mutant analyses will enable functional interactions between the affected proteins to be assessed. Genetic suppressor screens of *ppi1* and other translocon mutants are also likely to provide a means of identifying new and interesting loci. We hope that this study will provide the impetus for many, similar genetic studies of plastid protein import in the future.

5. Acknowledgements

This work was supported by grants from the US Department of Energy (ER13993; to J.C.) and Academia Sinica and the National Science Council of Taiwan (NSC 87-2312-B-001-018, to H.m.L.) and by an HFSP fellowship (to P.J.). J.C. is an associate investigator of the Howard Hughes Medical Institute (HHMI).

6. References

1. Cline, K. and Henry, R. (1996) Import and routing of nucleus-encoded chloroplast proteins, *Annu. Rev. Cell Dev. Biol.* **12**, 1-26.
2. Heins, J., Collinson, I., and Soll, J. (1998) The protein translocation apparatus of chloroplast envelopes, *Trends Plant Sci.* **3**, 56-61.
3. Pfanner, N., Craig, E.A., and Meijer, M. (1994) The protein import machinery of the mitochondrial inner membrane, *Trends Biochem. Sci.* **19**, 368-372.
4. Bernd, K.K. and Kohorn, B.D. (1998) *Tip* loci: Chlamydomonas nuclear suppressors that permit the translocation of proteins with mutant thylakoid signal sequences, *Genetics* **149**, 1293-1301.
5. Voelker, R., Mendel-Hartvig, J., and Barkan, A. (1997) Transposon-disruption of a maize nuclear gene, *tha1*, encoding a chloroplast SecA homologue: *in vivo* role of cp-SecA in thylakoid protein targeting, *Genetics* **145**, 467-478.
6. Settles, A.M., Yonetani, A., Baron, A., Bush, D.R., Cline, K., and Martienssen, R. (1997) Sec-independent protein translocation by the maize Hcf106 protein, *Science* **278**, 1467-1470.
7. Roy, L.M. and Barkan, A. (1998) A SecY homologue is required for the elaboration of the chloroplast thylakoid membrane and for normal chloroplast gene expression, *J. Cell Biol.* **141**, 385-395.

ISOLATION AND CHARACTERIZATION OF *CHLAMYDOMONAS* MUTANTS DEFICIENT IN THE PLASTID *YCF10* OPEN READING FRAME

N. ROLLAND[1-2], G. AMOROSO[3], D. BERNY-SEIGNEURIN[2], A.-J. DORNE[2], D. F. SÜLTEMEYER[3], J. JOYARD[2] and J.-D. ROCHAIX[1]

[1] *Departments of Molecular Biology and Plant Biology, University of Geneva, CH-1211 Geneva 4, Switzerland.* [2] *Laboratoire de Physiologie Cellulaire Végétale, URA 576 (CEA/CNRS/Université Joseph Fourier), Département de Biologie Moléculaire et Structurale, CEA-Grenoble, 38054 Grenoble-cedex 9, France.* [3] *Department of Biology, University of Kaiserslautern, Postfach 3049, D-6750 Kaiserslautern, Germany*

1. Introduction

The complete DNA sequences of a dozen plastid genomes have been determined and they have revealed the existence of approximately 120 genes that can be divided in three major groups. The first group (approximately 50 genes) corresponds to sequences required for plastid transcription and translation. The second group (approximately 40 genes) codes for components of the photosynthetic apparatus. The last group contains many open reading frames of unknown function (for review, see [1]). Some of these are species-specific and designated as ORFs. Those that are conserved in algae and in higher and lower plants have been designated *ycf* for "hypothetical chloroplast open reading frame". Chloroplast reverse genetics coupled with biophysical and biochemical analysis has become a powerful tool for analyzing the function of chloroplast-encoded proteins. Several of these ORFs of unknown function have been inactivated in *Chlamydomonas reinhardtii* and tobacco by biolistic transformation (for review, see [2]). This paper summarizes the characterization of a *Chlamydomonas ycf10*-deficient mutant. An unusual feature of this chloroplast gene is that it encodes a plastid envelope protein.

2. The *ycf10* open reading frame codes for a chloroplast envelope protein

The *ycf10* gene has been identified in all chloroplast genomes examined, with exeption of *Euglena* [3]. The *ycf10* gene codes for a putative polypeptide of approximately 230 amino acids in higher plants. This ORF was located upstream of *petA* (encoding apo-cytochrome *f*) in tobacco [4], rice [5], and *Marchantia* [6]. In *C. reinhardtii*, *ycf10* is cotranscribed with *atpA* and *psbI*. An additional non-*atpA* containing *ycf10* transcript has also been detected [7]. In pea, *ycf10* has been shown to be cotranscribed with *petA* [8,

J. H. Argyroudi-Akoyunoglou and H. Senger (eds.),
The Chloroplast: From Molecular Biology to Biotechnology, 143–148.

9]. The Ycf10 product was immuno-localized in the inner membrane of the pea chloroplast envelope [10, 11]. Accordingly, *ycf10* was renamed *cemA* (chloroplast envelope membrane). Two homologues were also detected in the cyanobacteria *Synechocystis sp.* PCC6803 (slr1596 or *cotA*; [12] and sll1685, Cyanobase). The *cotA* gene product was immunolocalized in the cytoplasmic membrane of *Synechocystis* [13].

3. The *Chlamydomonas ycf10* gene has unusual features

Cloning the *ycf10* gene from *C. reinhardtii* has revealed several unusual features, as compared to its higher plant homologues: localization in the chloroplast genome, size of the gene and presence of both 5' and internal additional sequences [14]. In *C. reinhardtii, ycf10,* located between. the *psbI* and *atpH* genes encodes a putative hydrophobic protein of 500 residues, which greatly exceeds the size of its higher plant homologues.

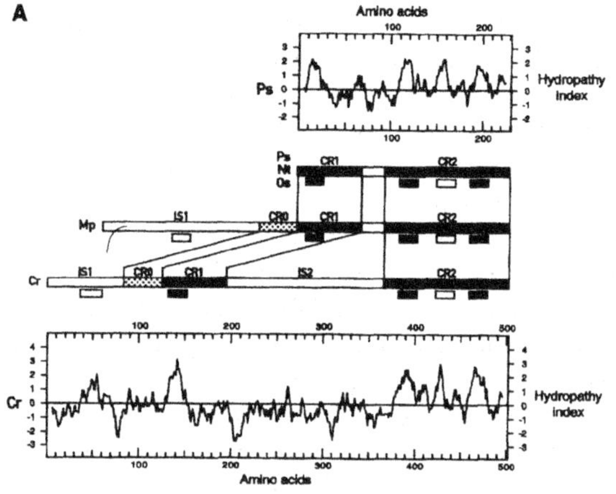

Figure. 1. Comparative analysis of the structure of Ycf10 of *C. reinhardtii* and plants. Cr, *C. reinhardtii*; Ps, *Pisum sativum*; Nt, *Nicotiana tabacum*; Os, *Oryza sativum*; Mp, *Marchantia polymorpha*. The two *Synechocystis* homologues present the structure of the protein from Mp (only N-terminal additional sequences). Gray regions (CR) represent conserved domains. White regions (IS) represent additional sequences. Boxes correpond to hydrophobic domains. Only black boxes are predicted to be transmembrane segments.

This unusual structure derives from the insertion of two additional sequences. The N-terminal sequence (130 residues) is partially conserved with a corresponding sequence found in the *M. polymorpha* and *Synechocystis* homologues. The second additional sequence (150 residues) is present only in *C. reinhardtii* and separates the conserved N and C termini (Figure 1). The size of the putative gene product is compatible with that of two of the four chloroplast encoded polypeptides found to be associated with the chloroplast envelope of *C. reinhardtii* [15]. The presence of additional sequences in algae, lower plants and cyanobacteria may confer additional functions to the protein or allow interaction with other chloroplast envelope components. Conservation of the N-terminal additional sequence in the gene product of one *Synechocystis* homologue has already been demonstrated [13]. Conservation of the additional sequences in the *C. reinhardtii ycf10* gene product remains to be demonstrated.

4. Disruption of the *Chlamydomonas ycf10* gene does not affect cell viability or photosynthetic reactions, but induces light sensitivity

To gain further insights into the function of *ycf10*, we have disrupted this chloroplast gene in the green unicellular alga *C. reinhardtii*, using biolistic transformation. Obtention of complete segregation of the mutated gene copy in the chloroplast genome (homoplasmic transformants) first demonstrated that the Ycf10 product is not essential for cell viability. The *ycf10*-deficient mutants were also found to grow both photoheterotrophically and photoautotrophically under low light (Figure 2), thereby revealing that the Ycf10 product is not essential for the photosynthetic reactions.

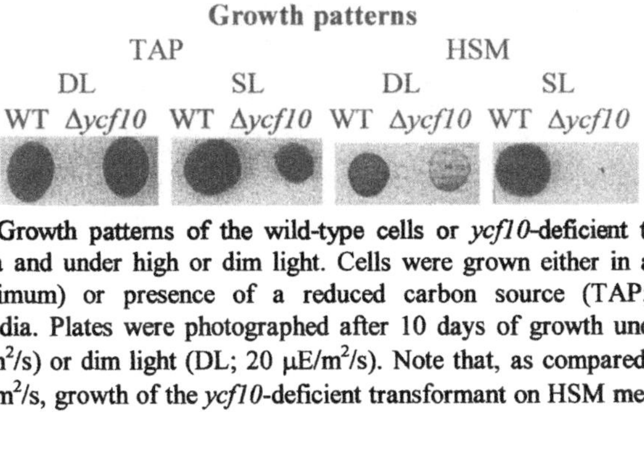

Figure 2: **A.** Growth patterns of the wild-type cells or *ycf10*-deficient transformant in different media and under high or dim light. Cells were grown either in absence (HSM; High-Salt-Minimum) or presence of a reduced carbon source (TAP; Tris-Acetate-Phosphate) media. Plates were photographed after 10 days of growth under strong light (SL; 100 $\mu E/m^2/s$) or dim light (DL; 20 $\mu E/m^2/s$). Note that, as compared to [14], where DL was 3 $\mu E/m^2/s$, growth of the *ycf10*-deficient transformant on HSM medium is already affected.

When tested under high light, the *ycf10*-deficient mutants did not grow photoautotrophically and barely photoheterotrophically (Figure 2). This light sensitivity could not be related to altered chlorophyll, carotenoid or quinone metabolism. The light sensitivity of the *ycf10*-deficient mutants was suppressed when growing the cells in the presence of spectinomycin, or when the *ycf10* deletion is combined with a deletion of *psbI*, which is known to limit accumulation of PSII. It was then concluded that *ycf10* is involved in a process essential for utilization and/or dissipation of photochemical energy and thus that *ycf10* function is essential only when the chloroplast electron transfer chain is highly active [14].

5. Light sensitivity of the *Chlamydomonas ycf10*-deficient mutants correlates with a reduced affinity of the CO_2 and HCO_3^- uptake systems for their substrates

Cyanobacterial mutants affected in CO_2 transport have been shown to contain mutations within the *cotA* gene which displays significant sequence identity to *ycf10* [12], thus raising the question whether Ycf10 may be required for CO_2 assimilation in the chloroplast of *Chlamydomonas.*

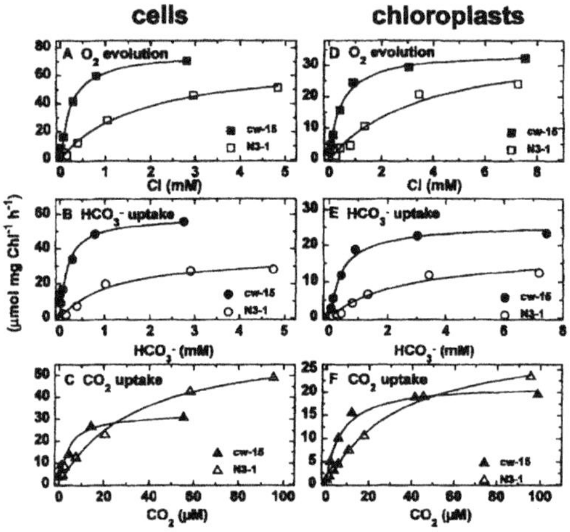

Figure 3. Substrate dependence of O_2 evolution, HCO_3^- and CO_2 uptake in intact cells and chloroplasts from wild-type (CW15) and the *ycf10*-deficient mutant (N3-1 = CW15, *ycf10::aadA*). Chloroplasts were purified from cell Wall-deficient strains (CW15).

Oxygraphic studies on *ycf10*-deficient *C. reinhardtii* mutants has revealed that loss of *ycf10* strongly limits the apparent photosynthetic affinity for Ci when mutant cells have reached the CO_2-compensation point. Furthermore, the deletion increases the sensitivity of mutants to inhibitors of carbonic anhydrases [14]. Mass spectrometric measurements with either whole cells or isolated chloroplasts from the transformants revealed that both the CO_2 and HCO_3^- uptake systems have a reduced affinity for their substrates (Figure 3, Table 1).

Spreitzer and Mets [16, 17] reasoned that *Chlamydomonas* mutants capable of photosynthetic electron transport but blocked in the dark reactions of photosynthesis would accumulate NADPH that could not be oxidized, and thus these mutants would die in the light. By using a dark selection, they selected the first *Chlamydomonas* mutant affected in the function of the large subunit of RUBISCO and demonstrated it's light sensitivity [16, 17]. Thus, a mutant affected in efficient uptake of inorganic carbon would present the same phenotype, but would be light sensitive only under high light intensities, *i.e.* only when the passive diffusion of CO_2 would become rate limiting.

The increased light sensitivity of the transformants appears to be a consequence of a limitation in photochemical energy utilization and/or dissipation which correlates with a decrease in CO_2-dependent photosynthesis and a reduced affinity of the CO_2 and HCO_3^- uptake systems for their substrates. These results suggest the existence of a *ycf10*-dependent system within the plastid envelope which promotes efficient inorganic carbon uptake into chloroplasts.

TABLE 1. Values for $K_{1/2}$ (Ci) (μM) for O_2 evolution, HCO_3^- and CO_2 uptake in cells for wild-type and the *ycf10*-deficient mutant. the numbers ± SD represent mean values from three separate experiments. CO_2 and O_2 gas exchanges were determined by Mass-spectrometric measurements.

Cells		**Wild-type**		***Δycf10***	
Growth medium		TAP	HSM	TAP	HSM
O_2 evolution	$K_{1/2}$ (Ci)	230 ± 37	738 ± 8.0	1642 ± 192	3300 ± 382
HCO_3^- uptake	$K_{1/2}$ (HCO_3^-)	168 ± 21	48 ± 6.0	1133 ± 186	1970 ± 232
CO_2 uptake	$K_{1/2}$ (CO_2)	3.8 ± 0.2	2.5 ± 0.3	32 ± 3.9	37.4 ± 5.2
Chloroplasts		**CW15**		**CW15, *ycf10::aadA***	
Growth medium		TAP		TAP	
O_2 evolution	$K_{1/2}$ (Ci)	457 ± 71		3731 ± 839	
HCO_3^- uptake	$K_{1/2}$ (HCO_3^-)	453 ± 64		2481 ± 771	
CO_2 uptake	$K_{1/2}$ (CO_2)	6.7 ± 0.8		34.3 ± 5.3	

6. Reduced affinity of the CO_2 and HCO_3^- uptake systems for their substrates in the *ycf10*-deficient mutants: a secondary effect ?

The cyanobacterial *cotA* product was recently demonstrated to be required for light-induced proton extrusion [13, 18]. Acidification of cyanobacterial periplasmic would accelerate conversion of HCO_3^- to CO_2 and thus, facilitate CO_2 diffusion in the cytoplasm of cyanobacteria. It was concluded that the inhibition of CO_2 transport in the mutant of cyanobacteria was a secondary effect of the inhibition of proton extrusion [13]. Several arguments would argue for this hypothesis, including the fact that high CO_2 concentrations cannot rescue the *Chlamydomonas ycf10*-deficient transformants.

Further genetic and biochemical experiments are clearly required to identify the exact function of the *ycf10* gene product. Species-specific functions cannot be excluded, since the protein of unicellular organisms contains additional sequences and since specific phenotype were observed in cyanobacteria and algae [13, 14]. Genetical and environmental conditions which can suppress the phenotype of the *Chlamydomonas ycf10*-deficient transformants may also help in understanding such a mechanism.

Finally, when considering the chloroplast envelope membrane and the control of photosynthesis, the Pi/TPi transporter is generally considered as the unique point of control of photosynthesis rate. A *ycf10*-dependent mechanism that modulate CO_2 uptake into the chloroplast, directly or indirectly, can also be considered as another good candidate for regulation of photosynthetic activity.

7. References:

1. Reardon, E.M. and Price, C.A. (1995) Plastid genes of three non-green algae are sequenced. *Plant Mol. Biol. Rep.* **13**, 320-326.1.
2. Rochaix, J.-D. (1997) Chloroplast reverse genetics: new insights into the function of plastid genes. TIPS., **2**, 419-425
3. Hallick, R.B., Hong, L., Drager, R.G., Faureau, M.R., Montfort, A., Orset, B., Spielmann, A. and Stutz, E. (1993) Complete sequence of *Euglena Gracilis* chloroplast DNA. *Nucleic Acids Res.* **25**, 3537-3544
4. Shinozaki, K. *et al.* (1986) The complete nucleotide sequence of the tobacco chloroplast genome: Its gene organisation and expression. *EMBO J.*, **5**, 2043-2049.
5. Hiratsuka, J. *et al.* (1989) The complete sequence of the rice (*Orysa sativa*) chloroplast genome. Intermolecular recombination between distinct tRNA genes accounts for a major plastid DNA inversion during the evolution of the cereals. *Mol. Gen. Genet.*, **217**,185-194
6. Ohyama, K. *et al.* (1986) Chloroplast gene organisation deduced from complete sequence of liverwort *Marchantia polymorpha* chloroplast DNA. *Nature*, **322**, 272-274.
7. Levy, H. Kindle, K.L. and Stern, D.B (1997) A nuclear mutation that affects the 3' processing of several mRNAs in *Chlamydomonas* chloroplasts. *The Plant Cell.* **9**, 825-836
8. Willey, D.L. and Gray, J.C. (1990) An open reading frame encoding a putative haem-binding polypeptide is co-transcribed with the pea chloroplast gene for apocytochrome *f. Plant Mol. Biol.*, **15**, 347-356.
9. Nagano, Y., Matsumo, R. and Sasaki, Y. (1991) Sequence and transcriptionnal analysis of the gene cluster *trnQ-zpfA-psaI*-ORF231-*petA* in pea chloroplasts. *Curr. Genet.*, **20**, 431-436.
10. Sasaki, Y., Sekiguchi, K., Nagano, Y. and Matsumo, R. (1993) Chloroplast envelope protein encoded by the chloroplast genome. *FEBS. Lett.*, **316**, 93-98.
11. Gray, J. (1996) Biogenesis of chloroplasts in higher plants. In: *Membranes: Specialized Functions in Plants.* M. Smallwood, J.P. Knox, D.J. Bowles, *eds.* pp. 441-458. *Bios. Scientific Publ.*
12. Katoh, A., Lee, K.S., Fukuzawa, H. Ohyama, K. and Ogawa, T. (1996a) *cemA* homologue essential to CO_2 transport in the cyanobacterium in *Synechocystis* PCC 6803. *Proc. Natl. Acad. Sci. USA,* **93**, 4006-4010.
13. Ogawa, T., Katoh, A. and Sonoda, M. (1998) Molecular mechanisms of CO_2 concentration and proton extrusion in cyanobacteria, in K. Satoh and N. Murata (eds.), *Stress responses of photosynthetic Organisms*, Elsevier Science B.V. pp. 181-196
14. Rolland, N., Dorne, A.-J., Amoroso, G., Sültemeyer, D.F., Joyard, J. and Rochaix, J.-D. (1997) Disruption of the plastid *ycf10* open reading frame affects uptake of inorganic carbon in the chloroplast of *Chlamydomonas. EMBO J.*, **16**, 6713-6726.
15. Clemetson, J.M., Boschetti, A. and Clemetson, K.J.(1992) Chloroplast envelope proteins are encoded by the chloroplast genome of *Chlamydomonas reinhardtii. J. Biol. Chem.* **267**, 19773-19779
16. Spreitzer, R.J. and Metz, L (1980) Non-mendelian mutation affecting ribulose-1-5-bisphosphate carboxylase structure and activity. Nature, **285**, 114-115
17. Spreitzer, R.J. and Metz, L (1981) Photosynthesis-deficient mutants of *Chlamydomonas reinhardtii* with associated light-sensitive phenotypes. *Plant. Physiol.* **67**, 565-569
18. Katoh, A., Sonoda, M., Katoh, H. and Ogawa, T. (1996b) Absence of light-induced proton extrusion in a *cotA*-less mutant of *Synechocystis sp.* strain PCC 6803. *J. Bacteriol.*, **178**, 5452-5455

STRUCTURE AND FUNCTION OF CEMA HOMOLOGUE (PXCA) IN CYANOBACTERIA

M. Sonoda[1], H. Katoh[2], A. Katoh[3], H. Ohkawa[1], W. Vermaas[4] and T. Ogawa[1,2]
[1]*Graduate School of Bioagricultural Sciences and* [2]*Bioscience Center, Nagoya University, Nagoya 464-8601, Japan,* [3]*Bioscience Department, Tsukuba University, Tsukuba 305-8572, Japan and* [4]*Center for the Study of Early Events in Photosynthesis, Arizona State University, Tempe, AZ 85287-1601, USA*

1. Introduction

The *cemA* (*ycf10*) gene codes for a chloroplast envelope membrane protein [1] and is conserved in higher and lower plants and in algae [2-8]. CemA in higher plants consists of 229 to 231 amino-acids [2-4] whereas that in liverwort (*Marchantia*) [5] and *Chlamydomonas* [6] is much larger and consists of 434 and 500 amino-acids, respectively. Recent sequencing of whole chloroplast genomes of *Porphyra* [7] and *Chlorella* [8] revealed that *cemA* in these algae encodes proteins of 278 and 264 amino-acids, respectively. The function of CemA is not known. Rolland et al [6] have constructed mutants by disrupting *cemA* in *Chlamydomonas*. They showed that the disruption of the gene led to increased light sensitivity and affected CO_2-dependent photosynthesis and inorganic carbon uptake.

We have isolated a homologue of *cemA* from *Synechocystis* PCC6803 as the gene which complemented mutants defective in CO_2 transport [9]. The gene, *pxcA* (formerly known as *cotA*), was also isolated from *Synechococcus* PCC7942 [10]. Mutants were constructed by disrupting *pxcA* in these two strains. In this paper, we show that the primary effect of *pxcA* disruption is inactivation of light-dependent proton extrusion and that the proton extrusion is PSII-dependent and is accompanied by an influx of protons. A possible role of PxcA-dependent proton exchange in pH and charge homeostasis is discussed.

2. Materials and Methods

The wild-type (WT) cells and *pxcA*$^-$ mutants of *Synechocystis* PCC6803 and *Synechococcus* PCC7942 and *psaAB*$^-$(PS I-less) mutant [11] were grown at 30°C in BG-11 medium [12] buffered with 20 mM N-Tris(hydroxy-methyl)methyl-2-aminoethanesulfonic acid (TES)-KOH at pH 8.0 during aeration with 3% (vol/vol) CO_2 in air. Glucose (5 mM) was added to the above medium for the growth of *psaAB*$^-$. Continuous illumination was provided by fluorescent lamps at 40 μmol photosynthetically active radia-

J. H. Argyroudi-Akoyunoglou and H. Senger (eds.),
The Chloroplast: From Molecular Biology to Biotechnology, 149–154.

tion/m^2s (400-700 nm) for *psaAB*⁻ cells which are sensitive to higher light intensity, and at 100 μmol/m^2s for the other strains.

Changes in pH of the cell suspension (3 ml) kept at 30 °C were monitored by a pH electrode with a meter (Inlar 423 and Delta 350; Mettler Toledo, Halstead Essex, UK). Details of this method and the methods for measuring uptake of CO_2 and nitrate have been described elsewhere [13, 14].

3. Results

3.1. PHYLOGENETIC TREE OF PXCA AND CEMA

The *pxcA* genes of *Synechocystis* PCC6803 and *Synechococcus* PCC7942 encode proteins of 440 and 427 amino-acids, respectively [9,10], which were close in size to CemA of *Marchantia* and *Chlamydomonas* (434 and 500 amino-acids, respectively) [5,6], but was much larger than CemA of higher plants (229-231 amino-acids) [2-4]. Although CemA proteins of *Porphyra* and *Chlorella* were smaller than PxcA and consists of 278 and 264 amino-acids, respectively [7,8], they showed higher homology to PxcA than CemA of *Marchantia* and *Chlamydomonas*. PxcA and CemA contain 4 membrane spanning domains in the C-terminal region [2-10]. The homology was high in this region but was low in the N-terminal region. Phylogenetic tree for PxcA of cyanobacteria and CemA of plants and algae, which was computed for the 230 amino-acid sequences in the C-terminal regions, suggested that PxcA and CemA have the same ancestor (Fig. 1).

In addition to *pxcA* (slr1596), another *cemA* homologue (sll1685) exists in *Synechocystis* genome [15]. This gene encodes a protein of 393 amino-acids which was less homologous to CemA than PxcA. The lower homology can be also seen from the phylogenetic tree in Fig. 1.

3.2. IDENTIFICATION AND LOCALIZATION OF PXCA

Western analysis using an antibody raised against *Synechocystis* PxcA (partial) fused to glutathione S-transferase indicated that a protein in the cytoplasmic membrane with an apparent molecular mass of 52 kDa cross-reacted with the antibody. No reacting band was

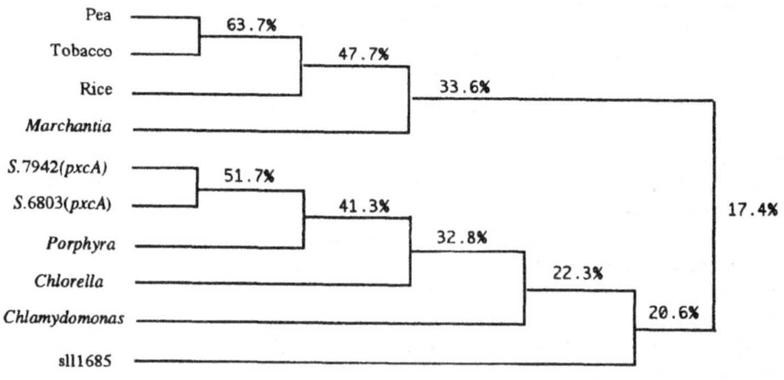

Figure 1. Phylogenetic tree for PxcA of cyanobacteria and CemA of plants and algae.

observed in the soluble fraction [16]. The antibody also reacted with the thylakoid membrane preparation. However, the cross reactivity was much less than with the cytoplasmic membrane and, therefore, is considered to be due to contamination of the cytoplasmic membrane. The 52-kDa band was not detected on the Coomassie brilliant blue staining profiles of the WT membranes. The amount of PxcA appears to be low.

3.3. GROWTH CHARACTERISTICS OF THE *pxcA*⁻ MUTANTS

Mutants were constructed by substituting *pxcA* in the WT cells of *Synechocystis* and *Synechococcus* with the omega fragment. Both mutants were unable to grow in BG11 medium (N- Na^+; ~17 mM Na^+) at pH 6.5 or, at any pH in a low Na^+ (L- Na^+; ~100 μM Na^+) under aeration with 3% (vol/vol) CO_2 in air [9,10]. The WT cells grew well in the pH range between 6.5 and 8.5 in BG11 medium but only at alkaline pH in the low Na^+ medium.

3.4. ABSENCE OF LIGHT-DEPENDENT PROTON EXTRUSION IN THE *pxcA*⁻ MUTANTS

Light-dependent proton extrusion, which has been observed with various cyanobacteria strains, was observed with the WT cells of *Synechocystis* and *Synechococcus*. Illumination of WT cells of *Synechocystis* resulted in an extrusion followed by an uptake of protons (Fig. 2, curve A). Similar profile was obtained for WT cells of *Synechococcus*. The light-dependent proton extrusion was not observed with the *pxcA*⁻ mutants (curves B and C). These results indicated that *pxcA* is involved in light-dependent proton extrusion.

A mutant constructed by disrupting another *cemA* homologue in *Synechocystis* (sll1685) showed the WT phenotype; the mutant extruded protons in the light and grew normally at pH 6.5 or in L-Na^+ medium. Another mutant constructing by inactivating *pxcA* and sll1685 genes showed the phenotype of the *pxcA*⁻ mutant.

3. 5. PHOTOSYSTEM II-DEPENDENT PROTON EXTRUSION

The proton extrusion was observed with WT cells in the presence of dimethylbenzoquinone (DMBQ; an electron acceptor from PS II) (curves A and B in Fig. 3) or with a PSI-deletion mutant(*psaAB*⁻) (curve C). The result indicated that linear electron transport *via* PS II is essential to this reaction and that the proton extrusion proceeds without PSI

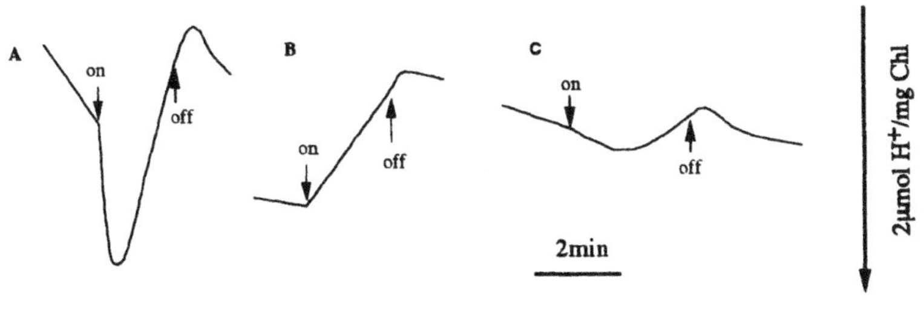

Figure 2. Net proton movements in cell suspensions of the WT (A) and *pxcA*⁻ mutant (B) of *Synechocystis* PCC6803 and *pxcA*⁻ mutant of *Synechococcus* PCC7942 (C). Light was switched on and off as indicated.

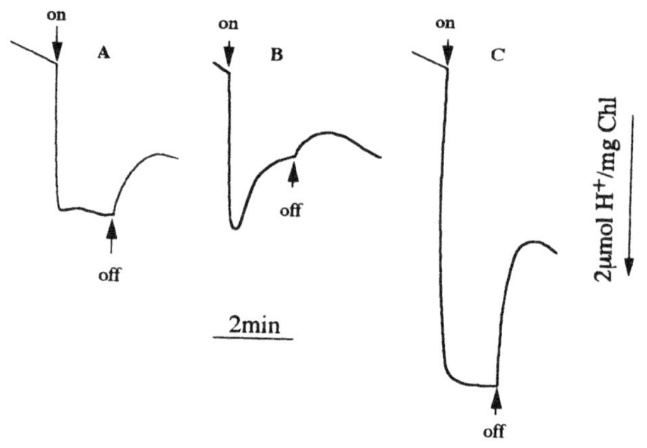

Figure 3. Net proton movement in the suspensions of WT Synechocystis (curveA) and Synechococcus (B) and *psaAB*⁻ (C) cells upon switching the light on and off. Cells were suspended in 0.2 mM Tes-KOH containing 15 mM NaCl. DMBQ was added prior to illumination for curves A and B.

reaction. Curves A and C in Fig. 3 indicate that the net proton extrusion does not proceed continuously in the light but ceases after a minute of illumination. This suggests that in the light both extrusion and influx of protons occur, reaching a stationary level where there is no net proton exchange. The influx of proton in the light was clearly observed with *Synechococcus* cells (curveB). After the light is turned off (causing proton extrusion to cease) proton influx continues for a short time until new steady-state level is attained. Thus, proton extrusion and proton uptake proceed simultaneously in the light and proton uptake continues for a short period when proton extrusion was ceased. Probably, *Δpmf* produced by proton extrusion is utilized for proton uptake.

3.6. EFFECT OF Na^+ AND pH ON THE UPTAKE OF CO_2 AND NO_3^-.

Protons are produced during the transport of CO_2; protons are consumed when NO_3^- is reduced to NH_4 *via* NO_2^-. Cells have a mechanism to maintain a homeostasis with respect to the intracellular pH and electroneutrality during these processes. To test whether the PxcA-dependent proton exchange is involved in maintaining this homeostasis, the uptake of CO_2 and NO_3^- was monitored as a function of the activity of proton exchange. For this purpose, the uptake of CO_2 and NO_3^- in WT and the *pxcA*⁻ strain was measured at pH 8.0 and 6.5 in the presence of normal concentration (15 mM) of NaCl (N-Na^+) or KCl with a low contaminating (~100μM) concentration of Na^+ (L-Na^+). At L-Na^+, the activity of NO_3^- uptake was very low in *pxcA*⁻ at pH 8.0 and was zero at pH 6.5 (lower columns in Fig. 4). At N-Na^+, no significant effect of *pxcA* inactivation was observed on CO_2 and NO_3^- uptake at pH 8.0 but CO_2-uptake activity was reduced significantly at pH 6.5 (upper and lower columns). No CO_2 uptake was observed in the mutant at pH 6.5 and L-Na^+. It is evident that the inactivation of *pxcA* strongly affected the CO_2 uptake under acidic conditions and the NO_3^- uptake at low Na^+ concentrations.

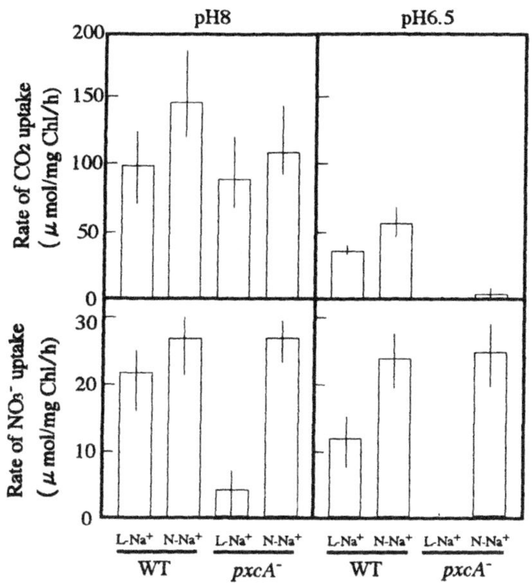

Figure 4. The rates of CO_2 and NO_3^- uptake in the WT and *pxcA*⁻ cells of *Synechocystis* at pH 8.0 and 6.5 in the presence of 15 mM NaCl (N-Na^+) or KCl (L-Na^+).

4. Discussion

Cells have a mechanism to keep their intracellular homeostasis against various environmental changes. PxcA-dependent proton exchange system may be involved in keeping intracellular pH and charge homeostasis. Another system involved in such homeostasis would be Na^+/H^+ antiport. The activity of both systems are dependent on Na^+ concentration in the medium. The hypothesis that the PxcA-dependent and Na^+/H^+ antiport-dependent proton exchange are involved in pH homeostasis may explain the growth characteristics of the WT and *pxc*- mutant. The ability of cells to keep their intracellular pH homeostasis depends on the presence of the system(s) and on Na^+ concentration in the medium. The ability would be in the following order:

WT(N-Na^+)>WT (L-Na^+], *pxcA*⁻ (N-Na^+) >*pxcA*⁻ (L-Na^+)

The highest ability will be attained by WT cells at N-Na^+ (~15 mM) and the lowest ability by the *pxcA*⁻ mutant at L-Na^+. This may explain why WT cells grow at wide range of pHs in BG11 medium and the mutant cells are unable to grow at any pHs in the L-Na^+ medium [9,10]. This model could be verified by studying the characteristics of disruption mutants of Na^+/H^+ antiporter genes. *Synechocystis* PCC 6803 possesses at least 5 Na^+/H^+ antiporter(-like) genes. We are now trying to construct such mutants.

There is no ATP-binding motif in PxcA, which suggested that PxcA is not a proton transporter driven by ATP. It is possible that PxcA plays a role in regulating or activating an H^+-ATPase or it could be a new type of proton pump. Further studies are in progress to answer these questions on the role of PxcA.

5. Acknowledgement

This study was supported by Grants-in Aid for Scientific Research (No. 09640767) from the Ministry of Education, Science, Sports and Culture of Japan and by grants from the New Energy and Industrial Technology Development Organization (NEDO) of Japan and from the Human Frontier Science Program.

6. References

1. Sasaki, Y., Sekiguchi, K., Nagano, Y.and Matsuno, R. (1993) Chloroplast envelope protein encoded by chloroplast genome. *FEBS Lett.* **316,** 93-98.
2. Willey, D.L. and Gray, J.C. (1990) An open reading frame encoding a putative haem-binding polypeptide is cotranscribed with the pea chloroplast gene for apocytochrome f. *Plant Mol. Biol.* **15,** 347-356.
3. Shinozaki, K., Ohme, M., Tanaka, M., Wakasugi, T., Hayashida, N., Matsubayashi, T., Zaita, N., Chunwongse, J., Obokata, J., Yamaguchi-Shinozaki, K., Ohto, C., Torazawa, K., Meng, B.Y., Sugita, M., Deno, H., Kamogashira, T., Yamada, K., Kusuda, J., Takaiwa, F., Kato, A., Tohdoh, N., Shimada, H. and Sugiura, M. (1986) The complete nucleotide sequence of the tobacco chloroplast genome: its gene organization and expression. *EMBO J.* **5,** 2043-2049.
4. Hiratsuka, J., Shimada, H., Whittier, R., Ishibashi, T., Sakamoto, M., Mori, M., Kondo, C. Honji, Y., Sun, C.R., Meng, B.Y., Li, Y.Q., Kanno, A., Nishizawa, Y., Hirai, A., Shinozaki, K. and Sugiura, M. (1989) The complete sequence of the rice (*Oryza sativa*) chloroplast genome: intermolecular recombination between distinct tRNA genes accounts for a major plastid DNA inversion during the evolution of the cereals. *Mol. Gen. Genet.* **217,** 185-194.
5. Ohyama, K., Fukuzawa, H., Kohchi, T., Shirai, H., Sano, T., Sano, S., Umesono, K., Shiki, Y., Takeuchi, M., Chang, Z., Aota, S., Inokuchi,H. and Ozeki, H. (1986) Chloroplast gene organization deduced from complete sequence of liverwort *Marchantia polymorpha* chloroplast DNA. *Nature* **322,** 572-574.
6. Rolland, N., Dorne, A.-J., Amoroso, G., Sultemeyer, D.F., Joyard, J. and Rochaix, J.-D. (1997) Disruption of the plastid *ycf10* open reading frame affcts uptake of inorganic carbon in the chloroplast of *Chlamydomonas. The EMBO J.* **16,** 6713-6726.
7. Reith M.E. and Munholland J. (1995) Complete nucleotide sequence of the *Porphyra purpurea* chloroplast genome. *Plant Mol. Biol. Rep.* **13,** 333-335.
8. Wakasugi, T., Nagai,T., Kapoor, M., Sugita, M., Ito, M., Ito, S., Tsudzuki, J., Nakashima, K., Tsudzuki, T., Suzuki, Y., Hamada, A., Ohta, T., Inamura, A., Yoshinaga, K.and Sugiura, M. (1997) Complete nucleotide sequence of the chloroplast genome from the green alga *Chlorella vulgaris*. The existence of genes possibly involved in chloroplast division. *Proc.Natl.Acad.Sci.U.S.A.* **94,** 5967-5972.
9. Katoh, A., Lee, K.S., Fukuzawa, H. Ohyama, K. and Ogawa, T. (1996) A *cemA* homologue essential to CO_2 transport in the cyanobacterium, *Synechocystis* PCC6803. *Proc. Natl. Acad. Sci. USA* **93,** 4006-4010.
10. Sonoda, M., Katoh, H., Ohkawa, H. and Ogawa, T. (1997) Cloning of *cotA* gene of *Synechococcus* PCC7942 and complementation of a *cotA*-less mutant of *Synechocystis* PCC6803 with chimeric genes of the two strains. *Photosynth. Res.* **54,** 99-105.
11. Shen, G., Boussiba, S. and Vermaas, W.F.J. (1993) *Synechocystis* sp. PCC6803 strains lacking photosystem I and phycobilisome function. *Plant Cell* **5,** 1853-1863.
12. Stanier, R.Y., Kunisawa, R., Mandel, M. and Cohen-Bazire, G. (1971)Purification and properties of unicellular blue-green algae (order *Chroococcales*). *Bacteriol. Rev. 35,* 171-205.
13. Katoh, A., Sonoda, M., Katoh, H. and Ogawa, T. (1996) Absence of light-induced proton extrusion in *cotA*-less mutant of *Synechocystis* sp. strain PCC6803. *J. Bacteriol.* **178,** 5452-5455.
14. Sonoda, M., Katoh, H., Vermaas, W., Schmetterer, G. and Ogawa, T. (1998) Photosynthetic Electron Transport Involved in the PxcA-Dependent Proton Extrusion in *Synechocystis* sp. PCC6803: Effect of *pxcA* Inactivation on CO_2, HCO_3^- and NO_3^- Uptake*J. Bacteriol.* in press.
15. Kaneko, T., Sato, S., Kotani, H., Tanaka, A., Asamizu, E., Nakamura, Y., Miyajima, N., Hirosawa, M., Sugiura, M., Sasamoto, S., Kimura, T., Hosouchi, T., Matsuno, A., Muraki, A., Nakazaki, N., Naruo, K., Okumura, S., Shimpo, S., Takeuchi, C., Wada, T., Watanabe, A., Yamada, M., Yasuda, M. and S. Tabata, S. (1996) Sequence analysis of the genome of the unicellular cyanobacterium *Synechocystis* sp. strain PCC6803. II. Sequence determination of the entire genome and assignment of potential protein-coding regions. *DNA Research* **3,** 109-136.
16. Sonoda, M., Kitano, K., Katoh, A., Katoh, H., Ohkawa, H. and Ogawa, T. (1997) Size of *cotA* and identification of the gene Product in *Synechocystis* sp. strain PCC6803. *J. Bacteriol.* **179,** 3845-3850.

IMPORT AND PROCESSING OF *E. COLI* EXPRESSED POLYPHENOL OXIDASE BY ISOLATED CHLOROPLASTS

S. KOUSSEVITZKY, E. NE'EMAN, E. HAREL*
Department of Plant Sciences, The Hebrew University,
Jerusalem 91904, Israel

1. Introduction

Polyphenol oxidases (PPOs) are copper containing enzymes which catalyze the oxidation of *o*-dihydroxyphenols [1]. The resulting, highly reactive, quinones undergo secondary reactions among themselves and covalently modify a variety of cellular constituents, including proteins [1]. Since PPO is believed to take part in protecting plants against herbivores and pathogens [1, 2] it is not surprising that the enzyme is sequestered in the thylakoid lumen, away from its substrates in the vacuole. PPO, the largest lumen protein described, is routed to its location in two steps. The 67 kDa precursor (pPPO) is processed by a stromal peptidase (SPP) to a 62 kDa intermediate (iPPO). The latter traverses the thylakoid and is converted by a thylakoid processing peptidase (TPP) to a 59 kDa mature protein [3, 4, 5]. It appears that translocation of iPPO across thylakoids may proceed by both the ΔpH- and Sec-dependent pathways (Koussevitzky, Ne'eman and Harel, unpublished observations).

We have over-expressed a tomato gene coding for pPPOA in *E.coli* and studied its import into chloroplasts, to enable its use in competition experiments designed to determine the mechanism of PPO translocation. Assembly of the enzyme in a phenol-free environment may enable determination of its genuine properties [6].

2. Materials and Methods

In vitro transcription and translation, isolation of plastids and sub-organellar fractions, import of protein percursors and the subsequent analysis were carried out as previously described [3, 4, 7]. The coding sequence of tomato ppoA [8] was amplified by polymerase chain reaction (PCR), subcloned in PCRTMII

J. H. Argyroudi-Akoyunoglou and H. Senger (eds.),
The Chloroplast: From Molecular Biology to Biotechnology, 155–160.

(Invitrogen, San Diego, CA), and, using the XbaI and SacI sites, transferred into pSP64poly(A)XmnI. The latter vector was formed by changing the EcoRI site of Promega's pSP64poly(A) to a XmnI site. The change was necessary to allow linearization of the vector, since the coding sequence of ppoA contains an EcoRI site. The transcription plasmids harboring Cab30 (from Lemna) and psaD (from spinach) were described previously [9, 10].

2.1. EXPRESSION IN *E. COLI* The coding region of ppoA was colned into pDS12-RBSII-2, over expressed and inclusion bodies purified as described [11]. Radioactively labeled pPPOA was prepared by adding a ^{35}S-[L-cysteine / L-methionine] mixture (Amersham) to bacteria grown on a sulfur deficient medium. pPPOA was solubilised using 8M urea , 8 mM DTT and diluted to 400 mM urea 3 mM DTT immediately before use.

3. Results and Discussion

pPPOA, a member of the tomato PPO gene family, accumulated in inclusion bodies when expressed in *E. coli* (Fig. 1). Only limited degradation of the accumulated pPPO was observed using antibodies prepared against the enzyme (Fig. 1, right panel). The accumulated pPPO was made import competent by dissolving in 8 M urea, 8 mM DTT and diluting upon addition to plastids (Fig. 2A). The time course of the import of radioactively labeled *E, coli* expressed pPPO closely resembled observations with the *in vitro* translated precursor [3].

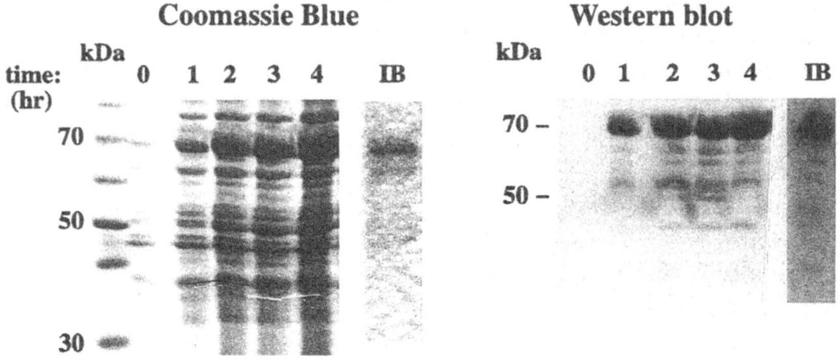

Figure 1. Overexpression of pPPO. The gene encoding pPPOA was cloned into pDS12/RBSII-2. *E. coli* cells containing the plasmid were grown in LB media to OD_{550}=0.45. IPTG was added to a final concentration of 4 mM and 20 μl samples were removed at 1 hour intervals. Cells were resuspended in buffer and denatured for 2 min at 95° C. IB - inclusion bodies (1 μg protein for the SDS-PAGE and 50 ng for the Westren blot). Antibodies against PPOwere detected using the ECL kit of Amersham.

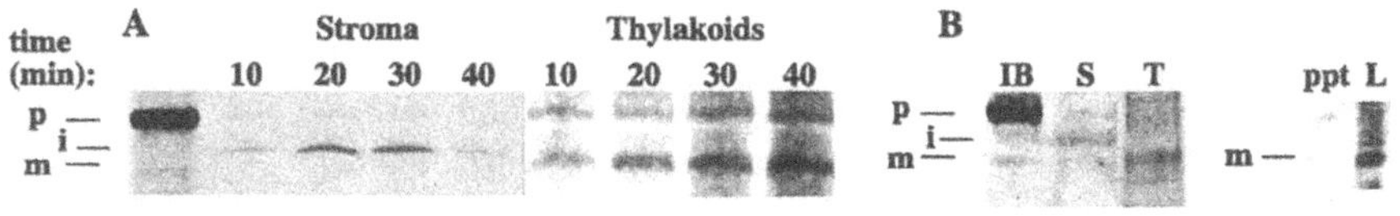

Figure 2. Import and processing of overexpressed pPPO. A. Inclusion bodies containing [^{35}S]pPPOA (500,000 cpm; 1.5 μg protein) were dissolved in 8 M urea, 8 mM DTT and diluted to 0.4 M urea, 3 mM DTT upon addition to import mixtures containing intact pea plastids (50 μg chlorophyll). Following incubation in light at 25° C for the indicated time, intact plastids were reisolated and fractionated into stroma and thylakoids. tp - 20,000 cpm of isolated inclusion bodies; precursor (p) intermediate (i) and mature (m) forms of PPO are indicated. B. Immuno-detection of imported and processed PPO species in stroma, thylakoids and (x2 concentrated) lumen (L) fractions using ECL. ppt - the membrane fraction precipitated after lysis of thylakoids and centrifugation to obtain the soluble lumen proteins fraction (L).

A transient appearance of the intermediate (iPPO) in the stroma was followed by accumulation of the mature protein in the thylakoid fraction (Fig. 2A). The three imported PPO species were also detected by antibodies (Fig. 2B), making use of the fact that pea does not contain PPO [3]. The PPO which accumulated in thaylakoids was found primarily in the lumen upon further fractionation (Fig. 2B). Analysis of competition between *in vitro* translated pPPO and relatively

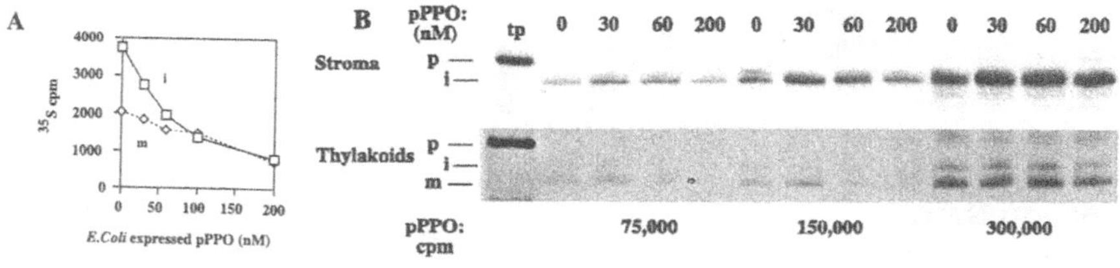

Figure 3. Competition between *E. coli*-expressed and *in vitro* translated pPPO.
A - The amount of the ^{35}S-labeled intermediate in the stroma (i_S) and the mature form in thylakoids (m_T) were plotted against the concentration of pPPO over-expressed in *E. coli*.
B - Competition at three levels of ^{35}S-labeled pPPO.

Figure 4. The effect of Cu^{2+} on the import of *E. coli*-expressed pPPO into chloroplasts.

low concentrations of the *E. coli* expressed precursor showed that the two behaved in the same manner in both routing steps - import into the stroma and translocation across the thylakoid (Fig. 3). Increasing the amount of labeled pPPO raised the relative level of its imported forms (Fig. 3B). This indicated that the apparent 'inhibition' by the over-expressed precursor was not due to poisoning of components of the translocation complex by a constituent in the inclusion bodies extract. Inhibition of over-expressed pPPO import by Cu^{2+} required higher concentrations than the ones used with the *in vitro* translated precursor [3, 4]. Furthermore, concentrations lower than 20 μM increased accumulation of mature PPO suggesting that translocation across the thylakoid might be assisted by Cu^{2+} (Fig. 4). Alternatively, Cu^{2+} might protect the protein

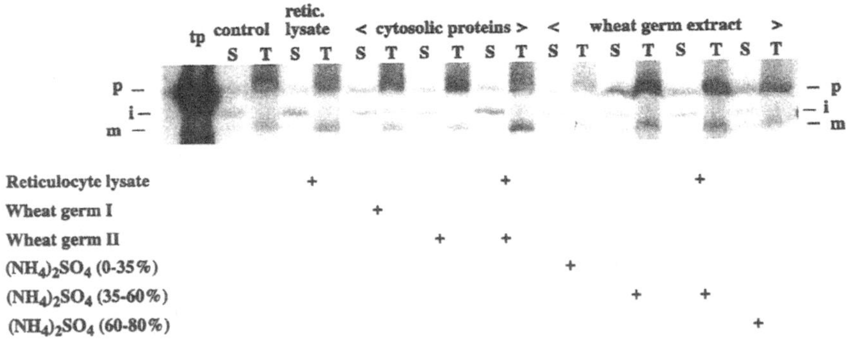

Figure 5. Imported was supported by chaperone-like factors. Reticulocyte lysate, wheat germ extract {from Promega (I) or prepared according to Chitnis *et al.* [12] (II)} or $(NH_4)_2SO_4$ fractions of cytosolic pea leaf proteins were added to the import mixture before the addition of pPPO.

against degradation. Import of *E. coli* expressed pPPO was improved by adding reticulocyte lysate, wheat germ extract or cytosolic proteins to the import mixture (Fig. 5). A combination of the first two was most effective and was therefor used in further experiments. The effect might be due to chaprone-like factors present in these preparations and might help to keep the precursor in an unfolded conformation.

Over-expressed pPPO competed with *in vitro* translated precursors of other nucleus-coded chloroplast proteins (Fig. 6). The import of pLHCPII was arrested at the envelope level, preventing processing by the stromal peptidase (SPP) (Fig. 6A). Competition with pPsaD was evident at both the envelope and interaction with thylakoids levels. Precursor which accumulated, when increasing pPPO concentartion, on envelope membranes fractionatng with thylakoids (Fig. 6B) could be removed by thermolysin treatment (not shown). Accumulation of processed PsaD could be noted in the stroma, indicating that processing by SPP is not a limiting step (Fig. 6B).

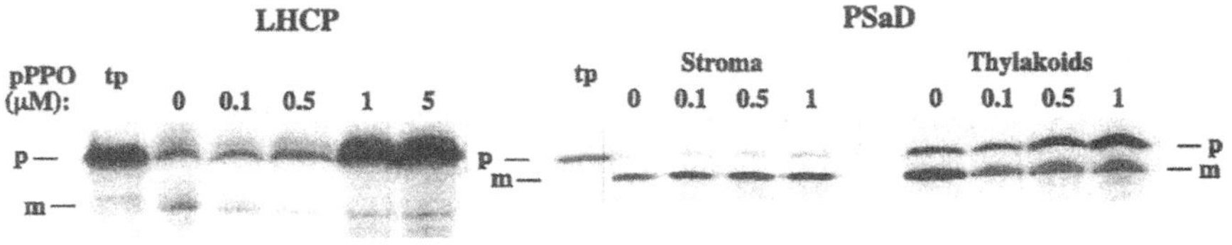

Figure 6. Competition between pPPO and thylakoid proteins. Various concentrations of pPPOA were added to import mixtures containing *in vitro* translated (500,000 cpm) pLHCPII (A) or pPsaD (B). Import was carried out for 30 min in light. tp - 20,000 cpm of *in vitro* translated precursor. Precursor (p) and mature (m) forms are indicated.

The course of *E. coli* expressed pPPO routing into the lumen resembles that of the *in vitro* translated precursor as shown for other nucleus-coded chloroplast proteins [13]. It can be made import-competent by dilution of the dissolving urea and addition of cytosolic proteins which probably contain chaperones. The bottle-neck in the import of over-expressed proteins appears to take place at the membrane (envelope and thylakoids) crossing steps. *E. coli* expressed pPPO
can therefor be reliably used in studying the mechanism of translocation of PPO across thyalkoids, while comparing and competing with other lumen proteins (S. Koussevitzky, P. Hynds, E. Ne'eman, C. Robinson and E. Harel,unpublished observations). Although mature PPOA accumulated in thylakoids to levels detectable by Western blots, no enzymatic activity was observed in chloroplasts

following import. It is possible that pea, which lacks active PPO, is unable to activate the imported polypeptide. This possibility is being examined using chloroplasts from PPO-null transgenic tomato plants in import assays.

4. References

1. Steffens, J.C., Harel, E., and Hunt, M.D. (1994) Polyphenol oxidase, in B.E. Ellis and H.A. Stafford (eds.), *Genetic Engineering of Plant Secondary Metabolism,* Plenum Press, New York, pp. 275-312.
2. Constable, P.C., Bergey, B.R., and Ryan, C.A. (1995) Systemin activates synthesis of wound-inducible tomato leaf polyphenol oxidase via the octadecanod defense signaling pathway. *Proc. Natl. Acad. Sci. USA* **92**, 407-411.
3. Sommer, A., Ne'eman, E., Steffens, J.C., Mayer, A.M. and Harel, E. (1994). Import, targeting, and processing of a plant polyphenol oxidase. *Plant Physiol.* **105**, 1301-1311.
4. Sommer, A., Ne'eman, E., Koussevitzky, S., Hunt, M.D., Steffens, J.C., Mayer, A.M. and Harel, E. (1995). The inhibition by Cu^{2+} of the import of polyphenol oxidase into chloroplasts. In P. Mathis (ed.) *Photosynthesis: from Light to Biosphere.* Kluwer Academic Publishers, Dordrecht, The Netherlands, Vol. III, pp. 827-830.
5. Koussevitzky, S., Ne'eman, E., Sommer, A., Steffens, J.C., and Harel, E. (1998). Processing of precursors of polyphenol oxidase by a chloroplast stromal peptidase: an unusual processing site. *J. Biol. Chem.*, in press.
6. Mayer, A. M. and Harel, E. (1979) Polyphenol oxidases in plants. *Phytochemistry* **18**, 193-215.
7. Yalovsky, S., Ne'eman, E., Schuster, G., Paulsen, H., Harel, E. and Nechushtai, R. (1992) Accumlation of a light-harvesting chlorophyll a/b protein in the chloroplast grana lamellae. *J. Biol. Chem.* **267**, 20689-20693.
8. Newman, S. M., Eannetta, N. T., Yu, H., Prince, J. P., De Vicente, M. C., Tanksley, S. D. and Steffens, J. C. (1993) Organisation of the tomato polyphenol oxidase gene family. *Plant Mol. Biol.* **21**, 1035-1051.
9. Kohorn, B. D., Harel, E., Chitnis, P. R., Thornber, J. P. and Tobin, E. M. (1986) Functional and mutatinal analysis of the light-harvesting chlorophyll a/b protein of thylakoid membranes. *J. Cell Biol.* **102**, 972-981.
10. Cohen, Y. and Nechushtai, R. (1992) Assembly and processing of subunit II (PsaD) precursor in the isolated photosystem I complex. *FEBS Lett.* **302**, 15-17.
11. Paulsen, H., Ruemler, U. and Rudiger, W. (1990) Reconstitution of pigment-containing complexes from Light-harvesting Chlorophyll a/b binding Protein over expressed in Escherichia coli. *Planta* **181**, 204-211.
12. Chitnis, P. R., Harel, E., Kohorn, B. D., Tobin, E. M., and J.P., T. (1986) Assembly of the precursor and processed light-harvesting chlrophyll a/b protein of Lemna into the light-harvesting complex II of barley etichloroplasts. *J. Cell Biol.* **102**, 982-988.
13. Cline, K, R Henry, L Changjiang and Y Jiangou (1993). Multiple pathways for protein transport into or across the thylakoid membrane. *EMBO J.* **12**: 4105-4114.

THE ROLE OF THE ENVELOPE IN ASSEMBLY OF LIGHT-HARVESTING COMPLEXES IN THE CHLOROPLAST: DISTRIBUTION OF LHCP BETWEEN CHLOROPLAST AND VACUOLES DURING CHLOROPLAST DEVELOPMENT IN *Chlamydomonas reinhardtii*

Laura L. Eggink, Hyoungshin Park and J. Kenneth Hoober
Department of Plant Biology and Center for the Study of Early Events in Photosynthesis, Arizona State University, Tempe, AZ 85287-1601

1. Introduction

The formation of thylakoid membranes during chloroplast development is crucial for support of the biosphere and remains a major question in plant cell biology. Biogenesis of functional photosystems involves assembly of light-harvesting complexes (LHCI and LHCII) and their association with PSI and PSII core complexes. We focused on assembly of LHCII as a means to probe for the context in which biogenesis of the membrane occurs. LHCPs, the apoproteins of LHCII, are the major proteins of the membrane and thereby provide a strong signal for analytical studies. These proteins are synthesized on cytoplasmic ribosomes as precursors, imported into the chloroplast, and processed by proteases in the stroma. The following are likely events leading to interaction of LHCPs with pigments. Photoreduction of protochlorophyllide (Pchlide) by light-dependent NADPH:Pchlide oxido-reductase (POR) produces chlorophyllide *a* (Chlide *a*). Binding of Pchlide to POR apparently causes release of Chlide [1]. The site at which this reaction occurs has not been established, but POR is an integral membrane protein in *C. reinhardtii* and possibly resides in the envelope. Fluorescence of Chlide *a* is quenched within seconds after release from the enzyme [2], which may result from interaction with xanthophylls in the membrane environment. Chlide *a* is esterified with geranylgeranyl (GG) diphosphate and normally reduced to the phytyl ester. During rapid greening of *C. reinhardtii y1* cells at 38°C, however, incorporation of Chl_{GG} into LHCII precludes substantial reduction of the side-chain [3]. Because lutein--the major xanthophyll in LHCII--is abundant in greening cells, assembly of LHCII under these conditions appears to be limited by synthesis of Chl. Some of the Chl *a* is oxidized to Chl *b*, as indicated by the same side-chain alcohol composition in both Chls, which stabilizes the complex. *In vitro* reconstitution experiments have shown that folding of LHCP requires the simultaneous presence of xanthophylls (lutein) and Chl [4]. Assembly *in vivo* can produce about 2 X 10^5 LHCII per min per chloroplast [5].

J. H. Argyroudi-Akoyunoglou and H. Senger (eds.),
The Chloroplast: From Molecular Biology to Biotechnology, 161–166.

2. Models

2.1. MOLECULAR MODELING THE ASSEMBLY OF LHCII

Extensive research has shown that excited states of Chl are quenched by carotenoids. We propose that lutein binds to Chlide *a* as it leaves the active site of POR, thereby protecting the cellular environment from the photosensitizing action of Chl(ide) *a* as it passes through the sequence indicated in the Introduction. We further suggest that folding of LHCP is initiated when a lutein-Chl pair impacts the protein, which is consistent with the requirement for both pigments during *in vitro* reconstitution. Lutein is active in assembly of LHCII whereas zeaxanthin, its isomer, is not [4]. Violaxanthin is also abundant in developing chloroplasts, yet is incorporated into LHCII to a lesser extent than lutein [8]. We examined structural models to explore whether specific interaction of active xanthophylls with Chl could occur. Shown in *Figure 1* are molecular models of zeaxanthin, lutein and Chl *a*. The 3' hydroxyl group of both xanthophylls potentially form ligands with Chl, although the planarity of the β ring of zeaxanthin requires a perpendicular (edge-on) approach to the Chl molecule. In contrast, lutein can interact while the ϵ ring is oriented parallel (face-on) with the chlorin ring. Thus, association of Chl with lutein is expected to be more stable.

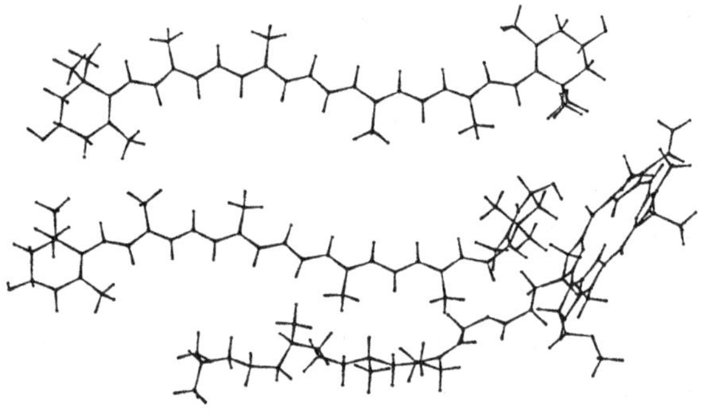

Figure 1. **Molecular models of zeaxanthin (*upper*), lutein (*middle*) and Chl *a*, shown without the central Mg^{2+} (*lower right*). The lutein-Chl pair illustrates that coordination of the hydroxyl group of lutein with Mg^{2+} can occur with parallel orientation of the ring systems. In contrast, coordination of the hydroxyl group of zeaxanthin requires a perpendicular orientation of the ring systems.**

2.2. ENVIRONMENT IN WHICH ASSEMBLY OF LHCII OCCURS

The two potential paths for translocation of LHCPs from their site of synthesis in the cytosol to thylakoid membranes were described as the "soluble intermediate" and the "membrane vesicle intermediate" models [6]. In the first model, LHCP precursors are processed in the stroma after complete translocation across the chloroplast envelope. This model is supported by experiments that demonstrated import of LHCPs into a soluble fraction of isolated pea chloroplasts. LHCPs were also integrated directly into thylakoid membranes in the presence of GTP and soluble chloroplast proteins including 54CP, a homolog of the 54 kD subunit of the signal recognition particle [7]. Any LHCP not in membrane should be recovered in the soluble fraction of the plastid. In the second model, LHCP precursors are translocated partially across the envelope, processed by stromal proteases, and interact with pigments within the inner envelope membrane. Thylakoid membranes then form by expansion of the envelope membrane, with subsequent release into the stroma of thylakoid domains as vesicles. These models predict different patterns of distribution of newly synthesized LHCPs during chloroplast development, which should be distinguishable by analysis of membrane density. For example, in the first model, the thylakoid pool of LHCP should increase by insertion of newly synthesized proteins, and the overall composition, and thus density, of the membrane should not change significantly as the membrane expands. Conversely, because the densities of envelope membranes are less than that of thylakoid membranes, the "membrane vesicle intermediate" model predicts that newly synthesized LHCPs would first be integrated into nascent thylakoid domains arising from the envelope, which would gradually increase in density as maturation of the membrane proceeds. Evaluation of these models requires bulk increases in the amount of membrane, which occurs only *in vivo*.

3. Results

3.1. ULTRASTRUCTURAL LOCALIZATION OF LHCPs

Electron micrographs of developing chloroplasts in degreened *C. reinhardtii* cells at 38°C showed a light-induced emergence of thylakoid-like membranes from the chloroplast envelope [9]. Within minutes after exposure of cells to light, the inner envelope membrane became immunoreactive with antibodies against LHCP, with no detectable labeling on the outer surface of the chloroplast. Continuous monitoring of LHCII accumulation showed that functional energy transducing systems formed within seconds, which functionally quenched fluorescence of newly synthesized LHCII [5]. PSI and PSII activities increased with no lag.

Also during the first few minutes of greening, an increasing amount of LHCPs were recovered in a non-membrane fraction that was not present in chloroplasts. ImmunoEM of greening cells detected an increase of LHCPs in cytoplasmic vacuoles [5]. Cells grown at 28°C in the light contained large, porous polyphosphate granules in these vacuoles (*Figure 2A*) that were highly reactive with antibodies against LHCPs (*Figure 2B*). Reduction in the cellular accumulation of Chl by removal of acetate from the medium, which produced a *chlorina*-like phenotype, or addition of norflurazon caused a shift in distribution of LHCPs toward granules [10]. Immunoblots after electrophoresis of proteins in purified granules showed that all detectable LHCPs were processed to the mature size [5]. These results suggested that precursors for the LHCPs in vacuoles had engaged the chloroplast import apparatus and were translocated sufficiently for the N-terminal transit sequence to be cleaved by stromal processing enzymes. In the absence of Chl synthesis, LHCPs were not detected in the plastid but rather in the cytosol and in vacuolar granules by immunoEM [10]. The kinetics of LHCII accumulation during greening, and the presence of LHCPs in vacuoles, suggested that assembly of LHCII and formation of functional thylakoid domains within the chloroplast envelope were necessary for LHCPs to be retained by the organelle.

Cryofixation techniques, developed to minimize ultrastructural artifacts, showed that membranes of the chloroplast envelope in green cells were tightly appressed over most of the chloroplast surface (*Figure 3A*). An intermembrane space, such as observed in micrographs after conventional fixation, was not detected. Such appression may be important for transfer of lipids, and possibly proteins, between the membranes [11]. Occasional protrusions of the outer envelope membrane enclosed material similar in appearance to stroma (*Figure 3B*). These protrusions seemed to be a means of transfer of chloroplast contents to vacuoles. Experiments with log-phase cells demonstrated substantial transfer of soluble, chloroplast-synthesized proteins to vacuoles (unpublished results), which apparently is a pathway for degradation of chloroplast material.

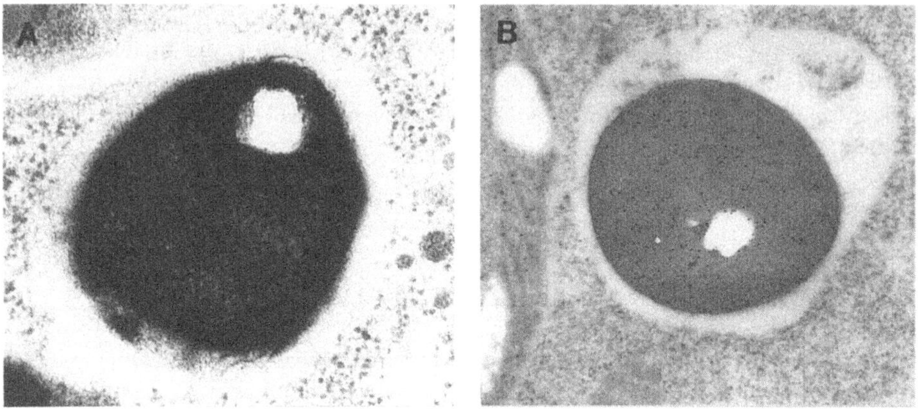

Figure 2. (A) Ultrastructure of a polyphosphate granule within a vacuole in a cryofixed cell of strain *cbn1-113 arg2*. (B) ImmunoEM detection of LHCPs in a vacuolar granule and adjacent chloroplast in a glutaraldehyde-fixed cell of strain *cw15 arg7A*. Cells were grown in the light at 28°C.

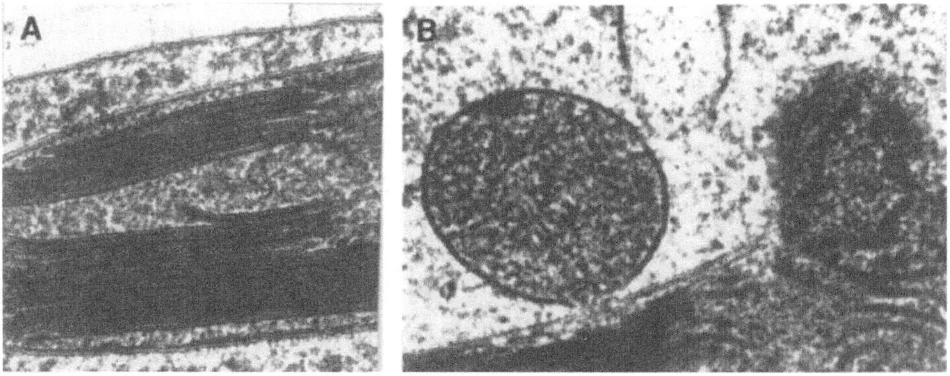

Figure 3. Micrographs of the chloroplast in cryofixed *cbn1-113 arg2* cells. (A) Image of a portion of the chloroplast in a green cell. Although this strain lacks Chl *b*, the chloroplast contained extensive grana. (B) Contents of protrusions of the outer membrane of the envelope were continuous with the stroma.

3.2. SUCROSE GRADIENT ANALYSIS

Newly synthesized Chl was incorporated into a continuously changing environment during greening. The bulk of the Chl was in LHCII that sedimented with large membrane fragments [9]. A likely source of such large membrane fragments at the beginning of greening was the chloroplast envelope. Integration of LHCPs into membranes was investigated by isopycnic sedimentation on 0.4 to 1.5 M sucrose gradients. This density range should encompass the densities of the envelope outer membrane (1.08 g cm^{-3}), inner membrane (1.13 g cm^{-3}) and thylakoid membranes (1.18 g cm^{-3}). Membranes in green or greening cells of *cw15* or MC9, a strain impaired in assembly of thylakoid membranes that was generated by insertional mutagenesis of *cw15 arg7A* cells [12], were fragmented by sonication. Samples were applied on the top or in 1.8 M sucrose at the bottom of the gradient, with similar results. Immunoblots after electrophoresis of proteins in each fraction of the gradient revealed that LHCPs were recovered mostly in membranes at the density of thylakoid membranes, as expected, but were also distributed throughout the density range (*Figure 4A*). Protease treatment of alternate fractions showed that LHCPs in the low density fractions were as protected from digestion as those in thylakoid membranes. Chl was distributed throughout the gradients, commensurate with the amount of LHCP. Addition of Triton X-100 to the low-density fractions increased the fluorescence of Chl *a* about 7-fold, similar to the effect of detergent on thylakoid membranes [9], which indicated association of LHCII with PSII core complexes. Dark-grown, yellow cells were pulse-labeled with [^{14}C]arginine for 5 min in the light at 38°C. Radioautograms after electrophoresis of proteins in fractions of gradients showed that labeled LHCPs were recovered in a broad band with densities of 1.14 to 1.18 g cm^{-3} (1.1 to 1.4 M), with membranes from MC9 shifted to slightly lower density, and also in membranes with much lower densities of 1.06 to 1.09 g cm^{-3} (0.5 to 0.7 M) (*Figures 4B* and *4C*). Significantly, the pellet fraction, containing vacuolar granules, from MC9 was much enriched in labeled LHCPs. Mature RbcS, but not RbcL, was also detected in the pellet, which indicated direct transfer of these proteins from the envelope to vacuoles (not shown).

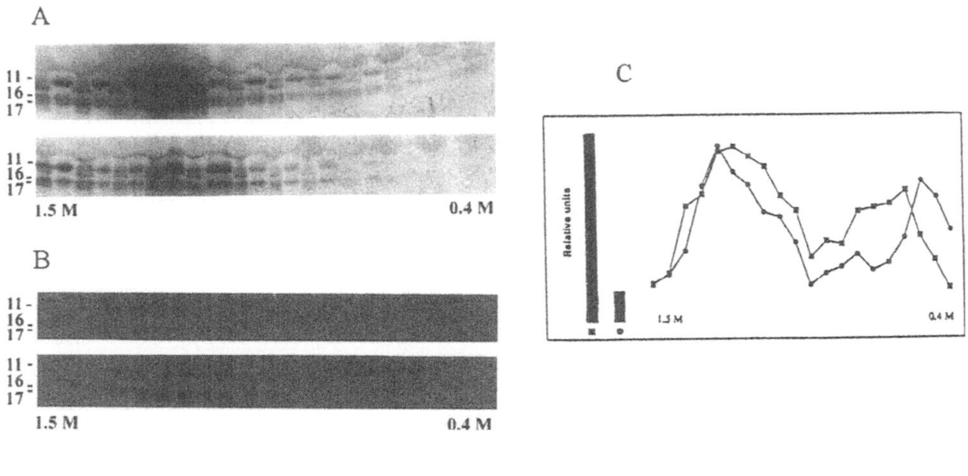

Figure 4. Analysis of membranes resolved with 0.4 to 1.5 M sucrose gradients containing 10 mM phosphate buffer, pH 7.0, and 1 mM EDTA. (A) Immunoblot analysis of the distribution of LHCPs (polypeptides 11, 16 and 17) after electrophoresis of proteins in each fraction. Alternate fractions were treated with 0.1 mg /ml thermolysin on ice for 1 h. (B) Radioautographic analysis of LHCPs synthesized in pulse-labeled cells. For (A) and (B), results with *cw15* are shown in the *upper* panel and with MC9 in the *lower* panel. (C) Graphical presentation of the distribution of newly-synthesized LHCPs shown in (B). Patterns were normalized with the maximal thylakoid fraction. The bars compare relative amounts of labeled LHCPs in the pellet fraction. (■), MC9; (●), *cw15*.

4. Conclusion

Our data show that newly synthesized LHCPs were distributed between membranes and vacuolar granules, with no detectable amount in a buffer-soluble fraction. Pulse-labeled LHCPs were recovered in thylakoid membranes, those intermediate in density between inner envelope and thylakoid membranes, and in low-density membranes. Chl fluorescence characteristics suggest that functional energy transducing systems were present in the low density membranes. Detection of LHCP-containing membranes with densities less than that of thylakoid membranes supported the "membrane vesicle intermediate" model, with the envelope as a site of assembly of LHCII. Consistent with this conclusion is the evidence for localization of reactions leading to Chlide *a* and thylakoid lipids in envelope membranes [13]. Formation of Chl-xanthophyll pairs in the envelope, which molecular modeling suggested is possible, would satisfy the requirement for both Chl and an active xanthophyll for assembly of LHCII. The greater recovery of processed, pulse-labeled LHCPs in granules in mutant MC9 pointed to a defect in integration of these proteins into thylakoid membranes. The apparent bifurcation of the targeting pathway for LHCPs indicate that events at the level of the envelope are critical to incorporation of the proteins into thylakoid membranes.

5. References

1. Reinbothe, S., Reinbothe, C., Lebedev, N., and Apel, K. (1996) PORA and PORB, two light-dependent protochlorophyllide-reducing enzymes of angiosperm chlorophyll biosynthesis. *Plant Cell* **8**, 763-769.
2. Eullaffroy, P., Popovic, R., and Frank, F. (1998) Changes in chlorophyll(ide) fluorescence yield induced by a short light pulse as a probe to monitor the early steps of etioplast phototransformation in dark-grown leaves. *Photochem. Photobiol.* **67**, 676-682.
3. Maloney, M.A., Hoober, J.K., and Marks, D.B. (1989) Kinetics of chlorophyll accumulation and formation of chlorophyll-protein complexes during greening of *Chlamydomonas reinhardtii y-1* at 38°C. *Plant Physiol.* **91**, 1100-1106.
4. Paulsen, H., Finkenzeller, B., and Kühlein, N. (1993) Pigments induce folding of light-harvesting chlorophyll *a/b*-binding protein. *Eur. J. Biochem.* **215**, 809-816.
5. White, R.A., Wolfe, G.R., Komine, Y., and Hoober, J.K. (1996) Localization of light-harvesting complex apoproteins in the chloroplast and cytoplasm during greening of *Chlamydomonas reinhardtii*. *Photosynth. Res.* **47**, 267-280.
6. Hoober, J.K., White, R.A., Marks, D.B., and Gabriel, J.L. (1994) Biogenesis of thylakoid membranes with emphasis on the process in *Chlamydomonas*. *Photosynth. Res.* **39**, 15-31.
7. Li, X., Henry, R., Yuan, J., Cline, K., and Hoffman, N.E. (1995) A chloroplast homologue of the signal recognition particle subunit SRP54 is involved in the posttranslational integration of a protein into thylakoid membranes. *Proc. Natl. Acad. Sci. USA* **92**, 3789-3793.
8. Pagano, A., Cinque, G., and Bassi, R. (1998) *In vitro* reconstitution of the recombinant photosystem II light-harvesting complex CP24 and its spectroscopic characterization. *J. Biol. Chem.* **273**, 17154-17165.
9. Hoober, J.K., Boyd, C.O., and Paavola, L.G. (1991) Origin of thylakoid membranes in *Chlamydomonas reinhardtii y*-1 at 38°C. *Plant Physiol.* **96**, 1321-1328.
10. Park, H., and Hoober, J.K. (1997) Chlorophyll synthesis modulates retention of apoproteins of light-harvesting complex II by the chloroplast in *Chlamydomonas reinhardtii*. *Physiol. Plant.* **101**, 135-142.
11. Rawyler, A., Meylan-Bettex, M., and Siegenthaler, P.A. (1995) (Galacto)lipid export from envelope to thylakoid membranes in intact chloroplasts. II. A general process with a key role for the envelope in the establishment of lipid asymmetry in thylakoid membranes. *Biochim. Biophys. Acta* **1233**, 123-133.
12. Wolfe, G.R., Park, H., Sharp, W.P., and Hoober, J.K. (1997) Light-harvesting complex apoproteins in cytoplasmic vacuoles in *Chlamydomonas reinhardtii* (Chlorophyta). *J. Phycol.* 33: 377-386.
13. Reinbothe, S., and Reinbothe, C. (1996) The regulation of enzymes involved in chlorophyll biosynthesis. *Eur. J. Biochem.* **237**. 323-343

CHARACTERIZATION OF THE PLASTID IMPORT REACTION OF THE PEA NADPH: PROTOCHLOROPHYLLIDE OXIDOREDUCTASE (POR).

H. ARONSSON[1], J. ALMKVIST[1], C. SUNDQVIST[1], M. P. TIMKO[2] and C. DAHLIN[1]
[1]Department of Plant Physiology, Göteborg University, Box 461, SE-405 30 Göteborg, Sweden. [2]Department of Biology, University of Virginia, Charlottesville, Virginia 22903 USA.

1. Introduction

NADPH: protochlorophyllide (POR) is a vital enzyme in the biosynthesis of chlorophyll where it catalyzes the reduction of Pchlide into Chlide in a light-dependent manner. POR is nucleus-encoded and imported into the plastids where it is found at the inner membranes. Together with its substrate and the co-factor NADPH it forms a ternary complex which is needed for catalytical activity. The anomaly of a decreasing POR level during active chlorophyll synthesis was cleared with the discovery of two different POR proteins, POR-A and POR-B, in barley and *Arabidopsis thaliana*. During greening, POR-A is negatively regulated by light both at transcriptional and proteolytical levels. In addition, the import of POR-A, but not POR-B, has been suggested to require Pchlide in order to be translocated into the plastid [1]. In this respect, POR-A differs from other known nucleus-encoded plastid proteins, and as it appears, this requirements represents a novel and exclusive import characteristic.
In pea, only one POR gene has been found [2] indicating that the situation for the regulation of POR import and accumulation is far from clear.
We here present a characterization of the import conditions of the pea POR, including the potentional role of Pchlide in the translocation step.

2. Procedure

2.1. PREPARATION OF PLASTIDS

Seedlings of pea (*Pisum sativum* L. cv. Kelvedon Wonder) were grown 7 days in a greenhouse before harvest. Purified plastids were obtained through differential centrifugation steps including use of a linear Percoll gradient as described [3].

J. H. Argyroudi-Akoyunoglou and H. Senger (eds.),
The Chloroplast: From Molecular Biology to Biotechnology, 167–170.

2.2. PROTEIN EXPRESSION

Transcription and translation were performed with Promegas wheat germ extract system with [^{3}H]-leucine and T7 and SP6 RNA polymerase for pPOR and pSS, respectively. Before use, translation mixtures were adjusted to 30 mM unlabelled leucine.

2.3. IMPORT ASSAY

Import was performed generally as described earlier [4] if not otherwise stated. To perform import assays without endogenous ATP (Mg-salt), plastids were put in darkness for 45 min in room temperature prior to the import reaction. Concentration of salts in the samples were controlled by filtration through Sephadex G-25 [5], followed by an addition of low conc. of ATP (final konc. 85 μM), high conc. of ATP (final konc. 5.0 mM) and NADPH (final konc. 2 mM). Translocations were also assayed with different levels of plastid-bound Pchlide. This was achieved in three different ways: 1, dark incubation of green plants, leading to a gradual etiolation of chloroplasts; 2, incubation of isolated chloroplasts with δ-aminolevulinic acid (ALA); 3, incubation of isolated chloroplasts with Pchlide solved in cholate buffer. Fluorescence emission spectra were recorded at -196°C with a SLM Aminco 8000C (SLM Instr., Urbana, Ill, USA) spectrofluorometer. The excitation wavelenght was 440 nm. Proteins were separated on 12.5 % SDS-PAGE gels. After electrophoresis, gels were prepared for fluorography and exposed to X-ray films.

3. Result and Discussion

To characterize the translocation pathway of POR across the envelope membranes, import was performed with radiolabelled pPOR into pea chloroplasts under different conditions. As seen in Table 1, the temperature regulation on import is dependent on the age of the plants used for plastid isolation. Plastids from 7 days old plants were more or less unaffected between 5°C and 35°C, whereas older plants (i.e. plastids) exhibited a narrower temperature range for optimum import efficiency.

The energy requirements of POR did not deviate from the the majority of other nucleus-encoded proteins. For instance, addition of 85 μM ATP to an ATP-depleted import mix will only permit the protein to bind to the plastid surface whereas 5 mM ATP allows the translocation into the plastids. Similar results were also obtained with the Rubisco small subunit (pSS; Fig. 1, lanes 5, 8, 10). However, the presence of NADPH slightly decreased binding and import of pPOR at 85 μM and 5 mM ATP, respectively. The NADPH was recently reported to inhibit protein import into glyoxysomes [6]. As an inhibitor of protein import into glyoxysomes it was not restricted to any special protein. In our experiments NADPH did not affect binding and import of pSS. If NADPH affects binding/unfolding properties of POR by interaction with the nucleotide-binding motifs, or acts more unspecifically is presently being investigated.

	Temperature				
Age of plant	5°C	15°C	25°C	35°C	45°C
7 days	+++	+++	+++	+++	-
11 days	+	+++	+++	++	-
15 days	+	++	++	+	-

Table 1. Import of pPOR at different temperatures into plastids isolated from plants of different ages.

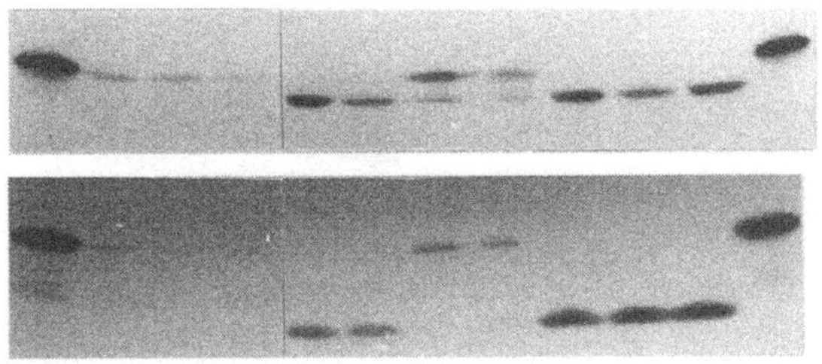

Figure 1. Fluorographs showing the import of pPOR and pSS into pea chloroplasts with or without endogenous ATP, lanes 6-10 and lanes 1-5, respectively. All import assays were performed in darkness, except for lane 10 which was performed in light as a control. TM: translation mixture. m: mature protein, pPOR 44 kDa, pSS 21 kDa.

To study substrate (Pchlide) requirement for the import reaction of the pPOR, *in vitro* import assays were performed with radiolabelled precursor protein into isolated plastids of pea. Several different strategies to alter the endogenous levels of Pchlide were used. First, by using etiolated pea chloroplasts, i .e. plastids isolated from greenhouse plants, grown in the dark for 4 days prior to plastid isolation. Second, isolated pea chloroplasts were incubated with ALA, an early precursor in the chlorophyll biosynthesis. Third, isolated pea chloroplasts were incubated with Pchlide. All treatments efficiently increased the relative amounts F636 and F657 indicating a build-up of plastid-bound Pchlide (Fig. 2). It is clear that the pPOR imported with almost equal efficiency into all plastids types containing different levels of Pchlide.

In conclusion, energy requirements and temperature regulation of pea POR import into pea chloroplasts resemble those of the majority of other investigated nucleus-encoded proteins. It is also clear that, in contrast to POR-A, import of the pea POR is not substrate-dependent.

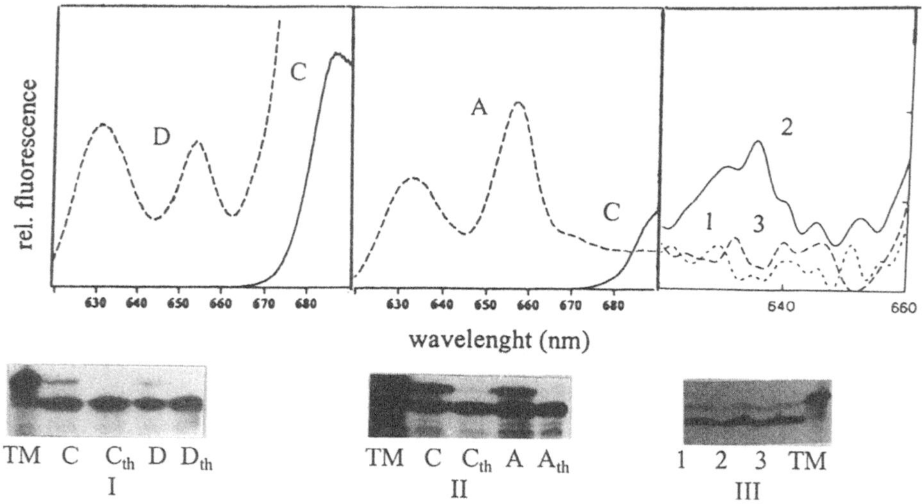

Figure 2. Import of pPOR with different levels of Pchlide. I, import with plastids from greenhouse grown plants either grown in light or in darkness 4 days prior to the assay. II, plastids incubated with ALA. III, plastids incubated with Pchlide solved in cholate buffer; 1, cholate and import buffer; 2, cholate and Pchlide; 3, import buffer. TM, translation mixture; C, control; D, dark-grown; th, thermolysin; A, ALA-incubated.

4. Acknowledgements

The authors wish to thank Nancy Ilenius for skilful technical assistance. This research was supported by the Helge Ax:son Johnsons Foundation and Carl Trygger Foundation. M. P. Timko was supported by funds from the US Department of Energy.

References

1. Reinbothe, S., Reinbothe, C., Holtorf, H. and Apel, K. (1995) Two NADPH:protochlorophyllide oxidoreductase in barley: evidence for the selective disappearance of PORA during the light-induced greening of etiolated seedlings. Plant Cell **7**: 1933-1940.

2. Spano, A. J., He, Z., Michel, H., Hunt, D. F. and Timko, M. P. (1992) Molecular cloning, nuclear gene structure, and developmental expression of NADPH-protochlorophyllide oxidoreductase in pea (*Pisum sativum* L.) Plant Mol. Biol. **18**: 967-972.

3. Dahlin, C., Sundqvist, C. and Timko, M. P. (1995) The *in vitro* assembly of the NADPH-protochlorophyllide oxidoreductase. Plant Mol. Biol. **29**: 317-330.

4. Dahlin, C. (1993) Import of nuclear-encoded proteins into carotenoid-deficient young etioplasts. Physiol. Plant. **87**: 410-416.

5. Olsen, L. J., Theg, S. M., Selman, B. R. and Keegstra, K. (1989) ATP is required for the binding of precursor proteins to chloroplasts. J. Biol. Chem. **264**: 6724-6729.

6. Pool, M. R., López-Huertas, E., Horng, J.-T. and Baker, A. (1998) NADPH is a specific inhibitor of protein import into glyoxysomes. Plant J. **15**: 1-14.

TOCOPHEROL BIOSYNTHESIS IN SENESCING CHLOROPLASTS - A MECHANISM TO PROTECT ENVELOPE MEMBRANES AGAINST OXIDATIVE STRESS AND A PREREQUISITE FOR LIPID REMOBILIZATION ?

B. CHROST, J. FALK, B. KERNEBECK, H. MÖLLEKEN
and K. KRUPINSKA
Institute of Botany, University of Cologne, Gyrhofstr. 15, 50931 Cologne, Germany

1. Abstract

Analyses of tocopherols in flag leaves of barley cultivated in a field revealed that both alpha- and gamma-tocopherol accumulate with the beginning of senescence. Correspondingly one of the genes expressed during barley leaf senescence is coding for a key enzyme of the biosynthetic pathway leading to tocopherols and plastoquinones. This enzyme, 4-hydroxyphenylpyruvate dioxygenase, catalyzes the formation of homogentisate from 4-hydroxyphenylpyruvate, a product of the shikimate pathway. It is hypothesized that tocopherols acting as lipophilic antioxidants play an important role in differential stabilization of plastid envelope membranes over thylakoid membranes which become degraded during senescence. A model emphasizing the important function of envelope membranes for senescence associated remobilization of thylakoid lipids is presented.

Keywords: senescence / oxidative stress / tocopherols / 4-hydroxyphenylpyruvate dioxygenase

2. Introduction

During its life span a leaf undergoes several phases of development. Senescence is the final stage of leaf development preceding death. It is well known that senescence is not a passive decay but is a highly regulated and genetically controlled process requiring the expression of specific genes [1]. During the phase of leaf senescence macromolecules, e.g. chlorophylls, proteins and nucleic acids, become degraded and the components are mobilized to other parts of the plant for the formation of new leaves and finally fruits and seeds. The maintenance of cellular compartmentation and order until the final stage of senescence is a prerequisite for the ordered sequence of the senescence associated remobilization processes. By ultrastructural analysis of senescing leaves it has been shown that the various membrane systems in a senescing cell have different stability [2]. While microsomal membranes including the ER and Golgi membranes are degraded early during senescence, mitochondrial membranes, nuclear membranes and the plasmamembrane seem to be rather stable until the ultimate stage of senescence. In case of plastids which during later stages of leaf development are undergoing a transition from chloroplasts to so-

J. H. Argyroudi-Akoyunoglou and H. Senger (eds.),
The Chloroplast: From Molecular Biology to Biotechnology, 171–176.

called gerontoplasts stability of the envelope membranes is higher than the stability of thylakoid membranes which are completely degraded during senescence [2].
In the present paper it is proposed that tocopherols play a key role in maintenance of subcellular compartmentation by regulating differential membrane integrity during senescence.

3. Material and Methods

Plant material. Spring barley (*Hordeum vulgare* L., cv. Scarlett) was grown in 1997 under field conditions. Flag leaves were collected and analysed from 18 June to 10 July. Sample collections were performed between 10.00 and 12.00 h.

Chlorophyll fluorescence measurements. Chlorophyll fluorescence parameters were measured directly in the field using a pulse amplitude fluorometer (Mini-PAM, Walz, Effeltrich, Germany) as described previously [3].

Determination of the relative chlorophyll content. The relative chlorophyll content was determined by a SPAD-502 chlorophyll meter (Minolta, Osaka, Japan). Mean values are based on 10 independent measurements, which were performed at a position in the middle of the flag leaf blade.

Determination of the relative tocopherol content. Tocopherols were extracted from 200 mg of frozen plant material as described [4]. HPLC analysis was performed with a Chromosorb Si 100 column and hexan as eluent. The separation has been monitored by a fluorescence spectrophotometer at excitation 290 nm and emission 340 nm.

Isolation and analysis of RNA. RNA of flag leaves was isolated and analyzed by northern blot hybridization as described earlier [3]. As a probe the cDNA fragment specific for the barley *hpd* gene [5] has been employed.

4. Results

In order to determine the onset of senescence in barley flag leaf blades, the efficiency of photosystem II and the chlorophyll content were determined after the flag leaves reached their final length. While chlorophyll content starts to decrease on 27 June the efficiency of photosystem II did not decrease until 1 July (Figure 1). Due to rainfall on 2 July the senescence related decline in F_V/F_M is reversed temporarily. After this point of time F_V/F_M declined dramatically. With regard to the course of F_V/F_M and chlorophyll content 27 June may be defined as the onset of senescence.

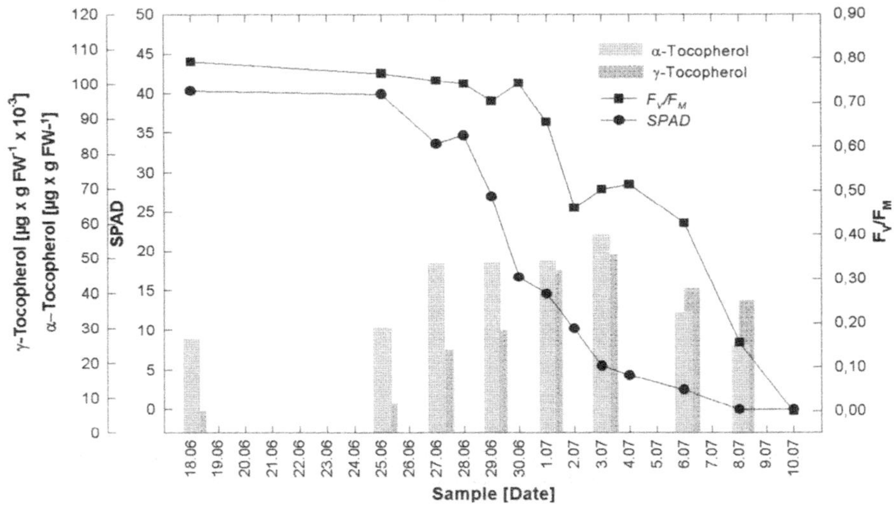

Figure 1. Changes in the efficiency of photosystem II (F_V/F_M), relative chlorophyll content (SPAD) and tocopherol levels of flag leaves during growth of barley under field conditions. Samples were collected after full emergence of the flag leaves.

Analysis of tocopherols in barley flag leaf blades showed that the content of alpha-tocopherol increases with the onset of senescence, has a peak on 3 July and declines only slightly during further senescence (Figure 1). Comparatively, a more dramatic increase of gamma-tocopherol occurs not until the onset of senescence. The increased level of gamma-tocopherol remains high during the senescence period of flag leaves. Ultrastructural analyses of leaf segments from flag leaves collected around the onset of senescence showed a dramatic increase in plastoglobuli content of chloroplasts (data not shown). A positive correlation of plastoglobuli and tocopherol content has been described before [6].

To investigate at which level the senescence associated changes observed in the contents of tocopherols are regulated, expression of the gene coding for the enzyme 4-hydroxyphenylpyruvate dioxygenase, a key enzyme of the tocopherol biosynthesis [7] was analyzed by northern blot hybridization of RNA from flag leaves with a cDNA specific for *hpd* gene [5]. It is evident that changes in the level of the *hpd* transcript positively correlate with the increase of tocopherols during senescence of flag leaves (Figure 2). Low levels of transcripts were present in mature flag leaves and increased markedly during senescence (Figure 2). Based on this result it is feasible that the senescence associated increase in the tocopherol content of barley leaves is regulated at least partly at the level of expression of the *hpd* gene.

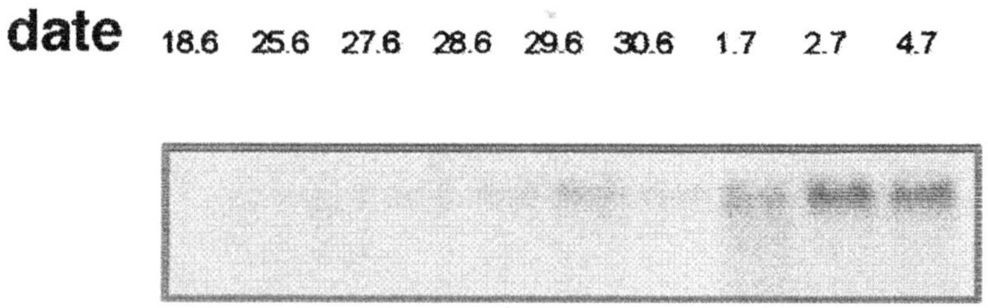

Figure 2. Northern blot analysis of *hpd* transcript levels in barley flag leaves during maturation and senescence under field conditions.

5. Discussion

Ultrastructural analysis of sections derived from senescent barley flag leaf blades revealed that the envelope membranes of plastids are well preserved even at a very late stage of senescence (unpublished data). Indeed integrity of plastid envelope membranes is a prerequisite for the remobilization of plastid lipids. At all stages which have been analysed plastids are in close contact to microbodies. It is well known that during senescence microbodies undergo a functional transition from peroxisomes to glyoxysome-like organelles. These organelles may play a key role in the remobilization of lipids and the use of fatty acids for beta-oxidation and gluconeogenesis [8].

With regard to their functions as lipophilic antioxidants and membrane stabilizers tocopherols might play a key role in the maintenance of the integrity of these membranes in senescing cells [9]. It is hypothesized that the prolonged stability of plastid envelope membranes over thylakoid membranes is due to their higher tocopherol content during senescence. Indeed it has been shown that senescing barley flag leaves contain more tocopherols than the mature leaves (Figure 1). So far informations on the intracellular distribution of the tocopherols are rare. Biochemical analysis of tocopherol biosynthesis has revealed that the main steps including the reaction catalyzed by the 4-hydroxyphenylpyruvate dioxygenase of this pathway are localized in chloroplasts [7].

In Figure 3 an hypothesis about the role of tocopherol biosynthesis for differential protection of envelope membranes is presented. Plastid membranes contain mainly galactolipids (MGDG, DGDG). Investigations on lipid catabolism in chloroplasts during senescence suggest that the decrease in the amount of galactolipids occurring during senescence is due to a reversal of the biosynthetic pathway [8]. According to this sequence of reactions galactolipids are converted to phosphatidylcholin (PC) via diacylglycerol (DAG). The first step in lipid catabolism is catalyzed by an alpha-galactosidase (αG) followed by either a galactolipid/galactolipid/galactosyltransferase (GGGT) or a beta-galactosidase (ßG) reaction (Figure 3). We have isolated a cDNA clone which shows high homology to known α-galactosidase genes and is specifically expressed during senescence (unpublished data). The diacylglycerol (DAG) may be exported and further metabolized

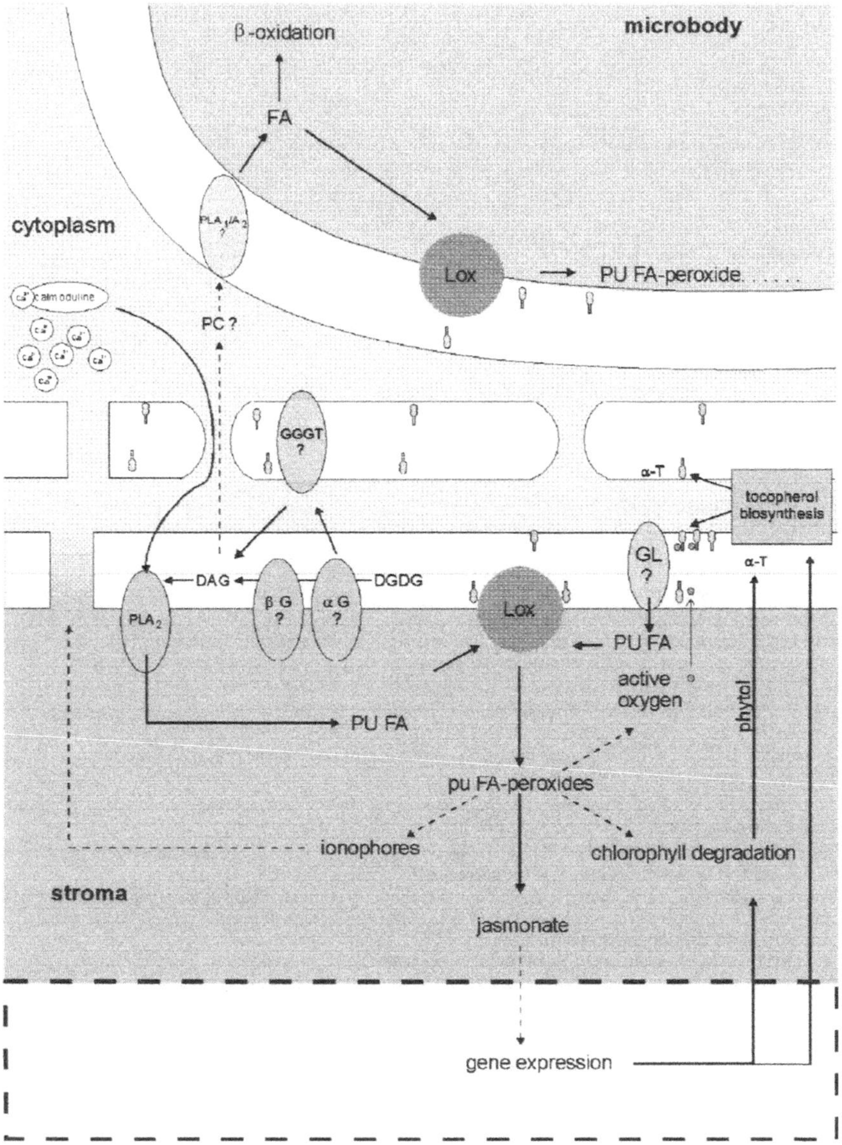

Figure 3.: Model explaining the reasons which might underly the differential stabilization of plastid and microbody envelope membranes during senescence (for details, see text).

outside of the plastid or may alternatively serve as a substrate for a putative phospholipase A_2 (PLA_2) activated by calcium ions [10]. Chloroplast lipids predominantly contain polyunsaturated fatty acids (PUFA) as linoleic and linolenic acid. If such fatty acids are released they may serve as substrates for a lipoxygenase (LOX). Fatty acid peroxides produced by the activity of this enzyme are precursors of jasmonate [11]. In view of this sequence of reactions PUFA released at early stages of senescence trigger the synthesis of the hormone-like substance jasmonate which is known to reprogram nuclear gene expression [11] (Figure 3). Several genes induced during senescence are indeed responsive to jasmonate, e.g. the *hpd* gene (data not shown).

In summary it is hypothesized that the increase in tocopherols during leaf senescence is a prerequisite for proper progression of senescence allowing efficient use of plastid thylakoid lipids for gluconeogenesis. Synthesis of tocopherols during senescence is likely to be regulated at the level of *hpd* gene expression by jasmonate which is produced as a result of initial lipid degradation in plastids.

6. References

1. Buchanan-Wollaston, V. (1997) The molecular biology of leaf senescence, J. Exp. Bot. **48**, 181-199.
2. Thomson ,W. W., Platt-Aloia, K A. (1987) Ultrastructure and senescence in plants, In Thomson, W. W., Nothnagel, E. A., Huffaker, R. C., (eds.) Plant senescence, Its Biochemistry and Physiology, American Society of Plant Physiologists, pp. 20-30.
3. Humbeck, K., Quast, S., and Krupinska, K. (1996) Functional and molecular changes in the photosynthetic apparatus during senescence of flag leaves from field-grown barley plants, Plant, Cell Environment **19**, 337-344.
4. Norris, S. R., Barrette, T. R., and DellaPenna, D. (1995) Genetic dissection of carotenoid synthesis in Arabidopsis defines plastoquinone as an essential component of phytoene desaturation, The Plant Cell **7**, 2139-2149.
5. Kleber-Janke, T., and Krupinska, K. (1997) Isolation of cDNA clones for genes showing enhanced expression in barley leaves during dark induced senescence as well as during senescence under field conditions, Planta **203**, 332-340.
6. Lichtenthaler , H. K. (1966) Verbreitung und Konzentration des α-Tocopherols in Chloroplasten, Ber. Dtsch. Bot. Ges. **79**, 111-117.
7. Schultz, G., Soll , J., Fiedler, E., and Schulze-Siebert, D. (1985) Synthesis of prenylquinones in chloroplasts, Physiol. Plant. **64**, 123-129.
8. Matile, P. (1992) Chloroplast senescence, In: Baker, N. R., and Thomas, H., (eds.) Crop Photosynthesis: Spatial and Temporal Determinants, Elsevier Science Publishers B.V., pp. 413-440.
9. Fryer, M. J. (1992) The antioxidant effects of thylakoid vitamin E (α-tocopherol), Plant Cell Environ. **15**, 381-392.
10. Leshem, Y. Y. (1992) Plant membranes: A biophysical approach to structure, development and senescence, Kluwer Acad. Publ., Dordrecht/Boston/London.
11. Loake, G. J. (1996) Jasmonates: global regulators of plant gene expression, In: Smallwood, M., Knox, J. P., and Bowles, D. J., (eds.) Membranes: Specialized functions in plants. BIOS Scientific Publishers, UK, pp. 215-230.

PIGMENT BIOSYNTHESIS

THE EFFECT OF LIGHT ON THE BIOSYNTHESIS AND FUNCTION OF NADPH-PROTOCHLOROPHYLLIDE OXIDOREDUCTASES (PORs) A AND B IN SEEDLINGS OF *ARABIDOPSIS THALIANA* AND *HORDEUM VULGARE*

Klaus Apel*, Ulrich Sperling and Gregory A. Armstrong
Swiss Federal Institute of Technology (ETH),
Institute of Plant Sciences (Plant Genetics),
Universitätstr. 2, 8092 Zürich, Switzerland

* Tel.:++41 1 632, 3821, Fax: ++41 1 632 12 39,
e-mail: klaus.apel@ipw.biol.ethz.ch

1. INTRODUCTION

Most oxygenic photosynthetic organisms have two options to complete the synthesis of chlorophyll. In a light-independent reaction the immediate precursor of chlorophyllide, protochlorophyllide (Pchlide), can be reduced to chlorophyllide through a Pchlide-reducing enzyme that consists of three subunits Ch L, Ch B and Ch N that in plants are encoded by chloroplast DNA (1). In addition to this light-independent enzyme they also contain a nuclear DNA-encoded light-dependent NADPH-Pchlide oxidoreductase (POR) (2, 3) (Fig. 1). In angiosperms, however, only one of these two routes is active. Through the loss of the light-independent enzyme angiosperms are no longer able to use this reaction during the night to avoid the accumulation of Pchlide, a photosensitizing pigment, that upon illumination may lead to severe photooxidative damage (Fig. 1). In order to compensate for the lack of the light-independent Pchlide-reducing enzyme angiosperms seem to have evolved at least two protection mechanisms that minimize the risk of photooxidative damage during the transition from the dark to the light. First, in contrast to other plants, some angiosperms – and also some of the gymnosperms – possess a second POR enzyme (2-6). It is tempting to speculate that angiosperms use this additional POR enzyme as part of a protection strategy against reactive oxygen species, that in particular during seedling development may lead to the rapid bleaching and death of etiolated seedlings after they have been exposed to light. Second, a feed-back inhibitory loop has been described in higher plants through which further synthesis of δ-aminolevulinic acid, a precursor common to all porphyrins, will be stopped once a critical level of Pchlide has been reached in the dark (7).

2. A COMPARISON OF POR A AND POR B

Upon illumination excited porphyrins, such as free Pchlide, may transfer the excitation energy directly to oxygen, thus leading to the formation of highly reactive singlet

J. H. Argyroudi-Akoyunoglou and H. Senger (eds.),
The Chloroplast: From Molecular Biology to Biotechnology, 179–184.

oxygen (8). Pchlide forms together with POR and NADPH a ternary photoactive complex (9) (Fig. 2). In this complex light is absorbed by Pchlide and the excitation energy is quenched during the reduction of the pigment. Thus, the second POR enzyme that has been discovered exclusively in higher plants may act as an additional light scavenger that reduces photooxidative damages that otherwise would occur during illumination of etiolated seedlings. In our own studies during recent years we have concentrated on elucidating this possible role of the second POR enzyme in angiosperm seedlings.

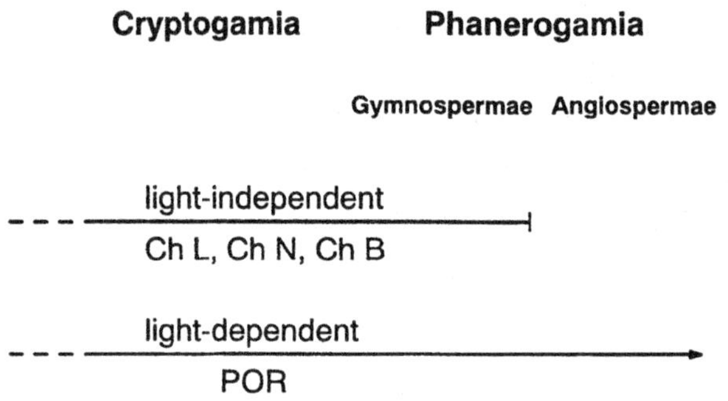

Figure 1
Light-dependent and light-independent reduction of Pchlide to Chlide in plants

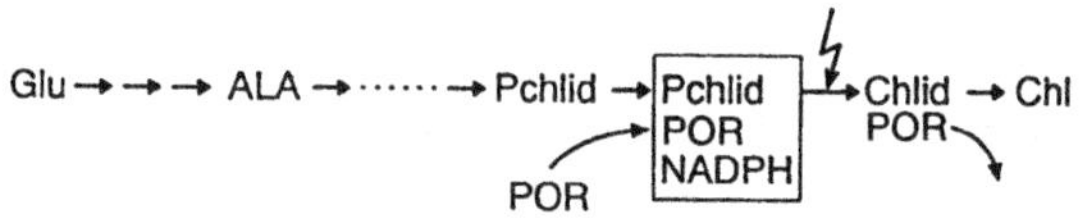

Figure 2
The formation of the ternary POR-NADPH-Pchlide complex in higher plants

In etiolated seedlings of monocots such as barley or wheat, large amounts of POR protein had been shown to accumulate within the prolamellar body of etioplasts (10, 11). Upon illumination the enzymatic activity of POR and the concentration of the enzyme protein were found to rapidly decline and at a time when chlorophyll accumulation had reached its maximum rate were hardly detectable anymore (12, 13). At the same time also the concentration of POR mRNA dropped, at least in the case of barley seedlings, below the limit of detection. This negative effect of light was mediated by the phytochrome photoreceptor (14). This behavior of POR was difficult to understand if indeed it was the only enzyme that could be used by plants to reduce Pchlide. Only recently it became possible to solve this apparent paradox. A second closely related POR protein (POR B) was discovered both in the dicot plant *Arabidopsis thaliana* (4) and in the monocot *Hordeum vulgare* (5), that differed in several ways from the previously described POR protein that accumulates in etiolated seedlings (Table 1). This latter enzyme was renamed POR A. While POR A appeared only transiently in both species during the initial phase of the light-induced greening of etiolated seedlings and was confined to the cotyledons and mesocotyl, respectively, POR B was constitutively expressed and was present in all photosynthetic tissues. These results suggest that in mature fully light-adapted plants the reduction of Pchlide to Chlide is catalyzed exclusively by POR B. While in both plant species the concentration of POR A mRNA rapidly declines during illumination of etiolated seedlings, POR B mRNA can still be detected in light-adapted plants (Table 1). In etiolated barley seedlings POR B makes up only a small fraction of total POR protein. In etiolated *Arabidopsis* seedlings, however, roughly equal amounts of both proteins are present. Two very similar proteins have also been discovered in *Pinus mugo* that are differentially affected by light (2). However, in etiolated seedlings of pea and cucumber only one POR protein has been found (15, 16).

TABLE 1 Comparison of NADPH-protochlorophyllide (Pchlide) oxidoreductases (PORs) A and B (*Arabidopsis thaliana** and *Hordeum vulgare***, resp.)

	POR A	**POR B**	**References**
Substrate	Pchlide	Pchlide	5, 9
Cosubstrate	NADPH	NADPH	5, 9
Gene localisation	Nucleus	Nucleus	4, 5
Mode of mRNA regulation	negative by light (Phytochrome)	constitutive**	2, 3, 4, 5
Tissue specificity	restricted to cotyledon* or to mesocotyl**	all photosynthetic tissues	2, 4
Precursor import into plastids	substrate-dependent**	constitutive**	2,5
Stability of ternary photoactive Pchlide-POR complex	stable in the dark**	stable in the dark**	18, 20
Stability of Chlide-POR complex	highly unstable**	highly unstable**	7, 8, 18, 20

3. THE SUBSTRATE-DEPENDENT IMPORT OF THE PRECURSOR OF POR A

Another rather remarkable difference between POR A and POR B was discovered when the import of precursor proteins of POR A and POR B into isolated plastids was studied. While the POR B precursor protein is translocated into isolated etioplasts and chloroplasts with equal efficiency, the POR A precursor is imported only into isolated etioplasts but not into chloroplasts. Subsequent studies revealed that the import of this precursor protein depends on the presence of the enzyme's substrate, Pchlide (17). Because the Pchlide concentration of plastids rapidly declines during illumination of etiolated seedlings, this pigment becomes soon a limiting factor for the import reaction (18). The competence of chloroplasts to import POR A can easily be restored by feeding δ-aminolevulinic acid – a common precursor of all porphyrins – to isolated chloroplasts in the dark which leads to the rapid accumulation of Pchlide (17, 18). As it has been shown previously that in etioplasts at least a fraction of the Pchlide is synthesized in the plastid's envelope (19), the substrate-dependent uptake of POR A seems to ensure the immediate assembly of the ternary photoactive POR-NADPH-Pchlide complex during the passage of the precursor protein through the envelope. In contrast to its free form, the complex-bound POR A protein is very stable and well protected against proteolytic attack (18, 20). At the same time, the accumulation of free Pchlide within the envelope may be prevented that otherwise upon illumination would lead to photooxidative damage. After photoreduction of Pchlide the resulting Chlide-POR complex is highly susceptible to proteolytic attack and is rapidly degraded to lower molecular weight fragments of approx. 2000-3000 Da (20). It is not known yet whether Chlide remains associated with one of the resulting degradation products and, if so, what role this complex may play during the subsequent translocation of chlorophyll(ide) to the growing thylakoid membrane.

4. A POSSIBLE FUNCTION OF POR A

As mentioned above, the differences between POR A and POR B suggest that POR A may represent a specialized enzyme form that has evolved in higher plants as part of an adaptive strategy to minimize photooxidative damage during illumination of etiolated seedlings. Ideally this hypothesis could be tested with a mutant plant that lacked a functionally intact POR A gene. In the absence of such a mutant an alternative approach has been used that exploits the highly selective, negative effect of light on POR A gene expression. In *Arabidopsis* seedlings grown under continuous non photooxidative far-red light, only POR B mRNA but not POR A mRNA can be detected (21). Furthermore, one of the two POR proteins is missing, while the second POR protein, most likely POR B, is still present in an enzymatically active form (21). When these far-red light-pretreated seedlings are transferred to white light, they do not accumulate chlorophyll as the etiolated control seedlings do when they are exposed to light. This result seems to suggest that the inability of the far-red light-pretreated seedlings to green may be caused by the apparent absence of POR A. However, the many pleiotropic effects that active phytochrome is known to have on these seedlings may preclude such a causal relationship between these two events. A transgenic approach was used to test the importance of POR A for the greening process. Complete cDNAs encoding either POR A or POR B were placed under the control of the constitutively active 35 S promoter of the cauliflower mosaic virus (CAMV) and this DNA construct was then transferred via *Agrobacterium tumefaciens* into *Arabidopsis*. When seedlings of such transgenic lines were grown under continuous far-red light, they synthesized large amounts of POR A or POR B, respectively. These seedlings were subsequently placed in white light of different intensities. Under very low light intensities of up to 10 $\mu Em^{-2} s^{-1}$ the POR-overexpressing seedlings accumulated similar amounts of chlorophyll to far-red light-treated control seedlings that were devoid of POR A. However, when the light intensity increased further the chlorophyll content declined in control seedlings and at an intensity higher than 125 $\mu Em^{-2} s^{-1}$ these seedlings contained virtually no detectable chlorophyll (22). In contrast transgenic seedlings that contained additional POR protein continued to accumulate large amounts of chlorophyll under these higher light conditions and reached a maximum of chlorophyll content that exceeded that to far-red pretreated wildtype control plants by more than 700-fold in some of the transgenic lines. Both POR forms gave a similar effect (22). These results demonstrate first of all that indeed one of the functions of the second POR form, that accumulates in etiolated seedlings and soon after the beginning of illumination declines in abundance again, is to protect the seedlings against photooxidative damage during the transition from the dark to the light. POR A is the most prominent protein in etioplasts and upon illumination rapidly degrades to form lower molecular weight fragments. It will be important in the future to determine the fate and possible function of these shorter peptide derivatives of POR A during the initial phase of rapid thylakoid assembly in developing chloroplasts. From what is known so far about the regulation of POR A synthesis and import, it must be a function that is only transiently required during the step from the dark into the light.

References

1 von Wettstein, D., Gough, S. and Kannangara, C.G. (1995) Plant Cell 7, 1039-1057
2 Forreiter, C. and Apel, K. (1993) Planta 190, 536-545
3 Fujita, Y. (1996) Plant Cell Physiol. 34, 411-421
4 Armstrong, G.A. et al. (1995) Plant Physiol. 108, 1505-1517
5 Holtorf, H. et al. (1995) Proc. Natl. Acad. Sci. USA 92, 3254-3258
6 Reinbothe, S. et al. (1996) Plant Cell 8, 763-769
7 Granick, S. (1959) Plant Physiol. Suppl. 34, 18
8 Foyer, C.H., Lelandais, M. and Kunert, K.J. (1994) Physiol. Plant. 92, 696-717
9 Griffiths, W.T. (1978) Biochem. J. 174, 681-692
10 Dehesh, K. and Ryberg, M. (1985) Planta 164, 396-399
11 Dehesh, K. et al. (1986) Planta 169, 172-183
12 Mapleston, E.R. and Griffiths, W.T. (1980) Biochem J. 189, 125-133
13 Santel, H.J. and Apel, K. (1981) Eur. J. Biochem. 120, 95-103
14 Apel, K. (1981) Eur. J. Biochem. 120, 89-93
15 Spano, A.J. et al. (1992) Plant Mol. Biol. 18, 967-972
16 Sundqvist, C. and Dahlin, C. (1997) Physiol. Plant. 100, 748-759

17 Reinbothe, S. et al. (1995) Plant Cell 7, 161-172
18 Reinbothe, S. et al. (1995) Plant Cell 7, 1933-1940
19 Joyard, J. et al. (1990) J. Biol. Chem. 265, 21820-21827
20 Reinbothe, C., Apel, K. and Reinbothe, S. (1995) Mol. Cell. Biol. 15, 6206-6212
21 Runge, S. et al. (1996) Plant J. 9, 513-523
22 Sperling, U. et al. (1997) Plant J. 12, 649-658
23 Batschauer, A. and Apel, K. (1984) Eur. J. Biochem. 143, 593-597
24 Holtorf, H. and Apel, K. (1996) Planta 199, 289-295
25 Reinbothe, C. et al. (1997) Proc. Natl. Acad. Scio. USA 94, 8890-8894

REDOX REACTIONS IN THE LAST STEPS OF CHLOROPHYLL BIOSYNTHESIS

W. RÜDIGER, H. KLEMENT, M. HELFRICH, A. TANAKA*, S. SCHOCH AND U. OSTER
Botanisches Institut der LMU München, Menzingerstr. 67, 80638 München, Germany
**The Institute of Low Temperature Science, Hokkaido University, Sapporo 0600819, Japan*

Abstract
The last steps of chlorophyll biosynthesis are formation of chlorophyllide *a* (Chlide *a*) from protochlorophyllide *a* (Pchlide *a*), esterification to chlorophyll *a* (Chl *a*) and formation of Chl *b*. Redox reactions are the reduction of ring D (Pchlide → Chlide), the oxidation of the methyl group at C-7 to a formyl group (Chl *a*→ Chl *b*), and the reverse reaction, reduction of the formyl group at C-7 to a methyl group (Chl *b* →Chl *a*). We investigated these redox reactions either with the enzymes that were (partially) purified from oat and barley etioplasts or with recombinant enzymes overexpressed in E. coli. Reduction of ring D, catalyzed by NADPH: protochlorophyllide oxidoreductase (POR), requires light. Several modified Pchlide derivatives were applied. The results allow a tentative model for the enzyme mechanism. The oxidation of Chl *a* derivatives to Chl *b* compounds with the recombinant chlorophyll *a* oxidase will be described here for the first time. The in vitro reaction will allow to investigate the requirement for cofactors.

Introduction

Redox reactions are widespread in nature and occur at some step in most if not all biosynthetic pathways. Here we will concentrate upon redox reactions which modify the electronic system of the Chl macrocycle such that the optical properties of the pigment are changed. This can be understood as an adaptation to particular functions during evolution. The Chl precursor Pchlide *a* shows only a weak absorption in the red and is therefore less suitable than Chl *a* if red light is used for photosynthesis (Fig. 1). The hypochromic shift of the Q_y band and the bathochromic shift of the Soret band, going from Chl *a* to Chl *b*, extends the overall absorption in the visible range if Chl *b* is present in addition to Chl *a*. Progress in the knowledge of two enzymatic steps, relevant to these spectral shifts, will be discussed here. These are the Phlide to Chlide transformation and the formation of Chl *b* from Chl *a* or from one of its precursors. The

J. H. Argyroudi-Akoyunoglou and H. Senger (eds.),
The Chloroplast: From Molecular Biology to Biotechnology, 185–190.

reverse redox reaction, the Chl *b* to Chl *a* transformation, will be discussed in an accompanying paper [1].

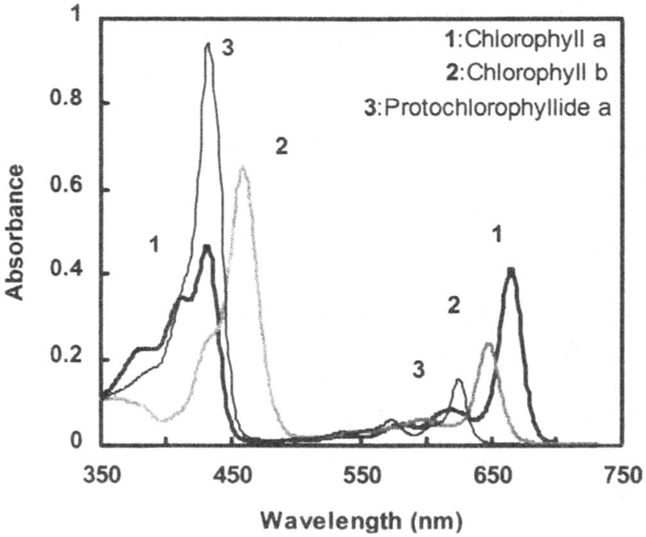

Figure 1. Absorption spectra of Pchlide *a*, Chl *a* and Chl *b* in acetone. The absorbance is shown for equimolar concentration of the pigments. The spectra were recorded with a diode array instrument (Tidas, J&M. Aalen) of pigments purified by HPLC (S. Schoch, unpublished results).

Results and Discussion

Pchlide to Chlide transformation

The well-known light-dependent transformation of Pchlide *a* to Chlide *a*, catalyzed by NADPH: protochlorophyllide oxidoreductase (POR), has been found in members of all groups of organisms that carry out oxygenic photosynthesis. The enzyme catalyzes a photobiochemical reaction, and our main interest deals with the role of light in this reaction. We purified pigment-free POR from etiolated oat seedlings and investigated the photoreaction with the natural substrate and several synthetic analogs of the substrate [2]. The enzyme was solubilized with octylglucoside, and the reaction was carried out at substrate (NADPH and Pchlide) and light saturation. The kinetics of the reaction are diffusion controlled under these conditions. Typical kinetics are shown in Fig. 2 in the form of difference absorption spectra. As can be seen, no other pigment (flavin, carotenoid) besides Pchlide is required for the reaction. When we studied the dependence of the kinetics on the concentration of Pchlide, we were surprised to detect substrate inhibition at ratios substrate: enzyme of about 2 or higher. This points to a second binding site of POR for Pchlide or its analogs.

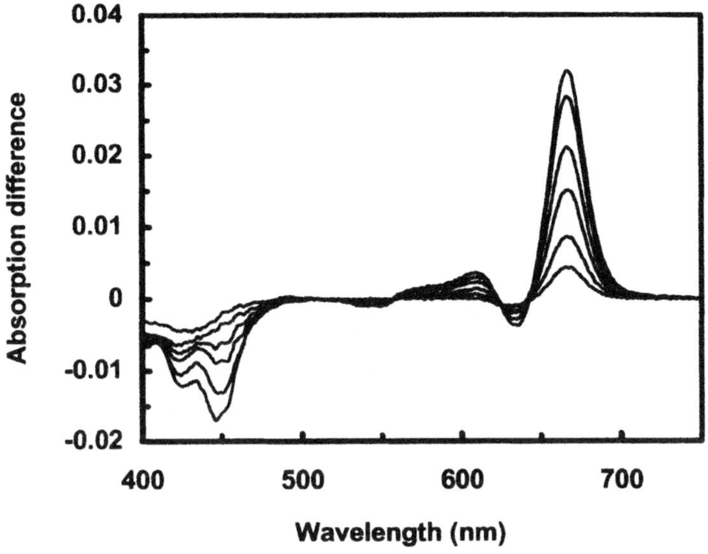

Figure 2. Absorption difference spectra for the enzymatic phototransformation of Zn Protopheide *a* to Zn Pheide *a*. POR was purified from oat etioplasts to a pigment-free state. For the reaction, saturating concentrations of NADPH and Zn Protopheide *a* were added. The mixture was then irradiated with white light (30·µmol·m^{-2}·s^{-1}) for 15s, 30s, 60s, 90s, 150s, 270s and absorption spectra were recorded at these time points as difference against the spectrum of the non-irradiated sample. The difference spectra show the decrease of the Q_y band of Zn Protopheide *a* (negative values at 626nm), the increase of the Qy band of Zn Pheide *a* (positive values at 650nm). The decrease of absorption at the Soret band is due to the smaller molar extinction coefficient of Zn Pheide *a* compared to Zn Protopheide *a*.

The application of several substrate analogs [2-4] allows now a conclusion on important structural features of compounds as suitable substrates for POR. Several modifications of the substituents at rings A and B are tolerated by the enzyme (R^1, R^2, R^3 in Fig. 3) whereas no structural change is allowed at ring D and the isocycling ring E (R^4, R^5, R^6, R^7 in Fig. 3). Since POR catalyses the hydrogenation of ring D, it is conceivable that changes in the structure of this ring are not tolerated. On the other hand, the strict structural and steric requirements for ring E are surprising. A possible explanation for this observation is the involvement of the isocyclic ring in the photoreaction. A model could be the photoenolization which is known in the photochemistry literature as Norrish type II reaction [5]. This reaction involves the formation of a bisradical structure via the triplet state. A different mechanism, starting with a proton attack at

C-18, has been proposed by Wilks and Timko [6] based on the observation that Tyrosin-275 is essential for the enzyme reaction. However, the three-dimensional structure of the enzyme-substrate complex and hence the exact position of Tyrosin-275 relative to Pchlide is not known. If Tyrosin-275 were close to the isocyclic ring, proton attack at the C-13^2 carbonyl group would lead to an enol-type structure and possibly also to a positive charge at C-17. Enolisation of ring E is facilitated in the native structure by two carbonyl groups, at C-13^1 and at C13^3. Both of these carbonyl groups must be present to form a suitable substrate structure. Additionally, the steric position of the methoxycarbonyl substituent at C-13^2 is important: this substituent must be below the macrocyclic ring (R^5 in Fig. 3) in the usual notation as in Fig. 3. The methoxycarbonyl substituent is probably important for correct binding to the enzyme. The transfer of the hydride ion from NADPH to C-17 of the substrate occurs from below the macrocycle so that the propionate side chain is above the macrocycle after the reaction, i. e. in trans position to the substituent at C-13^2. It remains to be shown how a presumed enol structure can facilitate the transfer of the hydride ion.

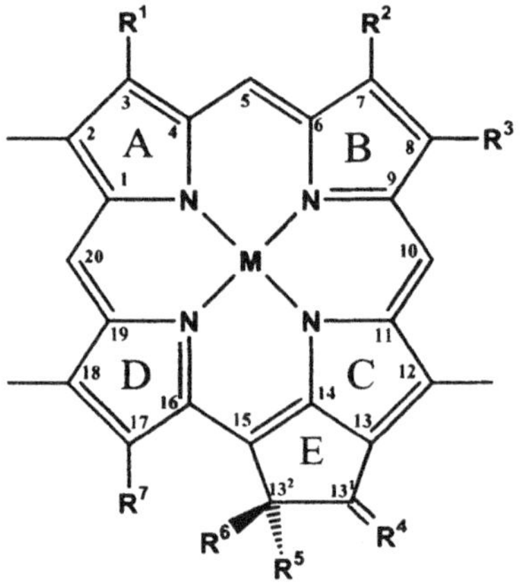

Figure 3. Structure of Pchlide and its analogs which were tested as possible substrates for POR. M=Mg or Zn, Substrates: R^1=vinyl, formyl or acetyl; R^2=methyl, hydroxymethyl or formyl; R^3=ethyl or vinyl; R^4=oxygen; R^5=methoxycarbonyl; R^6=proton; R^7=propionate. No substrates were compounds with either one of the following substituents: R^4=hydroxy or proton; R^5=hydroxy or proton; R^6=hydroxy or methoxy carbonyl; R^7=propionate methylester or acrylate.

Formation of Chlide b from Chlide a

It is known for a long time that etiolated plants, when exposed to light, form at first Chl *a* and only later Chl *b*. Early attemps to define the pathway of Chl *b* formation on the basis of indirect evidence are summarized by Rüdiger and Schoch [7]. An important step forward in the knowledge of Chl *b* biosynthesis was the finding that the oxygen of the formyl group of Chl *b* is derived from molecular oxygen, pointing to an oxygenase as the most likely class of enzymes for introduction of oxygen into the methyl group at ring B [8,9]. However, the enzyme was still not identified, neither was the proper substrate of the enzyme known, Chl *a* or one of its precursors.

Tanaka et al. [10] isolated Chl *b*-less mutants of *Chlamydomonas rheinhardtii* by insertion of either the nitrate reductase gene or the argininosuccinate lyase gene. Common to all mutants was a deleted region of genomic DNA which contained an open reading frame reminiscent of monooxygenases. Transformation of a DNA fragment containing this sequence complemented the Chl *b*-less phenotype. The enzyme encoded by this gene was named Chl *a* oxygenase (CAO).
We used the homologous CAO gene from *Arabidopsis thaliana* for expression in *E. coli* and for a test of enzyme activity. We found indeed production of Chlide *b*, analyzed as pheophorbide *b*, or ZnPheide *b* when we applied Chlide *a* or Zn-7^1-OH-Pheide *a*, respectively (Fig.4). The identity of the product was checked by HPLC (data not shown). The yield of the *b*-type pigment was higher , however, when we used the 7^1-OH-compound as substrate (Fig.4). This finding indicates that the 7^1-OH compound is an intermediate in the oxygenase reaction as proposed by Porra et al. [9].

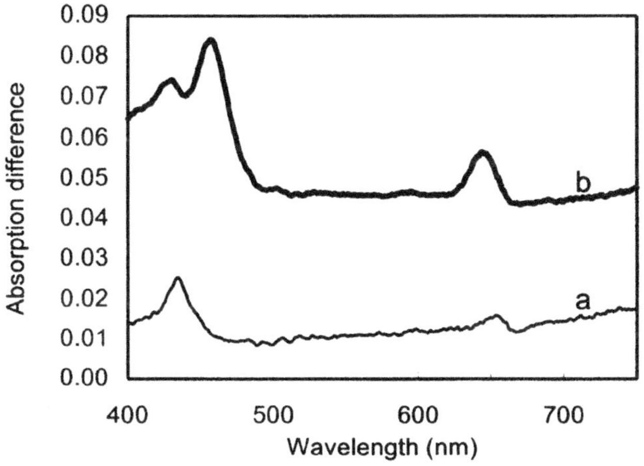

Figure 4. Absorption difference spectra of the *in vitro* oxygenation reaction of Chlide *a* to Chlide *b*, analyzed after removal of Mg (a) and of Zn 7^1-OH-Pheide *a* to Zn-Pheide *a* (b). The Chl *a* oxygenase gene of *Arabidopsis thaliana* was expressed in *E. coli* and incubated with the respective pigment for 120 min at 29°C. Formation of *b*-type pigment can be recognized by the increase of absorption around 450 and 650nm.

In contrast to the suggestion of these authors, no second enzyme is necessary for the second step of the reaction. We are presently working on the question whether first and second step of the oxygenase reaction require the same or different cofactors. Summarizing, we show here for the first time the enzymatic conversion of Chlide *a* to Chlide *b* via the 7^1-OH compound *in vitro*. The enzyme did not accept esterified Chl *a* under the applied conditions. Further experiments will show at which step of the Chl biosynthesis chain the oxygenation occurs.

Acknowledgement

This work was supported by the Deutsche Forschungsgemeinschaft, Bonn and the Fonds der Chemische Industrie, Frankfurt.

References

1. Scheumann, V., Schoch, S. and Rüdiger, W. (1998b) Cofactor requirement of the enzymatic reduction of chlorophyll b to chlorophyll a., this volume
2. Klement, H., Helfrich, M., Oster, U., Schoch, S. and Rüdiger, W. (1998) Pigment-free protochlorophyllide oxidoreductase from *Avena sativa* L.: purification and substrate specificity, submitted
3. Schoch, S., Helfrich, M., Wiktorsson, B., Sundqvist, C., Rüdiger, W. and Ryberg, M. (1995) Photoreduction of zinc protopheophorbide b with NADPH-protochlorophyllide oxidoreductase from etiolated wheat (Triticum aestivum L.), *Eur. J. Biochem.* **229**, 291-298
4. Helfrich, M., Schoch, S., Schäfer, W., Ryberg, M. and Rüdiger, W. (1996) Absolute configuration of protochlorophyllide a and substrate specificity of NADPH-protochlorophyllide oxidoreductase, *J. Amer. Chem. Soc.* **118**, 2606-2611
5. Wagner, P.J. and Park, B.-S. (1991) Photoinduced hydrogen atom abstraction by carbonyl compounds, *Org. Photochem.* **11**, 227-366
6. Wilks, H.M. and Timko, M. (1995) A light-dependent complementation system for analysis of NADPH: protochlorophyllide oxidoreductase: Identification and mutagenesis of two conserved residues that are essential for enzyme activity, *Proc. Natl. Acad. Sci. USA* **92**, 724-728
7. Rüdiger, W. and Schoch, S. (1988) Chlorophylls, in Goodwin, T.W. (ed.) *Plant Pigments*, Academic Press, London, pp.1-59
8. Schneegurt, M. A. and Beale, S. I. (1992) Origin of the Chlorophyll b formyl oxygen in Chlorella vulgaris, *Biochemistry* **31**, 11677-11683
9. Porra, R.J., Schäfer, W., Cmiel, E., Katheder, I. and Scheer, H. (1994) The derivation of the formyl-group of chlorophyll b in higher plants from molecular oxygen. Achievement of high enrichment of the 7-formyl-oxygen from $^{18}O_2$ in greening maize leaves, *Eur. J. Biochem.* **219**, 671-679
10. Tanaka, A., Ito, H., Tanaka, R., Tanaka, N.K., Yoshido, K. and Okada, K. (1998) Chlorophyll a oxygenase (CAO) is involved in chlorophyll b formation from chlorophyll a, *Proc. Natl. Acad. Sci USA* in press

COFACTOR REQUIREMENT OF THE ENZYMATIC REDUCTION OF CHLOROPHYLL *B* TO CHLOROPHYLL *A*

V. SCHEUMANN, S.SCHOCH AND W. RÜDIGER
Botanisches Institut der LMU München, Menzingerstr. 67, 80638 München, Germany

Introduction

There is only little known about the formation and degradation of chlorophyll *b* (Chl *b*). Ito *et al.* [1] suggested a chlorophyll cycle, in which Chl *b* is converted to Chl *a* and vice versa. This chlorophyll cycle can be important for the plants: during acclimation from low light to high light conditions the Chl *a* to Chl *b* ratio increases [2]. The reason is either a degradation of Chl *b* bound by the light harvesting complexes or a direct conversion of this Chl *b* to Chl *a*. Secondly, up to now only degradation products of Chl *a* were found in higher plants [3] and the key enzyme of Chl degradation, pheophorbide oxygenase, accepts only pheophorbide *a* (Pheide *a*) [4]. As a consequence, a separate degradation pathway for Chl *b* must exist in higher plants or Chl *b* must be reduced to Chl *a* before the action of Mg-dechelatase.
The reduction of Chl *b* to Chl *a* could be observed *in vitro* by addition of the chlorophyllide *b* (Chlide *b*) analogon zinc pheophorbide *b* (ZnPheide *b*) to isolated etioplasts of barley [5]. The cofactors of this two-step reduction of a formyl group to a methyl group are described here.

Materials and Methods

Etioplasts were prepared from primary leaves of 6 days old etiolated barley (*Hordeum vulgare* L.) as described by Scheumann *et al.* [6]. They were used either as intact etioplasts or lysed osmotically in reaction buffer (50 mM HEPES pH 8.0, 10 mM $MgCl_2$) without sorbitol. The etioplasts were incubated for 1.5 h at 28°C after addition of various supplements: about 1 nmol ZnPheide *b* in 5 µl acetone leading to a final concentration of 5% acetone, phytyl diphosphate (0.4 mM), ATP (10 mM), NADPH (2 mM), glucose-6-phosphate (16 mM), ferredoxin (Fd, 44 µM), glucose-6-phosphate dehydrogenase (0.01 units) and ferredoxin-$NADP^+$-oxidoreductase (0.01 units). All operations were performed under dim green save light.
Esterified and non-esterified pigments were analysed and quantified by HPLC using an analytical column filled with C-18 reverse phase silica gel (Shandon, Hypersil ODS

J. H. Argyroudi-Akoyunoglou and H. Senger (eds.),
The Chloroplast: From Molecular Biology to Biotechnology, 191–194.

5 µm). Peaks were detected with a diode-array detector (Tidas, J&M, Germany) and with a spectrofluorimetric detector (RF-551, Shimadzu). The zinc pheophorbides (see Fig. 1) were prepared as described previously [6].

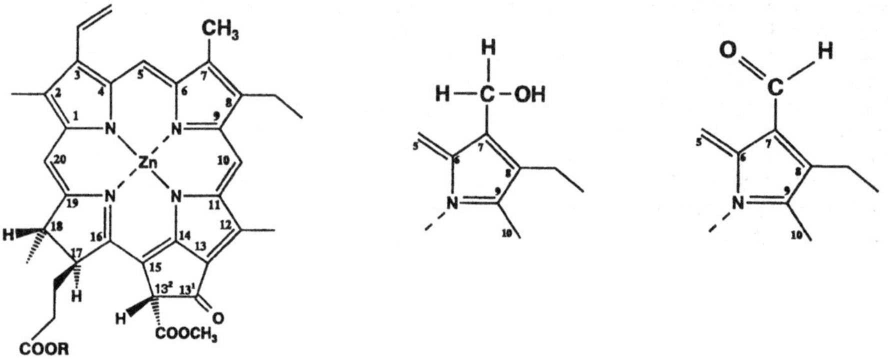

Figure 1. Chemical structure of zinc pheophorbide *a* (left, R=H). In zinc 7^1OH-pheophorbide *a*, the methyl-group at C-7 of the macrocycle is substituted by a hydroxymethyl group (middle), in zinc pheophorbide *b* by a formyl group (right). The corresponding zinc pheophytins are esterified with phytol (R=phytol).

Results and Discussion

The methyl group of the metal-free Pheide *b* can be reduced chemically in methanol with cyanoborohydride (BH_3CN^-). As products, 7^1OH-Pheide *a* and 7^1OCH_3-Pheide *a* could be observed. If zinc is present as a central atom, additionally ZnPheide *a* is formed under the same conditions. Zn-7^1OH-Pheide *a* and Zn-7^1OCH_3-Pheide *a* are just intermediates that can be further reduced [6], presumably with some unstable mesomeric intermediates. The central atom seems to be the crucial point because of its electronegativity: enzymatically the metal free Pheide *b* was not reduced at all [5].

In vitro, the first step of the enzymatic reduction of ZnPhe(ide) *b* to ZnPhe(ide) *a*, the reduction of ZnPhe(ide) *b* to Zn-7^1OH-Phe(ide) *a*, occurs in lysed etioplasts of barley and requires exogenous NADPH or NADH as reducing agent. ZnPheide *b*, phytyl diphosphate and different concentrations of NADPH were added to lysed and intact etioplasts of barley and incubated with or without addition of ATP. The esterified pigments were quantified (Fig.2). If lysed etioplasts were used, the amount of the product, Zn-7^1OH-Phe *a*, increases with higher concentrations of NADPH in the reaction buffer. In this system, exogenous ATP has no influence on the formation of Zn-7^1OH-Phe *a*. In contrast, with ATP about twice as much Zn-7^1OH-Phe *a* is formed compared to the assay without ATP if intact etioplasts were used instead. However, NADPH is only effective at low levels and the formation of Zn-7^1OH-Pheide *a* can not be enhanced by increasing concentrations of NADPH because NADPH can not penetrate the plastid's envelope membranes. So it seems like only the percentage of etioplasts that are broken during preparation or during incubation causes the difference of Zn-7^1OH-Pheide *a* formation in intact etioplasts with and without addition of NADPH.

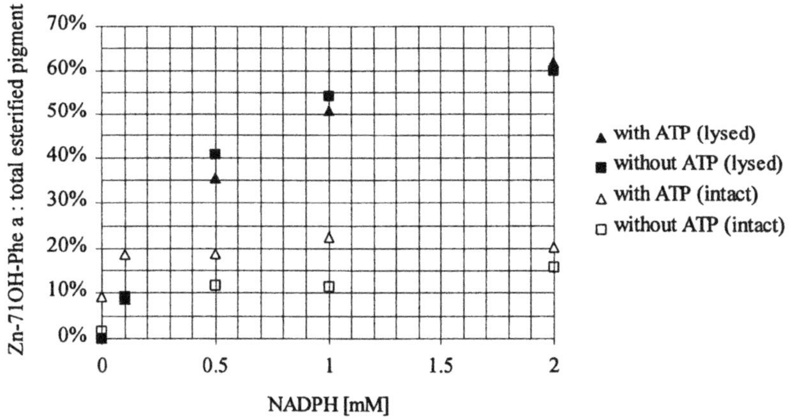

Figure 2. Formation of zinc 7[1]OH-pheophytin *a* in lysed and intact etioplasts depending on the concentration of NADPH. The etioplasts were incubated for 1.5 h at 28°C with zinc pheophorbide *b* and with or without 5.2 mM ATP. The esterified pigments were extracted and analysed by HPLC. The areas of the peaks were determined and the amount of the pigments was worked out after calibration with a standard.

In contrast to the first step of the reduction, the second step can not be performed with NADPH alone but requires reduced ferredoxin [7]. Ferredoxin and the enzyme ferredoxin-$NADP^+$-oxidoreductase are already present in etioplasts, but they are lost by dilution during lysis of etioplasts. If these substances are added exogenously to lysed etioplasts together with NADPH, glucose-6-phosphate and glucose-6-phosphate-dehydrogenase, ferredoxin was kept in the reduced state and Zn-7[1]OH-Phe(ide) *a* was reduced to ZnPhe(ide) *a*. In lysed etioplasts, this reaction was also independent of ATP. In contrast, ATP caused the formation of ZnPhe *a* if only ZnPheide *b* and phytyl pyrophosphate were added to intact etioplasts [7]. In this case ATP could be replaced by dihydroxyacetone phosphate. Maybe ATP serves for an intraplastidic transport of the substrate, because the enzymes of the reduction seem to be located in the inner plastid membranes [7]. Although the cofactor requirement indicate two different enzymes, we can not exclude the possibility that ferredoxin might be the reducing agent also for the first step *in vivo* and that only one enzyme catalizes both reaction steps.

Summary

The *in vitro* studies with lysed and intact etioplasts show the cofactor requirement of the enzymes responsible for the reduction of Chl *b* to Chl *a* via 7[1]OH-Chl *a* (Fig. 3). All reaction steps in lysed etioplasts are ATP-independent, however ATP has an influence on intact etioplasts. NADPH does not penetrate the envelope membrane of etioplasts and has no influence on intact etioplasts, but is absolutely necessary to perform the reduction in lysed etioplasts and to keep the reduced state of ferredoxin. Additionally, the reduction can only be performed if a central atom (Zn or Mg) is present in the tetrapyrrole.

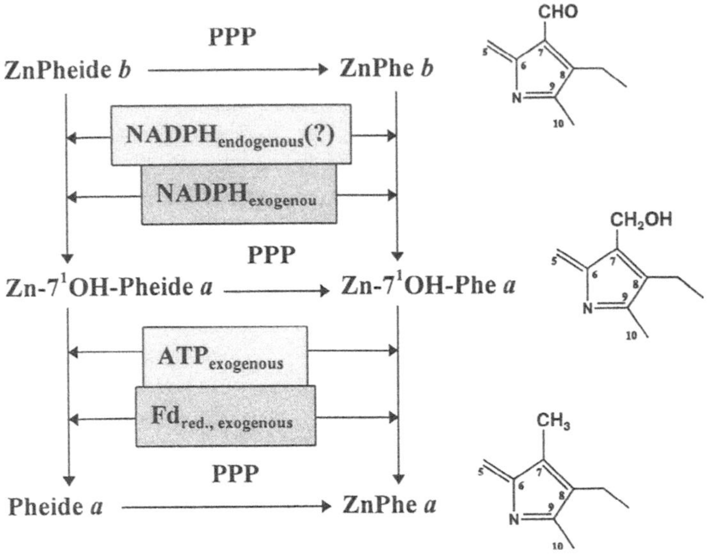

Figure 3. Formation of zinc pheophytin *a* from zinc pheophorbide *b in vitro*. The pigment can be reduced before or after esterification. The reducing agents required in [intact] and [lysed] etioplasts are marked between the arrows. PPP = phytyl pyrophosphate.

Acknowledgement

This work was supported by the Deutsche Forschungsgemeinschaft, Bonn and the Fonds der Chemische Industrie, Frankfurt.

References

1. Ito, H., Ohtsuka, T. and Tanaka A. (1996) Conversion of chlorophyll *b* to chlorophyll *a* via 7-hydroxymethyl chlorophyll. *J. Biol. Chem.* **271**, 1475-1479.
2. Lindahl, M., Yang, D.-H. and Andersson, B. (1995) Regulatory proteolysis of the major light-harvesting chlorophyll *a/b* protein of photosystem II by a light-induced membrane-associated enzymic system. *Eur. J. Biochem.* **231**, 503-509.
3. Matile, P. and Kräutler, B. (1995) Wie und warum bauen Pflanzen das Chlorophyll ab? *Chemie* **29**, 298-306.
4. Hörtensteiner, S., Vicentini, F. and Matile, Ph. (1995) Chlorophyll breakdown in senscent cotyledons of rape, *Brassica napus* L.: Enzymatic cleavage of phaeophorbide *a in vitro*. *New Phytol.* **129**, 237-246.
5. Scheumann, V., Ito, H., Tanaka, A., Schoch, S. and Rüdiger, W. (1996) Substrate specificity of chlorophyll(ide) *b* reductase in etioplasts of barley (*Hordeum vulgare* L.). *Eur. J. Biochem.* **242**, 163-170.
6. Scheumann, V., Helfrich, M., Schoch, S. and Rüdiger, W. (1996) Reduction of the formyl group of zinc pheophorbide *b in vitro* and *in vivo*: a model for the chlorophyll *b* to *a* transformation. *Z.Naturforsch.* **51c**, 185-194.
7. Scheumann, V., Schoch, S. and Rüdiger, W. (1998) Chlorophyll *a* formation in the chlorophyll *b* reductase reaction requires reduced ferredoxin. *(submitted)*.

EFFECTS OF NUCLEAR *y* MUTATIONS ON EXPRESSION OF PLASTID GENES REQUIRED FOR LIGHT-INDEPENDENT CHLOROPHYLL FORMATION IN *CHLAMYDOMONAS*

A.B. CAHOON and M.P. TIMKO
Department of Biology
University of Virginia
Charlottesville, Virginia 22903 USA

1. Introduction

During evolution two distinct mechanisms were established for the reduction of protochlorophyllide (pchlide) to chlorophyllide (chlide), a key step in the chlorophyll biosynthesis pathway. One mechanism, catalyzed by the enzyme NADPH: protochlorophyllide oxidoreductase (POR), is completely dependent upon light for its activity [1-2]. Light-dependent POR is present in cyanobacteria, green algae, most non-vascular and vascular plants, and is the only mechanism used for chlorophyll formation in angiosperms. The second mechanism, present in anoxygenic photosynthetic bacteria as well as cyanobacteria, non-vascular plants, ferns and gymnosperms, is capable of reducing pchlide to chlide in a light-independent manner [3]. All organisms containing this mechanism for pchlide reduction are capable of chlorophyll formation in the dark. While a significant amount of information is now available on the regulation of POR biosynthesis and activity [1], little is known about the enzyme mediating light-independent pchlide reduction (designated LIPOR), the factors that regulate its formation, and its requirements for catalytic function [3].

Previous studies have shown that the products of three chloroplast genes (designated *chl*L, *chl*N, and *chl*B) and at least seven nuclear loci (designated *y-1*, *y-5*, etc., for *yellow-in-the-dark* phenotype) are required for LIPOR activity in the green alga, *Chlamydomonas reinhardtii* [2]. Cells with mutations in any one of these three plastid-encoded genes [3-4] or containing one of the nuclear-encoded *y* mutations have identical "yellow-in-the-dark" phenotypes [5-6], reflecting a loss of chlorophyll formation and the accumulation of the biosynthetic precursor pchlide. When grown in the light, however, these same cells synthesize chlorophyll and achieve a wild-type phenotype, since they still have a functional light-dependent POR.

Although direct biochemical evidence is lacking, the products of the plastid-encoded *chl*L, *chl*N, and *chl*B genes are thought to constitute the subunits of an oligomeric LIPOR complex [2-3]. The role of the various *y* gene products in the process of light-independent pchlide reduction and their possible interaction with the *chl* gene products remains to be determined. In the present study, we have examined the expression of the plastid-encoded *chl*L, *chl*N, and *chl*B genes and the accumulation of their

J. H. Argyroudi-Akoyunoglou and H. Senger (eds.),
The Chloroplast: From Molecular Biology to Biotechnology, 195–200.

encoded proteins in light- and dark-grown wild-type and *y* mutant strains of *Chlamydomonas*. We show that the expression of at least one *chl* gene product, CHLL, may be negatively-regulated by light in wild-type cells and that its accumulation in the dark specifically blocked in the various *y* mutants.

2. Results and Discussion

A defining characteristic of all *y* mutants is their inability to reduce pchlide in the dark, but their continued capacity to synthesize and accumulate chlorophyll upon exposure to light. As part of our initial characterization of these strains, we determined the time course for pchlide accumulation in the various *y* mutants upon transfer from light to darkness and compared this to what occurs in wild-type cells. As shown in Figure 1, wild-type cells contained little or no pchlide when grown in the light or dark, since both the light-dependent and light-independent routes of chlorophyll formation are fully functional. In contrast, all of the *y* mutants accumulated pchlide upon transfer to darkness. Over the 96 hr time period following transfer into the dark, *y-6-1* and *y-7* showed the fastest rates of pchlide accumulation, with pigment levels reaching a maximum at 24 hr. However, the total amount of pigment accumulated per cell in these strains was low relative to the other *y* mutants. Strains *y-1a*, *y-8*, and *y-10* accumulated the highest levels of pchlide and these levels were maintained for longer periods of time. No significant pchlide accumulation was found in any *y* strain grown in HS medium, likely due to the limited availability of reduced carbon. Our findings are consistent with previous descriptions of these mutants [5-6] indicating that some of the strains have a more leaky phenotype than others.

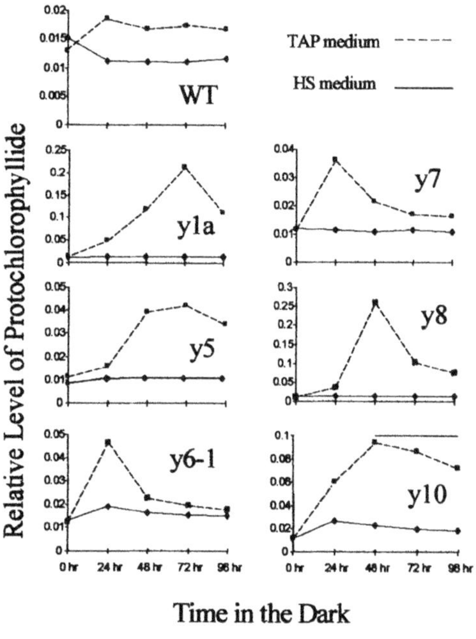

Figure 1. Protochlorophyllide accumulation in wild-type and mutant strains of *Chlamydomonas* during growth in the dark. Wild-type137c and *y* mutant strains of *C. reinhardtii* were grown in liquid culture on either Tris-acetate phosphate (TAP) or minimal (HS) medium at 24-28°C [4, 14]. After 3 days of growth under continuous white light, the cells were transferred to complete darkness (0 hr in the Figure) and pchlide accumulation was measured spectrofluorometrically [16] over the subsequent 96 hr time period.

Having shown that the various *y* mutants differed in their capacity to accumulate pchlide, we next examined whether any of these strains were altered in their ability to transcribe the plastid-encoded *chl*L, *chl*N, and *chl*B genes. As shown in Figure 2, *chl*L, *chl*N, and *chl*B transcripts were present in both dark- and light-grown wild-type cells, with steady-state levels of the *chl*L, *chl*N, and *chl*B messages ~2-4-fold greater in the light. The observed higher levels of *chl*L and *chl*N message in light-grown cells is

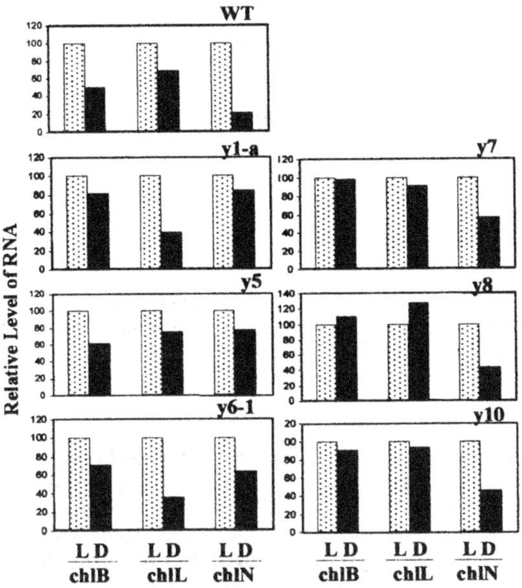

Figure 2. Relative abundance of *chl*L, *chl*N, and *chl*B transcripts in wild-type and *y* mutant strains. Total RNA was isolated from wild-type and mutant strains grown as described in Figure 1. RNA samples from dark-grown cultures were harvested 24-48 hr after transfer of the cultures to darkness. Transcript levels were quantitated by phosphorimaging following RNA gel blot analysis [4] and semi-quantitative reverse transcriptase PCR [13].

consistent with our previous studies [5] showing a slight increase in *chl*B transcript abundance in light- versus dark-grown wild type cells. By comparison to other plastid transcripts (e.g., *rbcL*), the abundance of all three messages is relatively low (data not shown). In general, the steady state levels of the *chl*L, *chl*N, and *chl*B transcripts in the various *y* mutants were similar to those seen in wild-type cells, with only a few minor differences distinguishing the various strains. Of particular note is the observation that *chl*L and *chl*B transcript levels were slightly elevated in dark-grown *y-7*, *y-8*, and *y-10* relative to wild-type. No differences were detected in the size of the *chl*L, *chl*N, or *chl*B transcripts present in the various *y* mutants compared to wild-type, suggesting that no gross defects in RNA maturation or splicing were present in these strains. RNA gel blot analysis did detect the presence of multiple bands hybridizing with the *chl*N probe, consistent with previous reports that this message is part of a polycistronic transcript [9].

Since only minor differences in transcript levels were observed, we were interested in knowing whether the levels of the CHLL, CHLN, and CHLB proteins were altered in the *y* mutants. Extracts of total cellular proteins were prepared from wild-type and mutant strains grown either in the light or darkness, the extracts were fractionated by SDS-PAGE, and the abundance of the CHL proteins analyzed by western blotting. As shown in Figure 3, the CHLL and CHLN antisera detected proteins bands of the appropriate size (~33 kDa and 67 kDa, respectively) in extracts of dark-grown wild-type cells. In contrast, little or no immunoreactive CHLL was detected in wild-type cells grown in the light and the levels of CHLN were substantially reduced. Only under prolonged exposure times could any CHLL protein be detected in extracts from light-grown cells. Antiserum directed against the β-subunit of CF_1-ATPase detected equivalent amounts of this protein in extracts from both dark- and light-grown cells, indicating that the loss of CHLL and CHLN was not the result of non-specific protein degradation. Unfortunately, no data is available for the CHLB protein at this time. Given the significant levels of *chl*L and *chl*N message present in light-grown wild-type cells, these data strongly suggest that the loss of CHLL and CHLN protein in these cells is either the result of a light-dependent decrease in protein synthesis (e.g., a block in translation initiation or elonga-

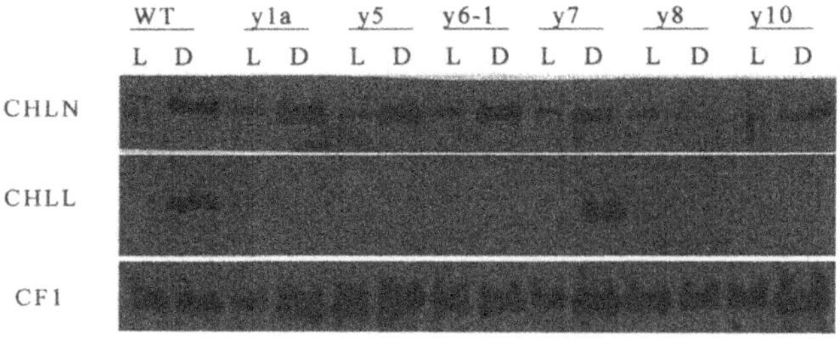

Figure 3. Western blot analysis of CHLL and CHLN protein levels in wild-type and *y* mutant strain.. Wild-type and mutant strains were grown either in the light or dark and extracts of total cellular proteins prepared as described in [5]. The extracts were fractionated by SDS-PAGE and CHLL and CHLN proteins detected by western blotting [4, 15] using polyclonal antisera prepared against bacterially synthesized CHLL and CHLN proteins and the chemiluminescent second antibody detection system (ECL, Amersham Corp). Antiserum directed against the β-subunit of CF_1-ATPase (kindly provided by Richard McCarty, Johns Hopkins University).

tion) or increased protein turnover (e.g., decreased stability and/or increased proteolysis).

In all but one case, the level of CHLN present in the mutants was similar to that observed in wild-type cells (i.e., higher levels of CHLN present in dark- versus light-grown cells) (see Figure 3). Only *y-8* showed any significant variation from this pattern, exhibiting a slightly decreased level of CHLN in the dark compared to wild-type. In contrast, a dramatically different effect was found on CHLL levels in the *y* mutants. Unlike wild-type cells, no CHLL was detected in the extracts of dark-grown *y* mutants, suggesting that the mutations present in these strains blocked CHLL formation or accumulation. The only exception was *y-7*. However, the immunoreactive CHLL present in this strain was slightly smaller than the 33 kDa protein found in wild-type cells, suggesting that it may be altered and/or nonfunctional. These results are interesting since they suggest that the inability to form CHLL (or in the case of *y-7*, the formation of a nonfunctional CHLL subunit), may be responsible for the inability of the various *y*-mutants to carry our light-independent chlorophyll formation. These results are also interesting since they suggest that the formation and accumulation of CHLN is not dependent upon the presence of CHLL. What factors are responsible for the loss of CHLL accumulation in the various *y* mutants and why *y-7* differs in its capacity for CHLL formation remain to be determined.

Figure 4. Western blot analysis of CHLL protein levels in wild-type and *y-7* mutant strains grown either in the light or dark in the presence or absence of acetate and CCCP. Total cellular proteins were prepared and fractionated by SDS-PAGE and the CHLL proteins detected by western blotting using the ECL chemiluminescent second antibody detection system [13, 15].

To explore further the mechanism by which light regulates the formation of CHLL, wild-type *Chlamydomonas* cells were grown in the light or dark on either TAP, HS, or HS medium containing acetate (HSA). When wild-type *Chlamydomonas* cells were grown photoautrophically (i.e., on acetate-free HS medium), higher levels of immunoreactive CHLL were found in light-grown cells compared to cells grown in the dark (Figure 4). Thus, under photoautotrophic growth conditions, wild-type cells synthesized and accumulated significant levels of CHLL in the light and maintained significantly lower levels of this protein in the dark. This is almost the inverse of what occurred when the cells wee grown in TAP medium. Wild-type cells growing in HS medium showed a loss of CHLL accumulation in the light and increased CHLL formation in the dark following the addition of acetate. In contrast, *y-7* showed no difference in its response to the presence or absence of acetate in the growth medium, and accumulated similar levels of CHLL in the dark in the presence or absence of an external carbon source. These findings suggest that the synthesis and/or accumulation of CHLL in wild-type cells may be controlled by the available levels of metabolically-derived energy and/or reductants provided via photosynthesis present in the cell. As shown in Figure 4, reducing photosynthetically derived ATP and/or reductants by treating cells with CCCP (carbonyl-cyanide-chlorophenyl hydrazone), an uncoupler of photosynthetic electron transport, resulted in increased CHLL accumulation in the light, only in the presence of acetate. The basis for this response is not known. One possibility is that formation of CHLL requires some minimal level of energy (either metabolically or photosynthetically derived), whereas degradation of the protein requires energy and/or reductants produced by photosynthesis. Thus, blocking photosynthetic electron transport would only partially affect CHLL formation, but might prevent activation of components for its turnover (e.g., an ATP-dependent protease).

3. Conclusions.

It is well documented that both light and development affect the translation of various chloroplast messages by influencing such processes as the recruitment of mRNAs onto polyribosomes, the assembly of translation initiation complexes, and the activation of RNA binding proteins required for either translation initiation or elongation [7-10]. Translation in turn can also affect message abundance and stability both directly and indirectly [9]. Light-regulated translation is also known to involve sequence specific nuclear-encoded RNA binding proteins, whose binding capacity can be modulated by ATP/ADP ratios, redox potential and/or the availability of photosynthetically-derived reductants [7, 11-12].

By analogy to previously described mechanisms [7, 10], we believe that in wild-type cells, CHLL and CHLN synthesis and accumulation may be regulated by one or more nuclear-encoded factors acting at a posttranscriptional level, most likely at the level of translation or protein turnover. Under our testable model, these factors function in a positive manner to promote CHLL and CHLN translation in the dark and are either missing or inactive in the light. Defects in these proteins or their ability to carry out their regulatory function result in a loss of CHLL formation in the dark in the various y-mutants. The ability to form CHLL (and CHLN) in the dark is modulated by the amount of metabolically-derived energy. Consistent with this, photoautotrophically growing wild-type cells transferred to the dark show lower levels of CHLL (and CHLN) expression than cells growing on media containing an external carbon source (acetate). Light-dependent activation of protein degradation machinery may also contribute to regulation of CHLL and CHLN levels. These processes are likely tied to light-dependent changes in redox potential and/or the availability of photosynthetically-derived energy or reduction potential. Thus, cells blocked in their ability to carry out photosynthetic electron transport would be expected to synthesize and accumulate CHLL, since translation would not be blocked and/or activation of light-dependent proteolytic activity (e.g., an ATP-dependent *Clp* protease) could not take place, if these processes were dependent upon energy or reductants derived from photosynthesis.

4. Acknowledgements

The authors wish to thank Nikolai Lebedev for his help with pigment analysis and Jean-David Rochaix (University of Geneva) for providing antiserum against CHLN. This work was supported by a grant to MPT from the U.S. Department of Energy.

5. References

1. Reinbothe, S. and Reinbothe, C. (1996) Regulation of chlorophyll biosynthesis in angiosperms, *Plant Physiol.* **111**, 1-7.
2. Timko, M.P. (1998) Pigment biosynthesis: Chlorophylls, Heme, and Carotenoids, in J.-D. Rochaix, M. Goldschmidt-Clermont, and S. Merchant (eds.), *Molecular Biology of Chlamydomonas: Chloroplasts and Mitochondria*, Kluwer Academic Publishers, The Netherlands, in press.
3. Fujita, Y. (1996) Protochlorophyllide reduction: a key step in the greening of plants, *Plant Cell. Physiol.* **37**, 411-421.
4. Li, J., M. Goldschmidt-Clermont, and Timko, M.P. (1993) Chloroplast encoded *chlB* is required for light-independent protochlorophyllide reductase activity in *Chlamydomonas reinhardtii*, *Plant Cell* **5**, 1817-1829.
5. Ford, C. and Wang, W.-Y. (1980) Three new *yellow* loci in *Chlamydomonas reinhardtii*, *Mol. Gen. Genet.* **179**, 259-263.
6. Ford, C. and Wang, W.-Y. (1980) Temperature sensitive *yellow* mutants of *Chlamydomonas reinhardtii*, *Mol. Gen. Genet.* **180**, 5-10.
7. Goldschmidt-Clermont, M. (1998) Coordination of nuclear and chloroplast gene expression in plant cells. *International Rev. Cytol.* **177**, 115-180.
8. Rochaix, J.D. (1996) Post-transcriptional regulation of chloroplast gene expression in *Chlamydomonas reinhardtii*, *Plant Molec. Biol.* **32**, 327-341.
9. Stern, D.B., Higgs, D.C., and Yang, J.J. (1997) Transcription and translation in chloroplasts, *Trends Plant Sci.* **2**, 308-315.
10. Mayfield, S.P., Yohn, C.B., Cohen, A., and Danon, A. (1995) Regulation of chloroplast gene expression, *Ann. Rev. Plant Physiol. Plant Mol. Biol.* **46**, 147-166.
11. Danon, A. and Mayfield, S.P. (1994) Light regulated translation of chloroplast messenger RNAs through redox potential, *Science* **266**, 1717-1719.
12. Danon, A. and Mayfield, S.P. (1994) ADP-dependent phosphorylation regulates RNA-binding *in vitro*: implications in light-modulated translation, *EMBO J.* **13**, 2227-2235.
13. Ausubel, F.M., Brent, R., Kingston, R.E., Moore, D.D., Seidman, J.G., Smith, J.A., and Struhl, K. (eds.)(1995) *Short Protocols in Molecular Biology, Third Edition*, John Wiley & Sons, Inc.
14. Harris, E.H. (1989) *The Chlamydomonas Source Book: A Comprehensive Guide to Biology and Laboratory Use*, Academic Press, Inc., San Diego, pp. 25-31.
15. Rochaix, J.D., Mayfield, S., Goldschmidt-Clermont, M., and Erikson, J. (1988) Molecular Biology of *Chlamydomonas*, in C.H. Shaw (ed.), *Plant Molecular Biology, A Practical Approach*, Oxford: IRL Press, pp. 253-275.
16. Lebedev, N.N., Djelepova, I.D., and Krasnovskii, A.A. (1991) Protochlorophyllide fluorescence in cells of the green alga *Chlamydomonas reinhardtii*. *Biofizika* **36**, 1022-1029.

CAN *PSILOTUM* AND/OR *GNETUM* SYNTHESISE CHLOROPHYLL IN DARKNESS?

J. Walmsley[1], H. Adamson[1,2], M. Wright[1], P. Wrench[1].
[1]Biological Sciences, Macquarie University, Sydney, N.S.W. 109 Australia
[2]Biological Sciences, University of Sydney, N.S.W. 2006, Australia

1. Introduction

Most plants including angiosperms can synthesise chlorophyll (Chl) in darkness as well as in light albeit at a lower rate [1]. For this, they require a light-independent reductase to catalyse the conversion of protochlorophyllide (Pchl(ide)) to chlorophyllide (Chl(ide)). Genes coding for individual sub-units of a such an enzyme have been identified in purple bacteria (*bch* L,*bch* N and *bch* B) and a wide variety of oxygenic photosynthetic organisms ranging from cyanobacteria to gymnosperms (*chl* L, *chl* N and *chl* B). However they have not been found in angiosperms. This could be a consequence of sequence divergence following gene transfer from the chloroplast to the nucleus. Alternatively, it might reflect gene deletion during evolution.

The experiments in this paper were prompted by the observation that *Psilotum*, (a fern ally) and *Gnetum*, (a gymnosperm) resemble angiosperms in failing to respond to light-independent Pchl(ide) reductase gene probes which elicit strong signals from a wide variety of lower plants [2,3]. Specifically we wanted to determine whether *Psilotum* and/or *Gnetum* , like many angiosperms, are capable of synthesising Chl in darkness.

2. Molecular Experiments

2.1 METHODS

Total DNA of Gnetum, Psilotum , Pinus and Ephedra was extracted with a minipreparation method (DNAeasy, Qiagen Australia). The presence or absence of *chl* L, *chl* B or *Chl* N homologous sequences was determined by (i) PCR amplification using pairs of 20mer primers with sequences corresponding to highly conserved regions of previously reported sequences of *chl* L, *chl* B and *chl* N[2,3,4] and (ii) non-rad Southern Blot hybridization methods using either an algal probe (*chl* L) or a gymnosperm probe (*chl* B or *chl* N). For Southern blot hybridization DNA was digested with Hind III, BamH 1 and EcoR V and electrophoresed on 0.8% agarose gels using a Tris-acetate-EDTA buffer system.

J. H. Argyroudi-Akoyunoglou and H. Senger (eds.),
The Chloroplast: From Molecular Biology to Biotechnology, 201–205.

Fluorescein labelled probes were prepared by standard methods (NEN RRPF Labeling kit, Dupont, Australia)

2.2 RESULTS

We could find no evidence of light-independent reductase genes in either *Gnetum* or*Psilotum*. When DNA of *Pinus* and *Ephedra* were used as controls PCR amplification and Southern blot hybridization experiments gave positive signals consistent with the presence of *chl* L and *chl* N. Negative results were obtained in PCR experiments with *Gnetum* amplified with primers for *chl* L (Fig.1a). Negative results were also obtained in Southern blot hybridization experiments with *Psilotum* probed for *chl* L (Fig. 1b) and *chl* N (results not shown).

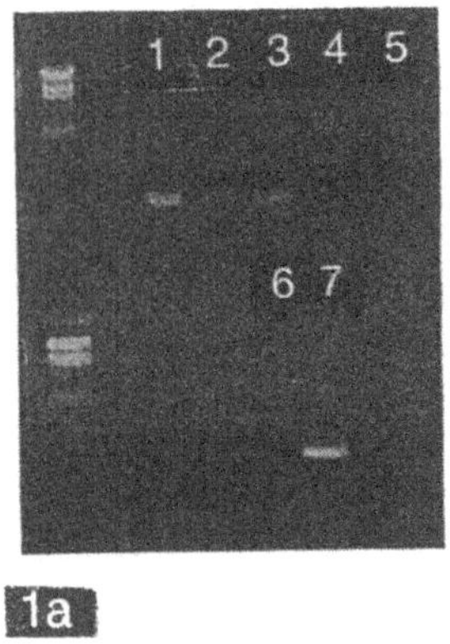

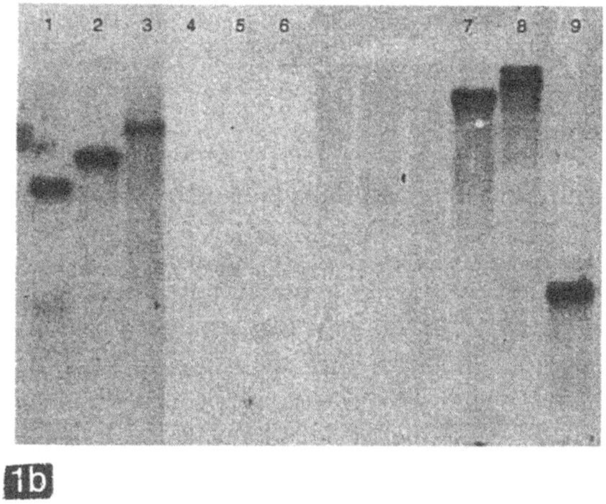

Fig 1(a) PCR amplification of genomic DNA obtained from *Marchantia* (lanes 1-3), *Gnetum* (lanes 4-5), *Pinus* (lane 6) and *Ephedra* (lane 7) using 20 mer primers with sequences corresponding to highly conserved regions of *Chl* L.

Fig 1(b) Southern blot analysis of genomic DNHA from *Pinus* (1-3), *Psilotum* (4-6) and *Ephedra* (7-9). The 3 lanes for each species represent restriction digests with BamH, Eco RV and Hind III respectively.

3. Physiological Experiments

3.1 METHODS

Individual branches of *Gnetum* (a 2m shrub) were enclosed in light-tight housings whose effectiveness in excluding light was confirmed each time they were used by enclosing a packet of barley seeds in moist vermiculite and testing the barley shoots produced during dark treatment for signs of Chl. Whole plants of *Psilotum,* growing in pots, were transferred from a glasshouse to the purpose built dark laboratory described in [1]. The tissue sampling strategy was dictated by the morphology of the plant. With *Gnetum*, initial (To) and final (T_D) Chl concentrations in each leaf were determined using matched circles of tissue removed with a cork borer from similar locations on either side of the mid-vein. With *Psilotum*, dichotomous apical branches of young main stems were used to determine initial and final Chl concentrations. Since these branches elongated significantly during dark treatment, they were marked and measured as required in order to calculate Chl dilution due to water uptake during dark treatment. Pigments were extracted in 85% acetone and quantified as described in [5]

3.2 RESULTS

Dark Chl synthesis in *Gnetum* was very obvious. The Chl concentration in 5 of the 6 leaves examined in the experiment summarised in Table 1 increased substantially during dark treatment (13 - 19%). This reflected increases in both Chl a and Chl b. Since the initial and final measurements were made on fully turgid leaf pieces the increases in Chl concentration cannot be attributed to a loss of turgor. The consistent decline in Chl a/b ratios also rules this out.

Dark Chl synthesis in *Psilotum* was also evident in some branches but not in all. In a typical experiment, eleven apical dichotomous branch pairs were examined before and after dark treatment. There was a substantial net gain in Chl content of 5 of them. This is shown in Table 2. Another 4 showed no change in Chl content and 2 sustained substantial loss (results not shown). Table 2 shows, for each apical dichotomous branch which accumulated Chl in darkness, the Chl a and Chl b concentration (ug per g FW) at time of transfer to darkness (T_O) and at the end of dark treatment (T_D (O)). It also shows the ***predicted*** Chl a and Chl b concentrations (T_D (P)). These were calculated taking into account the increase in fresh weight of the branches and ***assuming that no synthesis of Chl in darkness occurred during the experiment.***

In each case, the observed concentration exceeded the predicted concentration by a substantial amount (12 - 20%). The net gain in Chl content of the shoot which this implies can only be explained in terms of a light-independent Chl biosynthetic pathway. Also, in each case, the Chl concentration at the end of dark treatment is lower than at the beginning. This indicates that the apical dichotomous branches are growing faster than they are greening.

4. Discussion

To date, there is no evidence for *chl*L, chl N or chl B-type Pchl(ide) reductase genes in *Psilotum, Gnetum* or angiosperms. Nevertheless these plants can make Chl in the dark. This raises two questions. (1) Have the light-independent Pchl(ide) reductase genes been deleted from the chloroplast genomes of *Psilotum* and *Gnetum*? And (2) if so, have they been transferred to the nucleus, as has been postulated for angiosperms?

	Chl *a*			Chl *b*			Total Chl			Ratio *a/b*		
Sample	T_0	T_{11D}	*% change*	T_0	T_{11D}	*% change*	T_0	T_{11D}	*% change*	T_0	T_{11D}	*% change*
1	225	215	*-4*	84	100	*19*	309	315	*2*	2.68	2.15	*-20*
2	206	217	*5*	79	106	*34*	285	323	*13*	2.60	2.05	*-21*
3	233	246	*5*	82	107	*30*	315	353	*12*	2.84	2.30	*-19*
4	246	258	*5*	90	113	*26*	336	371	*10*	2.73	2.30	*-16*
5	223	253	*14*	88	113	*28*	311	366	*18*	2.53	2.20	*-13*
6	191	219	*15*	79	102	*29*	270	321	*19*	2.41	2.14	*-11*
Mean	**221**	**235**		**84**	**107**		**304**	**342**		**2.63**	**2.19**	
±SE	**± 8**	**± 8**		**± 2**	**± 2**		**± 10**	**± 10**		**± 0.06**	**± 0.04**	
P	*>0.05*			*< 0.01*			*< 0.05*			*< 0.01*		

Table 1. Changes in chlorophyll concentrations (μg Chl g FW^{-1}) and chlorophyll *a/b* ratios of matched leaf discs from an intact *Gnetum* plant at T_0 and after 11 days dark treatment. *P* is the probability that the difference between the values at T_0 and T_{dark} could have arisen by chance.

	Chl (µg per g FW)			Ratio	% Change	
	a	*b*	*a + b*	*a/b*	FW	length
T_0	236	90	326	2.6	18	29
$T_D(O)$	246	87	333	2.8		
$T_D(P)$	200	76	276			
% gain/loss	*+ 23*	*+ 14*	*+ 20*			
T_0	233	71	305	3.3	33	12
$T_D(O)$	217	53	270	4.1		
$T_D(P)$	175	53	229			
% gain/loss	*+ 24*	*0*	*+ 18*			
T_0	224	87	310	2.6	54	20
$T_D(O)$	174	51	225	3.4		
$T_D(P)$	145	56	201			
% gain/loss	*+ 20*	*- 9*	*+ 12*			
T_0	202	75	277	2.8	19	20
$T_D(O)$	203	57	260	3.6		
$T_D(P)$	170	63	232			
% gain/loss	*+ 20*	*- 10*	*+ 12*			
T_0	242	103	345	2.3	38	22
$T_D(O)$	218	61	279	3.6		
$T_D(P)$	175	75	250			
% gain/loss	*+ 25*	*- 19*	*+ 12*			

Table 2. Chlorophyll concentration of matched *Psilotum* apical branches at T_0 after 7 or 14 days dark treatment. T_D (*P*) was the predicted concentration calculated assuming no synthesis of chlorophyll in darkness (% net gain/loss = 100 x [$T_D(P)$ - T_D(O)]/ $T_D(P)$)

If the answer to (1) is 'yes' and (2) is 'no' then a second light-independent Pchl(ide) reductase widely distributed throughout the plant kingdom will be indicated.

5. References

1. Adamson, H.Y., Hiller, R.G., and Walmsley, J. D. (1997) Protochlorophyllide reduction and greening in angiosperms: a evolutionary perspective, *J. Photochem. Photobiol.* B. Biology **41**, 201-221.
2. Suzuki, J.Y. and Bauer, C.E. (1992) Light-independent chlorophyll biosynthesis: involvement of the chloroplast gene *chl* L (frxC), *Plant Cell,* **4**, 929-940.
3. Boivin, R., Richard, M., Beauseigle, D., Bousquet, J. and Bellemare, G. (1996) Phylogenetic inferences from chloroplasts *chl B* gene sequences of *Nephrolepis exaltata* (Filicopsida), *Ephedra altissima* (Gnetopsida) and diverse land plants, *Mol. Phylogenet. & Evol.*, **6**, 19-29.
4. Lidholme, J. and Gustafsson, P. (1991) Homologues of the green algal *gid* A gene and the liverwort frxC gene are present in the chloroplast genome of conifers, *Plant Mol. Biol.* **17**, 787-798.
5. Walmsley, J. and Adamson, H. (1989) Chlorophyll accumulation and breakdown in light-grown barley transferred to darkness: effect of seedling age, *Physiol Plant.* **77**, 312-319.

PROTOCHLOROPHYLLIDE AND POR IN THE *LIP1* MUTANT OF PEA

M. Seyyedi[1], M.P. Timko[2] and C. Sundqvist[1]

[1]Botanical Institute, Univ. of Göteborg, Box 461, SE-405 30 Göteborg, Sweden.
[2]Dept. of Biology, Univ. of Virginia, Charlottesville, VA 22901, USA

1. Introduction

The prolamellar bodies (PLBs) have a high content of the protochlorophyllide (Pchlide) reducing enzyme NADPH-protochlorophyllide oxidoreductase (POR). The genes for two different POR enzymes (PORA and PORB) have been isolated from barely and *Arabidopsis* [1, 2]. PORA is present mostly in dark-grown plants while PORB is more persistent and is suggested to be the main chlorophyll forming enzyme [3]. However, in pea only one type of POR seems to be present [4].

Pchlide is present in different spectral forms [5]. The Pchlide 633 and 657 nm fluorescence emission maxima in dark-grown plants varies between species and plant organs. Chlorophyllide (Chlide) is formed from Pchlide650-657 with a msec flash irradiation. When Pchlide fluorescing at 633 nm is transformed the Chl formation is slow and requires continuous light.

A pea mutant (*lip1*) with a light-independent photomorphogenesis, similar to the *det* and *cop* mutants of *Arabidopsis*, has been described by Frances et al. [6]. The *lip1* mutant has seedlings with short epicotyls and well developed leaves when grown in darkness. Etioplasts in dark-grown *lip1* plants lack typical PLBs.

The Pchlide650-657 form is regarded as a highly aggregated complex of Pchlide, NADPH and POR [7, 8] mainly localised to the PLBs. Sperling et al [9] have recently shown that overexpression of POR A or POR B in wild-type and a *cop* mutant of *Arabidopsis* led to an increased size of PLBs confirming at close correlation between POR and PLBs. As the etioplasts in the *lip1* mutant lack PLBs a comparison between the mutant and its wild-type is a potential way to elucidate the correlation of PLBs and the different forms of Pchlide and their possible functions during chloroplast development.

2. Materials and methods

Plant material: Peas (*Pisum sativum* L. cv. Alaska and *lip1* mutant) were grown as described in [5]. Unexpanded (Alaska) or expanded (*lip1*) leaves were used and all handling of the seedlings was done under dim green light.

J. H. Argyroudi-Akoyunoglou and H. Senger (eds.),
The Chloroplast: From Molecular Biology to Biotechnology, 207–212.

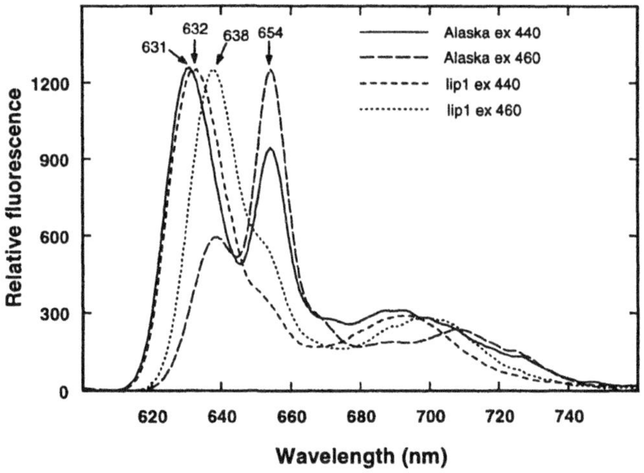

Figure 1. Fluorescence spectra (77 K) of dark-grown Alaska and *lip1* mutant peas with excitation wavelength 440 nm or 460 nm.

Irradiation: White fluorescent light (Osram L20, Warmton Warm White, 2 W m^{-2}) was used for continuous irradiation. The light source for flash irradiation was a Braun F 800 electronic photoflash with an average energy output of 160 J for 0.002 s. The cut leaves were irradiated with 2 flashes with 10 s in between.

Fluorescence spectroscopy: Fluorescence emission spectra were measured at 77 K using an SLM Aminco 8000C spectrofluorometer [5].

Protein extraction and determination: Protein extraction and determination was mainly as described in [10]. Fractionated proteins were electroblotted to PVDF membranes and POR polypeptides were detected using an antiserum containing polyclonal antibodies raised against POR from wheat, followed by a Goat anti-Rabbit IgG (H+L) Horseradish peroxidase conjugate (Biorad) and HRP color development reagent (Biorad).

Transmission electron microscopy: Samples for transmission electron microscopy was prepared as described in [10] and immunogold labelling was performed as described in [11], using Unicryl instead of Lowicryl.

3. Results

Alaska pea leaves grown in darkness were characterized by two evident fluorescence emission peaks of Pchlide at 631 nm and 654 nm when 440 nm light was used for excitation. Leaves of the *lip1* mutant had one fluorescence emission maximum at 632 nm. The fluorescence emission at 654 nm was evident only as a small shoulder (Fig. 1). The exitation with 460 nm light increased the fluorescence at 654 nm in Alaska leaves and promoted the fluorescence of a short-wavelength peak at 638 nm. The same changes

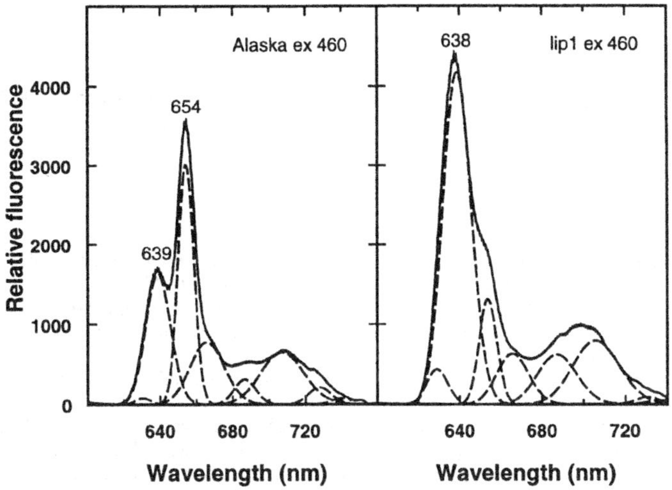

Figure 2. Gaussian resolution of fluorescence emission spectra (77 K) of Alaska and *lip1* leaves. Excitation wavelength 460 nm.

occurred also in *lip1* but no evident peak was formed at 654 nm. However, curve resolution into Gaussian components showed the presence of a 654 nm form also in *lip1* (Fig. 2)

Alaska pea leaves formed a considerable amount of Chlide with a fluorescence peak at 692 nm after flash irradiation (Fig. 3). During a following 30 min dark period the Chlide peak in the leaves shifted to 682 nm and a small peak at 652 nm indicated a regeneration of Pchlide phototransformable by a second flash irradiation. The Chlide formed in *lip1* leaves was diminutive but dominated of a short wavelength form with a maximum at 677 nm. During a following 30 min dark period the Chlide peak in the *lip1* leaves got a slightly reduced half-band width but showed no obvious peak shift. The small shoulder at 654 nm was present also after flash irradiation and no change occured during the 30 min dark period.

During continuous irradiation of the Alaska leaves a long-wavelength Chlide form was found after 30 s irradiation (Fig. 4). After 10 minutes this form had shifted to shorter wavelengths and the short wavelength Pchlide had more or less disappeared. In *lip1* leaves there was a formation of short wavelength Chlide already after 30 s irradiation and only a shoulder was seen at longer wavelengths. It can be noticed that a considerable amount of Pchlide was still left in the *lip1* leaves after 10 min irradiation.

Western blots indicated the presence of POR in leaves of dark-grown wild type and *lip1* mutant seedlings (Fig. 5). The leaf tissue was rich in POR but also the stem of the lipl mutant contained considerable amounts (results not shown). The levels of the enzyme both in Alaska and the *lip1* mutant were nearly stable during 6 hour of irradiation. Just one band at 36 kD was detectable on the Western blots in both Alaska and *lip1*.

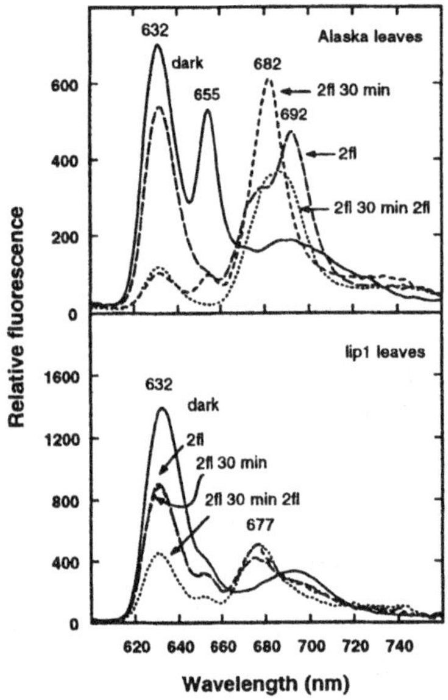

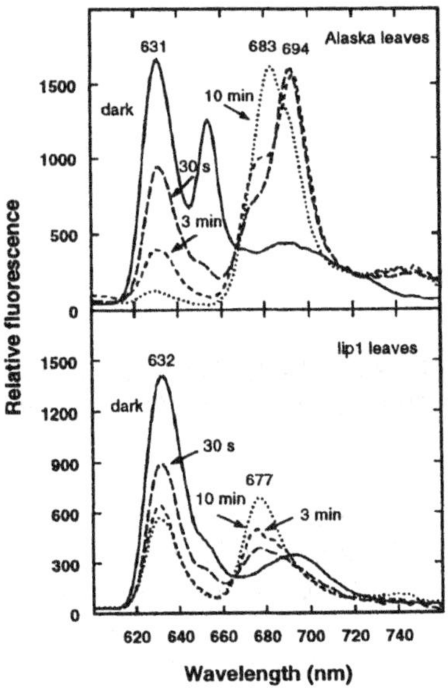

Figure 3 (left). Fluorescence spectra (77 K) of flash irradiated Alaska (A) and *lip1* (B) mutant pea leaves. The leaves were treated with a series of 2 flashes 30 min dark and 2 flashes. Samples were taken as indicated.

Figure 4 (right). Fluorescence spectra (77 K) of leaves of Alaska (A) and *lip1* (B) mutant pea irradiated for 30 s, 3 min and 10 min. Excitation wavelength 440 nm.

Electron micrographs showed that Alaska leaves contained regular PLBs. The *lip1* leaves had hardly any PLBs. Occasionally small remnants of PLBs could be found but they were never organized in a typical regular fashion. The immunolocalization using Unicryl embeded sections showed the presence of POR in the etioplasts of both Alaska and *lip1*. The content of POR seemed to be higher in *lip1* than in Alaska as revealed by number of gold particles per plastid (Table 1). In Alaska calculated per PLB area the number of particles was about double of that calculated per whole plastid area. In *lip1* the POR protein seemed dispersed along the thylakoids in the plastids or spread in the stroma.

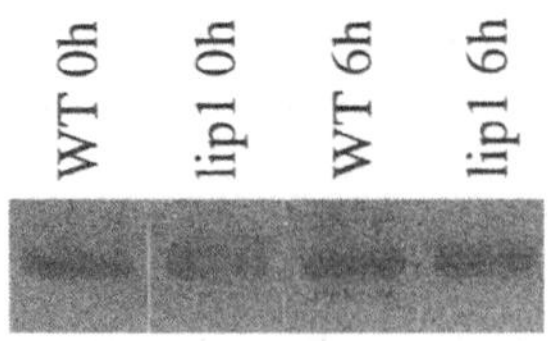

Figure 5. Immunological detection of POR in etioplast membranes of 7 day-old dark-grown or 6 h irradiated leaves of Alaska wild-type and *lip1* pea seedlings. Equivalent amounts of total protein were used for the different lanes.

Table 1. Statistics on gold labelling of POR in leaves of dark-grown wild-type (WT 0h) , *lip1* mutant (*lip1* 0h) and plants irradiated for 6 hours (WT 6h and *lip1* 6h).

Particles per plastid			
WT 0h	WT 6h	*lip1* 0h	*lip1* 6h
31	31	46	66

4. Discussion

Even when grown in complete darkness the *lip1* seedlings have several morphological characteristics of light-grown plants. Furthermore some light-regulated genes are expressed in dark-grown *lip1*. The *lip1* plants also show a fast chlorophyll accumulation when exposed to light [6, 12]. In the *lip1* etioplasts the Pchlide content is higher than in the Alaska etioplasts, but mostly short wavelength Pchlide forms are present (Fig. 1). Regular PLBs are absent but occasionally remnants of PLBs or PLB-like structures can be found. However, there are still considerable amounts of POR present in the *lip1* etioplasts. It appears that this POR has only a limited inclination to aggregate to form the long-wavelength Pchlide650-657 form. This can partly be an effect of the induced development of the plastid. In partially greened plants the reformed Pchlide is also typically characterized of short-wavelength forms. However, this implicates that the thylakoids in partially developed chloroplasts can hold POR and still remain in a dispersed form.

In *Arabidopsis* an increased Pchlide content lead to an increased photodestruction in white light. This effect was diminished by overexpression of POR A and POR B which also led to an increased Chl(ide) accumulation [13]. In *lip1* the Pchlide content is also increased considerably compared with the wild type. Nevertheless the *lip1* mutant is not photodestroyed when irradiated with white light but accumulate as much chlorophyll as the wild-type. The protection against photodestruction is probably a complex phenomenon where the activation of the phytochrome signal transduction pathway is one component.

The overexpression of POR A and POR B in dark-grown Arabidopsis leads to the formation of large PLBs. [9, 13]. In *lip1* a high amount of Pchlide is formed, short-wavelength forms dominates and an extensive thylakoid formation occurs. PLB formation is repressed as well as the formation of the long-wavelength form of Pchlide. The formation and accumulation of Pchlide and POR inside the etioplast seems, not in itself to induce the formation of PLBs or Pchlide650-657. The results presented here thus indicates that the formation of PLBs is a more complex phenomenon than earlier expected. It is not just an effect of the absence of light and an accumulation of precursors for chlorophyll biosynthesis. There are probaly other factor(s) present in non-phytochrome activated plants in darkness, which regulates the formation of the PLBs.

5. References

1. Holtorf, H. Reinbothe, S. Reinbothe, C. Bereza, B. and Apel, K. (1995) Two routes of chlorophyllide synthesis that are differentially regulated by light in barley (Hordeum vulgare L.), P. N. A. S. **92**, 3254-3258
2. Armstrong, G.A., Runge, S., Frick, G. Sperling, U. and Apel, K. (1995) Identification of NADPH:protochlorophyllide oxidoreductase A and B: abranched pathway for light-dependent chlorophyll biosynthesis in Arabidopsis thaliana. Plant Physiol. **108**, 1505-1517.
3. Runge, S., Sperling, U., Frick, G., Apel, K. and Armstrong, G. A. (1996) Distinct roles for light-dependent NADPH:protochlorophyllide oxidoreductases (POR) A and B during greening in higher plants, Plant J. **9**, 513-523.
4. Spano, A. J., He, Z., Michel, H., Hunt, D. F. and Timko, M. P. (1992) Molecular cloning, nuclear gene structure, and developmental expression of NADPH:protochlorophyllide oxidoreductase in pea (*Pisum sativum* L.), Plant. Mol. Biol. **18**: 967-972.
5. Böddi, B., Mc Ewen, B., Ryberg, M., and Sundqvist, C. (1994) Protochlorophyllide forms in non-greening epicotyls of darkgrown pea (Pisum sativum), Physiol. Plant. **92**, 160-170.
6. Frances, S., White, J.M., Edgerton, M.D., Jones, A.M., Elliott, R.C. and Thompson, W.F. (1992) Initial characterization of a pea mutant with light-independent photomorphogenesis, Plant Cell, **4**, 1519-1530.
7. Böddi, B., Lindsten, A., Ryberg, M. and Sundqvist, C. (1989) On the aggregational states of protochlorophyllide and its protein complexes in wheat etioplasts, Physiol. Plant. **76**, 135-143.
8. Wiktorsson, B., Engdahl, S., Zhong, L. B., Böddi, B., Ryberg, M. and Sundqvist, C. (1993) The effect of cross-linking of the subunits of NADPH-protochlorophyllide oxidoreductase on the aggregational state of protochlorophyllide, Photosynthetica, **29**, 205-218
9. Sperling, U., Franck, F., van Cleve, B., Frick, G. Apel, K. and Armstrong, G.A. (1998). Etioplast differentiation in Arabidopsis: Both PORA and PORB restore the prolamellar body and photoactive protochlorophyllide F655 to the cop1 photomorphogenic mutant, The Plant Cell **10**, 283-296.
10. Mc Ewen, B., Sundqvist, C. and Younis, S. (1994) Protochlorophyll(ide) forms in hypocotyls of dark-grown bean (Phaseolus vulgaris), Physiol. Plant. **90**, 396-407.
11. Ryberg, M. and Dehesh, K. (1986) Localiztion of NADPH-protochlorophyllide oxidoreductase in dark-grown wheat (*Triticum aestivum*) by immuno-electron microscopy before and after transformation of the prolamellar bodies, Physiol. Plant. **66**, 616-624.
12. Sundqvist, C., Seyyedi, M. and Timko, M.P. (1995) Chlorophyll formation in the *lip1* mutant of pea, in P. Mathis (ed.), *Photosynthesis from light to biosphere*, Kluwer Academic Publishers Dordrecht, vol III, pp. 1005-1008..
13. Sperling, U., van Cleve, B., Frick, G. Apel, K. and Armstrong, G.A. (1997) Overexpression of light-dependent PORA or PORB in plants depleted of endogenous POR by far-red light enhances seedling survival in white light and protects against photoxidative damage, The Plant Journal **12**, 649-658.

SUBTRATE SPECIFICITY OF OVEREXPRESSED BACTERIOCHLOROPHYLL SYNTHASE FROM *CHLOROFLEXUS AURANTIACUS*

S. SCHOCH, U.OSTER, K. MAYER, R. FEICK, AND W.RÜDIGER
Botanisches Institut der LMU München, Menzingerstr. 67 D-80638 München, Germany

Introduction

Two genes from *Chloroflexus aurantiacus* were reported to code for the protein synthase which catalizes the esterification of bacteriochlorophyllide to bacteriochlorophyll [1,2]. One gene, bchG, is located near genes coding for proteins associated with the bacterial reaction center, whereas the other, bchG2, is clustered with two genes coding for proteins associated with the antenna system. Lopez et al. [2] suggested, that the first one codes for the bacteriochlorophyllide *a* esterifying synthase and the second one codes for the synthase that esterifies the light-harvesting pigments, i.e. bacteriochlorophyllide *c*.

To prove this hypothesis we expressed the genes in *Escherichia coli* and tested both proteins for the synthase activity. For additional information on structure-function relationship, we studied the substrate specificity of these enzymes: we synthesized derivatives of naturally occurring (bacterio)chlorophylls and tested the esterification reaction with the two proteins BChG and BChG2.

Material and Methods

Chlorophylls and bacteriochlorophylls were extracted from plants and bacteria, respectively. The corresponding chlorophyllide and bacteriochlorophyllides were prepared either by the chlorophyllase reaction (to keep the central Mg ion) or by hydrolysis with trifluoracetic acid followed by zinc insertion according to Refs. 3 and 4. The synthesis of the pyro- and 3^1-hydroxyethyl- derivatives is described in Ref. 5. The enzyme assay was performed according to Ref. 4, using 100mM sodium carbonate to transfere the esterified Zn bacteriopheophytins into the hexane phase. Both bchG genes were cloned into the pT7-7 and the pQE60 vectors. For the pQE60 vector the restriction sites Nco I/ Hind III were used. The pT7-7 vector was cloned between Nde I/Hind III. Overexpression in E. coli was started using IPTG.

J. H. Argyroudi-Akoyunoglou and H. Senger (eds.),
The Chloroplast: From Molecular Biology to Biotechnology, 213–216.

Results and Discussion

The BChG protein was only active when it was overexpressed using the pT7-7 vector. On the other hand, the protein from the overexpressed bchG2 gene was more active with the pQE60 vector. For all the presented results the optimal vectors were used.

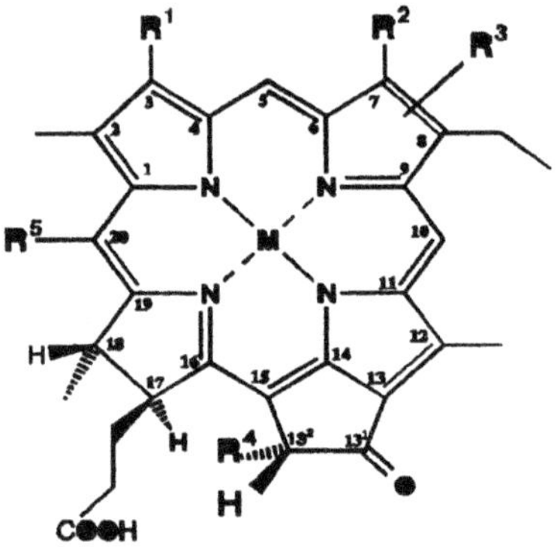

Name	M	R^1	R^2	R^3	R^4	R^5
Chlorophyllide a	Mg	$CH=CH_2$	CH_3	=	CO_2CH_3	H
Bacteriochlorophyllide a	Mg	CH_3CO	CH_3	-	CO_2CH_3	H
Pyrochlorophyllide a	Mg	$CH=CH_2$	CH_3	=	H	H
Zn-bacteriopheophorbide c	Zn	$CHOHCH_3$	CH_3	=	H	CH_3
Zn-bacteriopheophorbide d	Zn	$CHOHCH_3$	CH_3	=	H	H
Zn-bacteriopheophorbide e	Zn	$CHOHCH_3$	CHO	=	H	CH_3
Zn-3^1-hydroxyethyl-pyrobacteriopheophorbide a	Zn	$CHOHCH_3$	CH_3	-	H	H
Zn-3^1-hydroxyethyl-bacteriopheophorbide a	Zn	$CHOHCH_3$	CH_3	-	CO_2CH_3	H
Zn-3^1-hydroxyethyl-pheophorbide a	Zn	$CHOHCH_3$	CH_3	=	CO_2CH_3	H
Zn-3- acetyl-pheophorbide a	Zn	CH_3CO	CH_3	=	CO_2CH_3	H

Figure 1: Pigments used to test the synthase reaction with the proteins from the overexpressed bchG and bchG2 genes.

Both proteins were separately used for the synthase reaction with the pigments shown in Fig. 1. Substrates of the chlorophyll synthase of higher plants, such as chlorophyllide *a* or zinc pheophorbide *a* were accepted neither by the synthase BChG nor by BChG2. BChG synthase esterified only bacteriochlorophyllide *a* and Zn-3-acetyl-pheophorbide *a*. This result clearly shows, that the BChG enzyme specifically recognizes the 3-acetyl group (R^1 = CH_3CO) but accept both: the oxidized (R^3 = double bond, chlorin-type) and reduced (R^3 = single bond, bacteriochlorin-type) ring, respectively (Fig. 1).

The best substrates for the BChG2 synthase were Zn-bacteriopheophorbide *c*, *d* and *e*. These compounds are chlorin-type pigments having the pyrolyzed isocyclic ring (lacking the methoxycarbonyl group at C-13^2) and the hydroxyethyl group at C-3^1. All the other pigments listed in Fig 1 were not accepted as substrates, Zn-3-acetyl-pheophorbide *a* excepted, which was esterified to a small extent. This finding shows, that only chlorin-type pigments with oxygen-containing side chain at C-3 were essential for the BChG2 enzyme. The pyrolyzed isocyclic ring and the hydroxyethyl group at C-3^1 are possibly not essential, but pigments owing all these attributes together are strongly favoured as substrates.

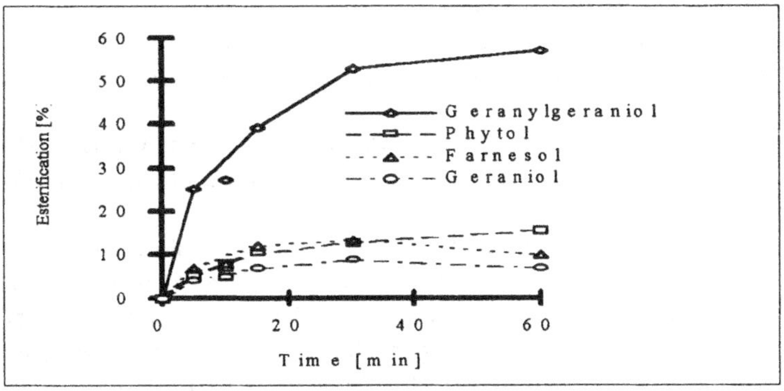

Figure 2: Kinetics of the esterification of Bchlide *a* with the diphosphates of the indicated alcohols with the synthase BChG

The time course of the esterification of bacteriochlorophyllide *a* with the diphosphates of several alcohols, using the BChG protein, is shown in Fig. 2. Geranylgeranyl diphosphate was more than three times more effective than the other alcohols, including phytyl diphosphate. After about 20 to 30 minutes the reaction slows down. The analogous kinetics of the synthase BChG2 with Zn-bacteriopheophorbide *c* shows, that the geranylgeranyl diphosphate is also the most effective alcohol followed by the diphosphates of phytol, farnesol and geraniol. Neither the BChG nor BChG2 enzyme accept the diphosphates of the cetyl- and stearyl-alcohols (data not shown), although bacteriochlorophyll *c* esterified with cetyl alcohol was isolated from *Chloroflexus aurantiacus* [6].

Summary

1.-The synthase BChG recognizes the 3-acetyl group of the tetrapyrrole pigments, but does not discriminate between the chlorin- or bacteriochlorin-type pigment.

2.-The synthase BChG2 seems to be specific for the chlorin-type tetrapyrrole with an oxygen-containing side chain at C-3. The enzymic reaction is greatly enhanced if the substrate contains specifically the hydroxyethyl group at C-3^1 and lacks also the methoxycarbonyl group at C-13^2 (pyro-type).

3.- Geranylgeraniol is the most preferred esterifying alcohol for both enzymes.

Acknowledgements

We thank H. Scheer for supplying us with bacteriochlorophyll *c*, and D. Steensgaard, A. R. Holzwarth and K. Schaffner for providing us with bacteriochlorophyll *d* and *e*. The work was supported by the Deutsche Forschungsgemeinschaft (SFB 184)

References

1. Niedermeier et al. (1994) EMBL accession no. Z34000
2. Lopez, J.C., Ryan S., and Blankenship R.E. (1996) Sequence of the *bch*G Gene from *Chloroflexus auriantiacus*: Relationship between Chlorophyll Synthase and Other Polyprenyltransferases, *J. Bacteriol.* **178**, 3369-3373
3. Schoch, S. Helfrich, M., Wiktorsson, B., Sundqvist, C., Rüdiger, W., and Ryberg, M. (1995) Photoreduction of zinc protopheophorbide *b* with NADPH-protochlorophyllide oxidoreductase from etiolated wheat (*Triticum aestivum* L.), *Eur.J.Biochem.* **229**, 291-298
4. Helfrich, M., Schoch, S., Lempert, U., Cmiel E., and Rüdiger W. (1994) Chlorophyll synthetase cannot synthesize chlorophyll *a'*, *Eur.J.Biochem.* **219**, 267-275
5. Struck, A., Cmiel, E., Katheder, I., Schäfer W.,and Scheer H. (1992) Bacteriochlorophylls modified at position C-3: long-range intramolecular interaction with position c-13^2, *Biochim.Biophys.Acta* **1101**, 321-328
6. Fages, F., Griebenow, N., Griebenow, K., Holzwarth, A.R., and Schaffner, K. (1990) Characterization of Light-harvesting Pigments of *Chloroflexus aurantiacus*. Two New Chlorophylls: Oleyl (Octadec-9-enyl) and Cetyl (Hexadecanyl) Bacteriochlorophyllides-*c*, *J. Chem. Soc. Perkin Trans. 1* **1990**, 2791-2797

CAROTENOIDS IN HIGHER PLANTS AND ALGAE

Susanne Römer

Lehrstuhl für Physiologie und Biochemie der Pflanzen
Universität Konstanz
D-78434 Konstanz, Germany

Carotenoid Biosynthesis

Carotenoids are tetraterpenoids which are synthezised from the common precursor isopentenyl diphosphate. The first dedicated step is the formation of phytoene via phytoene synthase [1]. Subsequent desaturation reactions lead to the production of lycopene. Cyclization of lycopene by the action of either β - or ε -cyclase (alone or in combination) results in the formation of ß- or α-carotene, respectively [2]. Addition of different oxygen-containing side groups gives rise to the various xanthophylls. The three β-carotene derived xanthophylls zeaxanthin, antheraxanthin and violaxanthin can be interconverted and are constituents of the xanthophyll cycle.

Function of Carotenoids in Photosynthesis

In photosynthesis carotenoids serve as antenna pigments broadening the absorption cross section of the photosystems [3]. One of the most important features, however, is their protective function due to the ability to quench excess light energy and to act as scavengers of radicals and activated oxygen species [4]. In addition, a role in the stabilization of the thylakoid membrane by the xanthophyll zeaxanthin and a structural function of lutein for oligomerization of light harvesting complexes during assembly of the photosynthetic apparatus have been suggested [5,6].

Localization of Carotenoids in the Photosynthetic Apparatus

Carotenoids are essential components of pigment-protein complexes. Whereas β-carotene is mainly located in the core complexes of photosystem I and II, the xanthophylls violaxanthin and neoxanthin as well as lutein are predominantly found in the light harvesting system [7]. The carotenoid composition can change in response to environmental and light conditions. The formation of zeaxanthin from violaxanthin under high light stress via the so-called xanthophyll cycle is a well described phenomenon [8,9].

J. H. Argyroudi-Akoyunoglou and H. Senger (eds.),
The Chloroplast: From Molecular Biology to Biotechnology, 217–223.

Carotenoid Genes

Since the pioneering work unraveling the carotenoid gene clusters of *Rhodobacter capsulatus* and *Erwinia uredovora* many other carotenoid genes have been identified from cyanobacteria, algae and higher plants using different cloning approaches [10-12]. Also genes encoding enzymes of the xanthophyll cycle have been isolated [13,14]. We have recently isolated a partial clone for β-carotene hydroxylase from tobacco after screening of a cDNA library (kindly provided by Dr. B. Grimm, IPK Gatersleben) with a PCR derived probe of β-carotene hydroxylase from *Arabidopsis thaliana*. Fig. 1 shows selected regions of a sequence alignment of all known β-carotene hydroxylases including the tobacco one. The conserved histidine motif 2 is marked (a). It is apparent that the plant sequences are very well conserved (b). The identity of the tobacco amino acid sequence compared to the pepper ones is greater than 90%.

	(a) histidine motif 2 HXXHH		(b) highly conserved region HDGLVH
ALCABCH	WGW**H**KS**HH**EE	ALCABCH	IYFIL**HDGLVH**QRWP
AGROBCH	WGW**H**KS**HH**EE	AGROBCH	IYFVL**HDGLVH**QRWP
FLAVOBCH	WGW**H**KS**HH**DE	FLAVOBCH	IYYFL**HDGLVH**GRWP
ERWHERBCH	WRW**H**ES**HH**TP	ERWHERBCH	LYFLV**HDGLVH**QRWP
ERWUREBCH	WGW**H**LS**HH**EP	ERWUREBCH	LYFMV**HDGLVH**QRWP
CABECARH	WHM**H**ES**HH**RP	CABECARH	AYMFV**HDGLVH**KRFP
CABCH2	WHM**H**ES**HH**KP	CABCH2	AYMFV**HDGLVH**KRFP
TOBACBCH	WHM**H**ES**HH**KP	TOBACBCH	AYMFV**HDGLVH**KRFP
ARABCH	WNM**H**ES**HH**KP	ARABCH	AYMFV**HDGLVH**KRFP
	* * ***		* .****** *.*

Fig. 1 Sequence alignment of all kown β-carotene hydroxylases
β-carotene hydroxylases from *Alcaligenes* (ALCABCH), *Agrobacterium* (AGROBCH), *Flavobacterium* (FLAVOBCH), *Erwinia* (ERWHERBCH, ERWUREBCH), pepper (CABECARH, CABCH2), tobacco (TOBACBCH), *Arabidopsis* (ARABCH)

Regulation of Carotenoid Biosynthesis

The knowledge of gene sequences now provides the basis for investigation of the regulation of the carotenoid biosynthesis. In this context, it is worth mentioning that several carotenoid promoters have been identified. Two promoters from pepper, namely the fibrillin promoter of a gene whose product is involved in carotenoid storage in chromoplasts and the CCS (capsanthin-capsorubin synthase) promoter have been shown to be also functional in another plant species (tomato). In addition, a positive light effect on the regulation of these promoters was observed which most likely is mediated by photooxidation [15].

Approaches to Study Carotenoid Biosynthesis and Function

I. Analysis of Mutants

Carotenoid mutants represent useful tools for investigation of carotenoid function and have also facilitated the identification of certain genes. Tab. 1 gives a summary of algal and higher plant mutants which exhibit defects in different steps in the carotenoid biosynthetic pathway. Three groups of mutants have been isolated to date: one in which early steps in the carotenoid biosynthesis are blocked resulting in a lack of carotenes and xanthophylls (and the accumulation of carotenoid precursors such as neurosporene or lycopene), a second one in which the formation of α-carotene derived carotenoids is impaired and a third one in which the synthesis of β-carotene derived xanthophylls is inhibited at different points.

Scendesmus mutant C-6E which does not contain any carotenoids clearly demonstrates that carotenoids are essential for photoprotection since this strain can only be cultivated in heterotrophic medium in the dark and suffers photooxidative damage when it is transfered to light [16]. The *lut* mutants of *Arabidopsis thaliana* (defect in ε-cyclization) proved the presence of two independent cyclases [2]. Analysis of the xanthophyll cycle mutants showed that neoxanthin, violaxanthin and antheraxanthin are not absolutely necessary for survival [25]; but failure to convert violaxanthin into zeaxanthin results in lower levels of non-photochemical quenching [23]. Taken together the data strongly emphazise the importance of a functional xanthophyll cycle in order to modulate and optimize absorbance and /or quenching of light energy.

Table 1

Group I: Mutants accumulating no carotenoids or carotenoid precursors

Mutant	Source	Carotenoid composition	Effect	Cause	Ref
C-6E	*Scenedesmus*	no carotenoids	highly photo-sensitive		16
C-6D dark-grown	*Scenedesmus*	carotenoid pre-cursors	light sensitive only PSI	defect in production of suitable isomers for cyclization	17,18
C-6D illuminated		normal carotenoid composition	functional PS		

PS, photosynthesis; PSI, photosystem I; psy, phytoene synthase

Group II: Mutants with defects in the synthesis of lutein or ε-cyclization

Mutant	**Source**	**Carotenoid composition**	**Effect**	**Cause**	**Ref**
C-2A'-34	*Scenedesmus*	no lutein, increased amount of zeaxanthin	less LHC trimers	defect in ε-cyclase	21
lor 1	*Chlamydomonas*	no lutein, increased amount of zeaxanthin	normal phenotype	defect in ε-cyclase	20
lut1	*Arabidopsis*	no lutein, accumulation of zeinoxanthin	increase in β,β-carotenoids	defect in ε-ring hydroxylation	19
lut2	*Arabidopsis*	no formation of α-carotene	increase in β,β-carotenoids	defect in ε-cyclase	19

Group III: Mutants with defects in the synthesis of xanthophyll cycle pigments

Mutant	**Source**	**Carotenoid composition**	**Effect**	**Cause**	**Ref**
npq1	*Chlamydomonas*	no conversion of violaxanthin to zeaxanthin	less npq	defect in violaxanthinde-epoxidase	22
npq1	*Arabidopsis*	as for *Chlamydomonas*	less npq		23
npq2	*Chlamydomonas*	high levels of zeaxanthin, no conversion in violaxanthin		defect in zeaxanthin-epoxidase	22
C-2A'-67,1	*Scenedesmus*	no violaxanthin, no antheraxanthin, no neoxanthin	increase in zeaxanthin	defect in zeaxanthinepoxidase	26
npq2	*Arabidopsis*	as for *Chlamydomonas*			23 24

npq, non photochemical quenching, LHC, light harvesting complex, chlb, chlorophyll b

Double mutants

Mutant	Source	Carotenoid composition	Effect	Cause	Ref
C-2A'-67,3b	*Scenedesmus*	only β-carotene and zeaxanthin	limited synthesis of chlb and LHC	defect in zeaxanthin-epoxidase and ε-cyclase	26
npq1/lor1	*Chlamydomonas*	no lutein, no conversion of violaxanthin	highly photosensitive, no npq	defect in ε-cyclase and vde	20

vde, violaxanthin deepoxidase

II. Employment of Transgenic Plants

Progress in molecular biology and plant transformation techniques made transgenic plants a powerful tool to elucidate specific aspects of carotenoid biosynthesis. Several approaches have been made to express carotenoid genes in higher plants chloroplasts. Homologous expression of the fruit-specific phytoene synthase (psy1) under constitutive 35S CaMV promoter led to the formation of lycopene also in vegetative tissues and resulted in a dwarf phenotype of strongly expressing transformants. The latter effect was due to reduced levels of gibberellins (GA) [27]. Heterologous expression of a bacterial carotenoid gene was successfully achieved using the phytoene desaturase gene (*crtI*) of *Erwinia uredovora* fused to the transit sequence of the pea small subunit of Rubisco for import into the chloroplast. Expression of the transgene caused changes in carotenoid composition and confered resistance to the bleaching herbicide norflurazone which targets the plant phytoene desaturase [28, Römer, unpublished results]. Using a viral vector tobacco has also been transformed with the carotenoid gene encoding capsanthin-capsorubin synthase. The transfected plants exhibited a yellow leaf phenotype which was shown to be a consequence of the accumulation of capsanthin-capsorubin, a ketocarotenoid responsible for the red colour of pepper fruits. This pigment was functionally incorporated in the chloroplast [29].

Outlook

Future research will certainly focus on the regulation of carotenoid biosynthesis since more and more information about genes and their promoters becomes available. Nevertheless, the structure and requirements of the formation of active pigment-protein complexes will still be a matter of investigation. Genetic engineering allows modification of pigment composition and/or introduction of additional traits. The last point also raises the need to increase our knowledge about carotenoid storage.

References

1. Bramley P.M. (1997) Isoprenoid metabolism, in P.M. Dey and J.B. Harborne (eds.), *Plant Biochemistry*, Academic Press, San Diego, London, Boston, pp. 417-437.
2. Cunningham F.X., Pogson B., Sun Z., McDonald K.A., DellaPenna D., and Gantt E. (1996) Functional analysis of the β and ε lycopene cyclase enzymes of *Arabidopsis* reveals a mechanism for control of cyclic carotenoid formation, *Plant Cell* **8**, 1613-1626.
3. Siefermann-Harms D. (1985) Carotenoids in photosynthesis. I. Location in photosynthetic membranes and light-harvesting function, *Biochim. Biophys. Acta* **811**, 325-355.
4. Young A.J. (1991) The photoprotective role of carotenoids in higher plants, *Physiol. Plant.* **83**, 702-708.
5. Sarry J.-E., Montillet J.L., Sauvaire Y., Havaux M. (1994) The protective function of the xanthophyll cycle in photosynthesis, *FEBS Letters* **353**, 147-150.
6. Plumley F.G. and Schmidt G.W. (1987) Reconstitution of chloropyll a/b light harvesting complexes. Xanthophyll-dependent assembly and energy transfer, *Proc Natl. Acad. Sci. (USA)* **84**, 146-150.
7. Lee A., I-c and Thornber J.P. (1995) Analysis of the pigment-protein complexes from barley (*Hordeum vulgare*). The xanthophyll cycle intermediates occur mainly in the light-harvesting complexes of photosystem I and II, *Plant Physiol.* **107**, 565-574.
8. Demmig-Adams B., and Adams III W.W. (1996) Xanthophyll cycle and light stress in nature: uniform response to excess direct sunlight among higher plant species, *Planta* **198**, 460-470.
9. Pfündel E., and Bilger W. (1994) Regulation and possible function of the violaxanthin cycle, *Photosynth. Research* **42**, 89-109.
10. Armstrong G.A., and Hearst J.E. (1996) Genetics and molecular biology of carotenoid pigment biosynthesis, *FASEB J.* **10**, 228-237.
11. Bartley G.E. and Scolnik P.A. (1995) Plant carotenoids: pigments for photoprotection, visual attraction, and human health, *Plant Cell* **7**, 1027-1038.
12. Sandmann G. (1994) Carotenoid biosynthesis in microorganisms and plants, *FEBS Letters* **223**, 7-24.
13. Bugos C.R. and Yamamoto H.Y. (1996) Molecular cloning of violaxanthin de-epoxidase from romaine lettuce and expression in *Escherichia coli*, *Proc. Natl. Acad.* Sci. *(USA)* **93**, 6320-6325.
14. Marin E., Nussaume L., Quesada A., Gonneau M., Sotta B., Hugueney P., Frey A., and Marion-Poll A. (1996) Molecular identification of zeaxanthin epoxidase of *Nicotiana plumbaginifolia*, a gene involved in abscisic acid biosynthesis and corresponding to the ABA locus of *Arabidopsis thaliana*, *EMBO J.* **15**, 2331-2342.
15. Kuntz M., Chen H.C., Simkin A.J., Römer S., Shipton C.A., Drake R., Schuch W., and Bramley P.M. (1998) Upregulation of two ripening-related genes from a non-climateric plant (pepper) in a transgenic climateric plant (tomato), *Plant J.* **13**, 351-361.

16. Römer S., Senger H., and Bishop N.I. (1995) Characterization of the carotenoidless strain of *Scenedesmus obliquus*, mutant C-6E, a living photosystem I model, *Bot. Acta* **108**, 80-86.
17. Humbeck K., Römer S., and Senger H. (1989) Evidence for an essential role of carotenoids in the assembly of an active photosystem II, *Planta* **179**, 242-250.
18. Humbeck K. (1990) Photoisomerization of lycopene during carotenogenesis in mutant C-6D of *Scenedesmus obliquus*, *Planta* **182**, 204
19. Pogson B., McDonald K.A., Truong M., Britton G., and DellaPenna D. (1996) *Arabidopsis* carotenoid mutants demonstrate that lutein is not essential for photosynthesis in higher plants, *Plant Cell* **8**, 1627-1639.
20. Niyogi K.K., Björkman O., and Grossman A.R. (1997) The roles of specific xanthophylls in photoprotection, *Proc. Natl. Acad. Sci.* USA **94**, 14162-14167
21. Bishop N.I., Urbig T., and Senger H. (1995) Complete separation of the β,ε- and β,β- carotenoid biosynthetic pathways by a unique mutation of the lycopene cyclase in the green alga *Scenedesmus obliquus*, *FEBS Letters* **367**, 158-162.
22. Niyogi K.K., Björkman O., and Grossman A.R. (1997) *Chlamydomonas* xanthophyll cycle mutants identified by video imaging of chlorophyll fluorescence quenching, *Plant Cell* **9**, 1369-1380.
23. Niyogi K.K., Grossman A.R., and Björkman O. (1998) *Arabidopsis* mutants define a central role for the xanthophyll cycle in the regulation of photosynthetic energy conversion, *Plant Cell* **10**, 1121-1134.
24. Tardy F., and Havaux M. (1996) Photosynthesis, chlorophyll fluorescence, light-harvesting system and photoinhibition resistance of a zeaxanthin-accumulating mutant of *Arabidopsis thaliana*, *J. Photochem. Photobiol.* **34**, 87-94.
25. Hurry V., Anderson J.M., Chow W S., and Osmond C.B. (1997) Accumulation of zeaxanthin in abscisic acid-deficient mutants of *Arabidopsis* does not affect chlorophyll fluorescence quenching or sensitivity to photoinhibition *in vivo*, *Plant Physiol.* **113**, 639-648.
26. Bishop N.I. (1996) The β,ε-carotenoid, lutein is specifically required for the formation of oligomeric forms of the light harvesting complex in the green alga, *Scenedesmus obliquus*, *J. Photochem. Photobiol.* **6**, 279-283.
27. Fray R.G., Wallace A., Fraser P.D., Valero D., Hedden P., Bramley P.M., and Grierson D. (1995) Constitutive expression of a fruit phytoene synthase gene in transgenic tomatoes causes dwarfism by redirecting metabolites from the gibberellin pathway, *Plant J.* **8**, 693-701.
28. Misawa N., Yamano S., Linden H., de Felipe M.R., Lucas M., Ikenaga H., and Sandmann G. (1994) Expression of an *Erwinia uredovora* phytoene desaturase gene confers not only multiple resistance to herbicides interfering with carotenoid biosynthesis but also alters xanthophyll metabolism in transgenic plants, *Plant J.* **6**, 481-489.
29. Kumagai M.H., Keller Y., Bouvier F., Clary D., and Camara B. (1998) Functional integration of non-native carotenoids into chloroplasts by viral-derived expression of capsanthin-capsorubin synthase in *Nicotiana benthamiana*, *Plant J.* **14**, 305-315.

REGULATORY MECHANISMS IN PHOTOSYNTHETIC UNIT BIOGENESIS/TURNOVER

ABUNDANCE OF PHOTOSYSTEM I PROTEINS IN CYANOBACTERIA AND CHLOROPLASTS

JUN SUN, DONALD A. HECK, WU XU, VAISHALI P. CHITNIS, PARAG R. CHITNIS
Department of Biochemistry and Biophysics
Iowa State University, Ames IA 50014 USA

1. Introduction

The biosynthesis of photosynthetic complexes encompasses the coordination of diverse processes with the potential for many levels of regulation. Collectively, complex assembly involves the expression of chloroplast and nuclear genes, the targeting of subunits to their proper locations within the chloroplast, and the proper organization and assembly of the subunits and cofactors into functional complexes [5]. Degradation of proteins is another crucial process that is regulated by internal and extrinsic factors. The abundance of functional photosystem I (PSI) can be affected by altering any one of these processes. Different environmental conditions and various mutations have been characterized that affect the abundance of PSI complexes.

The PSI complex of cyanobacteria and chloroplasts is a multisubunit membrane-protein complex that catalyzes electron transfer from reduced plastocyanin to oxidized ferredoxin [4]. X-ray crystallographic analysis of the PSI complex from *Synechococcus elongatus* has been used to propose a model for the three dimensional structure at 4-Å resolution [16]. The PsaA and PsaB subunits form a heterodimeric core of the complex that harbors ~89 chlorophyll *a* molecules, the primary reaction center P700 and a chain of electron acceptors A_0, A_1, and F_X. PsaC binds the terminal electron acceptors, F_A and F_B, each a [4Fe-4S] cluster. The PsaC, PsaD, and PsaE subunits are located on the stromal side of the PSI complexes. The remaining subunits of cyanobacterial PSI are integral membrane proteins [4]. In this paper, we describe how the transmembrane architecture and external factors regulate PSI abundance in cyanobacteria and plants.

2. Intrinsic factors

2.1. Role of the small subunits of PSI

Studies on subunit-deficient strains of cyanobacteria have shown that assembly of the core subunits of PSI is not dependent on the presence of other PSI proteins [4]. In contrast, most other proteins require the presence of the core complex for their assembly. PsaF is a major exception. It can be found in the membranes of the PsaA-PsaB-less mutants of the cyanobacteria [18]. Thus the assembly of PSI proteins involves both sequential and parallel addition of different subunits. Assembly of PSI is

J. H. Argyroudi-Akoyunoglou and H. Senger (eds.),
The Chloroplast: From Molecular Biology to Biotechnology, 227–232.

a highly coordinated process and may be controlled by multiple mechanisms [5]. In contrast, disassembly of PSI and degradation of its protein subunits remain largely unexplored. The lack of progress in studying PSI degradation is mainly due to its stability. Slow turnover of the PSI complexes causes experimental difficulties in determining half lives of the PSI proteins and in monitoring accumulation of degradation intermediates [6]. The architecture of PSI may provide clues to its extraordinary stability. The peripheral subunits PsaD and PsaE may provide physical protection to the redox centers on the stromal side of membranes whereas the integral membrane subunits may protect the redox chain from the lipid-soluble oxidants.

We used the PsaD-PsaE-less DE mutant strain of the cyanobacterium *Synechocystis* sp. PCC 6803 to investigate the molecular mechanisms of PSI destruction (Q. Xu, V. P. Chitnis and P. R. Chitnis, unpublished results). When the cells were grown at 38°C, the levels of many, but not all, PSI subunits reduced rapidly in the DE strain. PSI proteins show differences in their degradation under various conditions. Some PSI proteins in the DE strain were more susceptible to *in vivo* proteolysis than the others. For example, the PsaF levels remained constant whereas the PsaA-PsaB levels decreased when the DE cells were exposed to high temperature. Degradation of some PSI subunits in the DE strain is a light-regulated process. The PsaB and PsaC subunits were protected under darkness. In addition, photosynthetic electron transfer regulated PSI degradation of some PSI proteins, possibly by affecting the redox pool. Therefore, many mechanisms exist for degradation of different PSI proteins. The machinery and pathway(s) for disassembly of the PSI complex and degradation of its proteins are not yet elucidated.

2.2. Importanceofcofactors

PSI contains chlorophyll *a*, β-carotenes, phylloquinone and Fe-S clusters. Binding to some of these cofactors is one way to stabilize the complex. Chlorophyll is an important determinant for the level of PSI in the membranes. It is required for the stability of proteins and also for gene expression (see section 3). PSI contains at least 89 chlorophyll a molecules [16]. Site-directed mutagenesis has revealed roles of specific chlorophyll molecules in PSI assembly. The P700 chlorophyll molecules are coordinated by a pair of conserved histidyl residues. Replacement of these residues affects PSI abundance in the cells.

To investigate the roles of phylloquinones in PSI, we generated mutations in two cyanobacterial genes that are homologous to the *menA* and *menB* genes of *E. coli* [14]. Products of the *menA* and *menB* genes catalyze naphthoate synthesis and isoprenyl transfer in biosynthesis of menaquinones, respectively. These cyanobacterial mutant strains lack phylloquinones in the thylakoid membranes. These mutants contain properly assembled PSI complexes. Therefore phylloquinones are not required for the assembly of PSI. Function of β-carotenes in imparting stability to PSI has not been examined.

2.3. Role of transmembrane architecture

Hydrophobic interactions and packing of transmembrane domains are the major stabilizing forces in membrane proteins. We examined the published and unpublished Western blotting results for the site-directed mutants of *Synechocystis* sp. PCC 6803 and *Chlamydomonas reinhardtii*. Several conclusions can be drawn from the comparison of these results. First, amounts of the PSI complexes in the membranes are reduced greatly when the residues that surround redox centers are mutated. Similarly, most mutations in phylloquinone pocket dramatically decrease the amount of PSI. Second, disruptive changes in the transmembrane regions are more detrimental than the conservative changes. For example, the Y667F, H651C, L522V and C565S mutants in the PsaB protein of *Synechocystis* sp. PCC 6803 have more levels of PSI in the membranes than the Y667A, H651L, L522P and C565H mutants (Wu and Chitnis, unpublished results). Third, mutations in the residues that are closer to the lipid-water interface decrease PSI abundance less than the mutations in the interior of the lipid bilayer. Systematic mutations of amino acids in the interior of the *m* transmembrane helix abolish the ability of PSI proteins to accumulate in the membrane (Figure 1). In contrast, replacement of the residues that are closer to the exterior of the lipid bilayer allow accumulation of PSI proteins in the membrane.

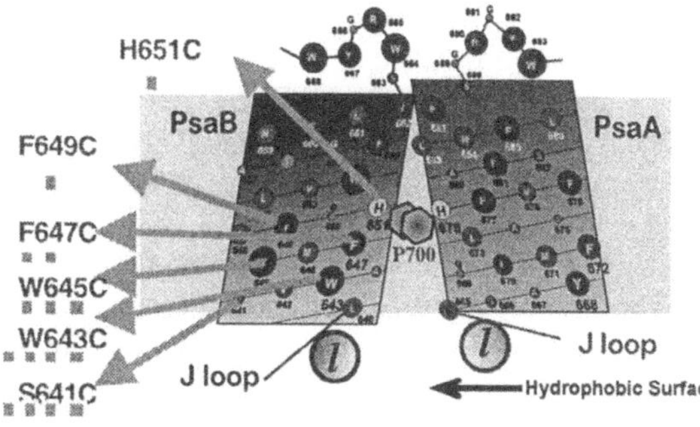

Fig. 1. **Mutations in the m-helix of PsaB.** The dots below mutations indicate the level of PSI in the membranes. Four dots is the wild type level.

We also generated deletion mutations in the H extramembrane loop of PsaB (Sun and Chitnis, unpublished results). The PSI proteins failed to accumulate in the membranes if the deletions were close to the transmembrane helices. In contrast, deletion of the middle regions of the loop had small effect on the accumulation of PSI in the membranes. Therefore, packing of transmembrane helices and cofactor-binding are two crucial intrinsic factors that determine PSI stability.

3. Regulation of the *psaA-psaB* operon in chloroplasts and cyanobacteria.

The *psaA* and *psaB* genes code for the core proteins of PSI and are located in an operon in the chloroplast genome of higher plants. The synthesis of PS I core proteins during the light-induced chloroplast development is regulated at the translational or posttranslational levels [10]. The stability of the chloroplast mRNAs can be regulated through polyadenylation in the 3'-untranslated region (3'-UTR) [11]. Also, nuclear-encoded translational activators that interact with the 5'-untranslated region of the mRNAs for chlorophyll proteins is required for their translation [8]. The light-induced chlorophyll biosynthesis triggers the accumulation of the chlorophyll proteins PsaA and PsaB by enhancement of chlorophyll apoprotein stability [9].

Expression of PSI genes in cyanobacteria has not been studied in detail. The transcripts for the *psaA-psaB* operon accumulate in the dark [19]. To investigate how physiological and environmental factors regulate PSI levels in cyanobacterial membranes, we grew *Synechocystis* cells under different culture conditions and measured levels of chlorophyll *a* and PS I proteins (Table 1). Both parameters varied according to growth conditions, the highest levels being in cells grown under autotrophic growth condition (AG). Therefore, genetic background, the presence of glucose and light intensity affected accumulation of PSI proteins. To study the role of 3' UTR in the accumulation of PS I proteins, we made constructions in which a chloramphenicol resistance cartridge was inserted at different positions. The p3C+ DNA contained the resistance cartridge just after the stop codon of *psaB* whereas pBC+ contained the cartridge about 200 bp downstream. These DNAs and the wild type gene were transformed into three recipient strains: wild type, a strain with deletion of the 3'-region of *psaB* gene grown under HL condition (LKC14) and a strain with the same deletion grown under LAHG condition (ΔB). As shown in Table 2, the accumulation of PS I proteins, again showed similar trends as the accumulation of chlorophyll *a*, with a decreasing order of WT>pBC+>p3C+ and WT>ΔB>LKC14. Therefore, 3' UTR, genetic background and environmental conditions have profound effect on the abundance of PSI in cyanobacteria.

TABLE 1. Accumulation of chlorophyll and PS I proteins under different growth conditions.

Strains	Conditions**			
	AG	HM	HL	LAHG
Wild type	5.66/ +++++*	4.02/ ++++	3.76/ ++++	1.76/ +++
Recovered WT	1.52/ ++	1.59/ +	0.73/ (+/-)	0.20/ -

** autotrophic growth under medium light (AG), heterotrophic under medium light intensity (HM), low light intensity (HL), and light-activated heterotrophic growth (LAHG)

TABLE 2 The accumulation of chlorophyll and PS I proteins in transformants with different constructions and different recipient strains.

Recipient strains	Construction		
	WT	pBC+	p3C+
Wild type (WT)	4.71/ +++++*	4.03/ ++++	2.52/ +++
ΔB	4.05/ ++++	3.67/ ++++	2.11/ +++
LKC14	3.19/ ++++	2.45/ +++	1.51/ ++

*: the numbers indicates the chlorophyll *a* content in an unit of mg/OD_{730}/l; the *plus/minus* symbols indicates the abundance of PS I proteins accumulated in the thylakoid membrane immunodetected by antibody against the PsaC protein.

4. Mutants defective in PSI abundance

Characterization of mutants is an effective way to identify genes that are involved in regulating the abundance of PSI in the membranes. Mutants with reduced activity or level of PSI have been identified in plants, *Chlamydomonas*, and *Synechocystis* (Table 3). The most common of these mutants show a high chlorophyll fluorescence (hcf) phenotype. These mutants have a reduced amount or complete absence of one or more of the photosynthetic complexes. These mutants have defects at different levels of gene expression. The maize mutant hcf38 has a deficit in mRNA processing resulting in a two to three-fold reduction in PSI polypeptide [1]. The *Arabidopsis* mutant hcf109 contains a lesion in a nuclear gene that specifically causes a decrease in mRNA stability

TABLE 3. Mutants affecting photosystem I abundance.

SPECIES	MUTANT	LESION	PSI ABUNDANCE
Synechosystis	BP26[a]	missense mutation in *btpA*	substantially reduced PSI level
Chlamydomonas	B4[b]	unknown	no properly assembled *psaA* mRNA
	ycf3-less, *ycf4*-less[d]	inactivation of *ycf3* and *ycf4* genes	loss of assembly of PSI complex
Arabidopsis	*hcf109*[e]	alteration of gene for the stability of operon transcripts	loss of PsaC subunit resulting in failure of PSI complex to assemble
Maize	*hcf2*[f]	unknown	reduced PsaA-B subunits
	hcf18, 43, 44, 47, 49, 50, 104, 113, 122[f-h]	unknown	loss or reduction of PSI complex and/or PSI subunits
	hcf38[f]	unknown	reduced *psaAB* mRNA and subunits
	hcf101[h]	unknown	loss of PSI
	hcf106[f,i]	defective gene in the ΔpH targeting pathway	reduced PSI complex
	tha1, tha2[j]	SecA homologue; alteration in SecA targeting pathway	reduced translocation of PsaF across the thylakoid membrane
	Psa1-4[h]	unknown	loss of PSI complex
Tobacco	*ycf3*[k]	inactivation of *ycf3* gene	reduced PSI subunits

References: [a][2], [b][7], [c][20], [d][3], [e][12], [f][1], [g][13], [h][13], [i][17], [j][21], [k][15]

of the *ndhH* operon that contains the *psaC* gene [12]. Many mutants affect targeting of proteins to thylakoid membranes. For example, the maize mutant *hcf106* mutation disrupts the transport of polypeptides via the ΔpH pathway resulting in a substantial reduction of photosynthetic complexes including PSI [17]. Other mutations have been shown to affect the formation of stable PSI complexes. Inactivation of the *ycf3* and *ycf4* genes results in a loss of PSI stability in *Chlamydomonas* and tobacco. In both species, disruption of one or both genes reduces PSI complex assembly without any effects on the transcription or stability of mRNAs for the PSI subunit. A novel gene affecting the assembly of PSI complexes has been discovered recently in *Synechosystis* [2]. A lesion in the *btpA* gene results in a reduction of functional PSI without significantly affecting any other photosynthetic complexes.

To identify the genes that regulate PSI levels in higher plants, we investigated maize mutants that are defective in the level of photosynthetic apparatus. The zebra necrotic and zebra stripe mutants of maize are phenotypically characterized as having leaves with necrotic lesions (zebra necrotic) or bands of light green tissue (zebra stripe). Mutant tissues from both phenotypes contain diminished amounts of photosynthetic membrane proteins. The leaves of the high chlorophyll fluorescence (hcf) mutants hcf44 and hcf47 contain reduced levels of chlorophyll, decreased PSI activity and lower levels of PSI polypeptides. Characterization of these mutants is expected to lead to discovery of additional genes that influence PSI levels.

To conclude, stability of the PSI complex is dependent on the protein architecture, cofactor interactions and genetic background. We have identified some of the physiological, environmental and genetic components that affect levels of PSI in the thylakoid membranes of cyanobacteria and higher plants.

5. Acknowledgements

This work was supported in part by grants from NSF and USDA to PRC. Journal Paper No. J-18049 of the Iowa Agriculture and Home Economics Experiment Station, Ames, Iowa, Projects No. 3416 and 3496, and supported by Hatch Act and State of Iowa funds.

6. References

1. Barkan, A., Miles, D., Taylor, W.C. (1986) *EMBO J.* **5**, 1421-7.
2. Bartsevich, V.V., Pakrasi, H.B. (1997) *J Bioll Chem* **272**, 6382-7.
3. Boudreau, E., Takahashi, Y., Lemieux, C., Turmel, M., Rochaix, J.D. (1997) *EMBO J.* **16**, 6095-104.
4. Chitnis, P.R. (1996) *Plant Physiol.* **111**, 661-669.
5. Chitnis, P.R. (1997). in A.S. Raghavendra (ed), *Photosynthesis: A comprehensive Treatise*, pp. 58-71. Cambridge University Press.
6. Chitnis, P.R., Nelson, N. (1992). in J. Argyroudi-Akoyunoglou (ed), *Chloroplast Biogenesis*, pp. 285-290. Plenum Press, New York.
7. Choquet, Y., Goldschmidt-Clermont, M., Girard-Bascou, J., Kuck, U., Bennoun, P., Rochaix, J.-D. (1988) *Cell* **52**, 903-913.
8. Girard-Bascou, J., Pierre, Y., Drapier, D.A. (1992) *Curr. Genet.* **22**, 47-52.
9. Kim, J., Eichacker, L.A., Rudiger, W., Mullet, J.E. (1994) *Plant Physiol* **104**, 907-16.
10. Klein, R.R., Mullet, J.E. (1987) *J. Biol. Chem.* **262**, 4341-4348.
11. Kudla, J., Hayes, R., Gruissem, W. (1996) *EMBO J.* **15**, 7137-7146.
12. Meurer, J., Berger, A., Westhoff, P. (1996) *Plant Cell* **8**, 1193-1207.
13. Neuffer, M.G., Coe, E.H., Wessler, S.R. (1997). *Mutants of maize.* Cold Spring Harbor Laboratory Press, Plainview, NY.
14. Reatégui, R., Johnson, W., Xu, W., Zybailov, B., Vassiliev, I., Shen, G., Golbeck, J.H., Chitnis, P.R. (1998) In: J. Garab (ed) Proceedings of International Congress on Photosynthesis, pp. In press. Kluwer Academic, Budapest (1998).
15. Ruf, S., Kossel, H., Bock, R. (1997) *J. Cell Biol.* **139**, 95-102.
16. Schubert, W.D., Klukas, O., Krauß, N., Saenger, W., Fromme, P., Witt, H.T. (1997) *J Mol Biol* **272**, 741-69.
17. Settles, A.M., Yonetani, A., Baron, A., Bush, D.R., Cline, K., Martienssen, R. (1997) *Science* **278**, 1467-1470.
18. Shen, G., Bryant, D.A. (1995) *Photosynthesis Res.* **44**, 41-53.
19. Smart, L.B., McIntosh, L. (1991) *Plant Moll Bioll* **17**, 959-71.
20. Takahashi, Y., Goldschmidt-Clermont, M., Soen, S.-Y., Franzen, L.G., Rochaix, J.-D. (1991) *EMBO J.* **10**, 2033-2040.
21. Voelker, R., Mendel-Hartvig, J., Barkan, A. (1997) *Genetics* **145**, 467-478.

DE NOVO SYNTHESIS OF CHLOROPHYLL A TRIGGERS ASSEMBLY OF REACTION CENTERS IN BARLEY ETIOPLASTS

Bernd Müller and Lutz A. Eichacker
Botanisches Institut der LMU München, Menzingerstr. 67, 80638 München, Germany

1. Introduction

Many investigations have been carried out to study the assembly of chlorophyll binding proteins (Chl-P) of photosystem I and II. Measurements of PSI and PSII activity after illumination of etiolated barley did not produce clear results [1,2,3,4,5,6] In this work, sucrose density gradients of solubilized inner etioplast membranes were utilised to measure assembly of the Chl-P as a shift in the molecular mass of the *de novo* stabilized apoproteins. In order to define the molecular mass fractions of the sucrose gradient we first isolated Chl-P complexes from chloroplasts. Protein complexes were identified by silver staining of SDS-PAGs, Western analysis and spectroscopy. In etioplasts, assembly was kinetically followed by radiolabeling of *de novo* synthesized proteins in the absence and presence of increasing concentrations of Chl. When the endogenous concentration of Chl was increased over a threshold value, stabilization of all Chl *a* binding proteins (Chl *a* -P) was observed. When the assembly kinetics of the Chl-P were followed for 80 min *in vitro*, Chl-P of PSI (P700) assembled in a complex of about 300 kDa, whereas Chl-P of PSII (D1,D2, CP43, CP47) assembled in a complex of about 100-150 kDa. Interestingly, the Chl *a* -P of P700 started to assemble after 20 minutes Chl synthesis whereas the Chl *a* -P of PSII were degraded in a high amount in parallel to the assembly of P700. We therefore conclude a Chl-transfer from *de novo* stabilized and assembled Chl *a* -P of PSII to the *de novo* synthesized Chl *a* -P of P700 [7]. The presence of Chl within the *de novo* assembled complexes was verified by spectroscopic analysis of the sucrose gradient fractions and by isolation of the pigment. We therefore conclude that in etioplasts isolated from dark-grown barley seedlings *de novo* synthesis of Chl triggers not only stabilization of Chl-P, but also assembly of Chl-P to reaction center particles of PSI and PSII.

2. Material and Methods

Barley (*Hordeum vulgare* L., var Steffi) seeds were planted in moist vermiculite and grown for 4.5 days at 25°C in a light-tight growth chamber located in a dark room. Intact etioplasts were immediately isolated in a dark room with green safety light ($<10nE/m^2s$) as described [8]. Conditions for Chl and Chl *a* -P synthesis in plastids (1.4×10^7 plastids/75µl final volume) were obtained in a buffer mixture at 25°C

J. H. Argyroudi-Akoyunoglou and H. Senger (eds.),
The Chloroplast: From Molecular Biology to Biotechnology, 233–236.

containing 50 mM HEPES/KOH (pH 8.0), 2.0 mM ATP, 0.2 mM GTP, 7 mM magnesium acetate (pH 7.0) 118 mM potassium acetate (pH 7.0), 10 mM DTT, 100 μM of each amino acid (except methionine) and 98 MBq of [35S] methionine (spec. act. 3.7×10^{13} Bq/mmol, Amersham). Synthesis of Chl and *Chl a* -P in etioplasts was induced through the addition of geranylgeraniol-pyrophosphate GGPP (Sigma) (0 - 20nmol/4.2×10^{7} plastids) and illumination with white light (500μE/m^2s) for 10 seconds. The assays were preincubated in the dark with ^{35}S-methionine for 15 minutes. To start the reaction the assays were incubated for 0, 5, 10, 20, 40 and 80 minutes in in the light (50μE/m^2s) by adding GGPP and lincomycin (Sigma) at 25°C. Translation was stopped by spinning down the plastids (4000×g, 1 min, 4°C), removing the supernatant and the intact plastids were lysed in RNase- free lysis medium (10 mM Tris·HCl pH 6,8, 20 mM KCl, 10mM $MgCl_2$). The membranes were spun down and washed twice in the lysis-medium and the first supernatant after lysis was immediately frozen in liquid N_2. For separation of thylakoid-bound complexes, membranes of 1×10^{8} plastids were gently resuspended in lysis-medium and solubilized by adding a three fold volume of 2% (w/v) n-dodecyl-β-D-maltoside (DM, Biomol, Hamburg) in the dark on ice for 40 minutes. The solubilized membranes were centrifuged (21000×g, 4 °C) and the supernatant was loaded on a linear sucrose-gradient of 10 ml (0.1-1.0 M sucrose in lysis-medium and 0.06% (w/v) DM) and spun in the ultracentrifuge for 16 hours at 160000×g at 4°C in a RPS 40-T rotor. After centrifugation, the sucrose-gradient was divided in 10 fractions of equal volume (1ml) from bottom to top. The pellet was resuspended in 1 ml lysis-medium containing 0.06% (w/v) DM. Absorbance- and fluorescense spectras were immediately recorded of each fraction and 500μl of each fraction were precipitated with 10% trichloroacetic acid for 20 minutes.

SDS-PAGE - The precipitated proteins were collected by centrifugation (21000×g, 30 min, RT), the pellet dried in the speedvac and afterwards resuspended in SDS-solubilization buffer (2% SDS, 20% sucrose, 66mM DTT, 66mM Na_2CO_3). Radiolabeled proteins from soluble and membrane fractions were solubilized in solubilisation-buffer and samples were loaded onto 12.5% SDS-PAGs containing 4M urea and detected by autoradiography. The quantification of the autoradiograms was carried out using a phosphoimager (Fuji) and the TINA 2.09g software.

3. Results and Discussion

3.1 A CHL A TRESHOLD VALUE IS REQUIRED FOR STABILIZATION OF CHL A APOPROTEINS

To determine the amount of *de novo* synthesized chlorophyll corresponding to the stabilization of *de novo* synthesized Chl *a*- binding proteins (P700, CP47, CP43, D2 and D1) we added increasing amounts of GGPP (0, 0.5, 1, 1.5, 2, 2.5 and 3 nmol/1.4×10^{7} plastids) to the translation-assay. Due to the isolation-procedure of etioplasts the endogenous GGPP is lost. This fact permitted us to investigate the dependence of stability of *de novo* synthesized Chl *a* -proteins on chlorophyll-synthesis. Simultaneously we carried out a nonradioactive assay under the same conditions as the radioactive assay to determine the amount of *de novo* synthesized Chl *a* corresponding to the added GGPP by acetone/hexane extraction. Interestingly we could determine a

Chl *a* treshold referring to the stabilized Chl *a* -proteins. Only at the maximum of chlorophyll-synthesis (more than 2 nmol GGPP/1.4×10^{7}plastids) Chl *a*- binding proteins could be protected against proteolysis (Fig. 1). This indicates that *de novo* synthesized Chl *a*-P (except D2) can only be stabilized by a saturated chlorophyll synthesis.

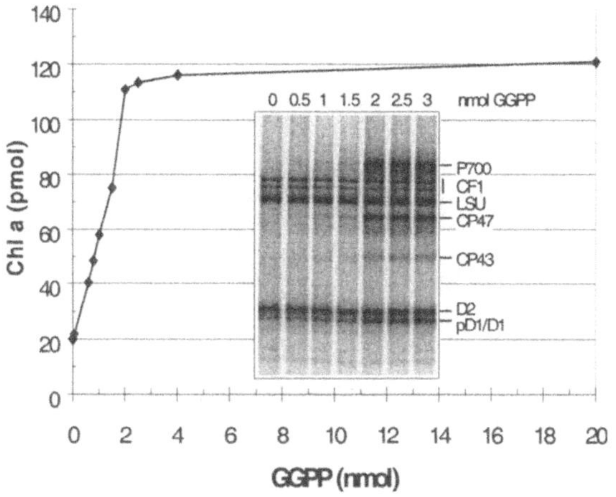

Figure 1: Stabilization of Chl *a* binding proteins by *de novo* synthesis of Chl *a* Intact barley etioplasts preilluminated with white light for 10 seconds (500 µE/m^{2}xs^{-1}) were pulse-labeled in the dark with ^{35}S-methionine in the presence of exogenously offered GGPP (0, 0.5, 1, 1.5, 2, 2.5 and 3.0 nmol) for 20 minutes. After a 30 minute chase with L-methionine etioplasts were lysed and the amount of esterfied chlorophyll per sample was determined by acetone/hexane extraction (Scheumann, V.; Ito, H.; Tanaka, A.; Schoch, S.; Rüdiger, W. (1996) Eur. J. Biochem. **242**, 163-170). Membrane-proteins were loaded on a SDS-PAG on the basis of equal plastid number (5×10^6 plastids/lane) and detected by autoradiography.

3.2 A STABILIZATION/DESTABILIZATION CYCLE OF PSII PROTEINS PRECEEDS POSTTRANSLATIONAL ASSEMBLY OF P700

To test the ability of the *de novo* synthesized Chl *a* -binding proteins (P700, CP47, CP43, D2 and D1) to assemble into of PSII and PSI we utilized sucrose gradient centrifugation of solubilized membrane proteins to follow the molecular mass shift due to the formation of the photosynthetic complexes. We therefore investigated different points of time (0, 5, 10, 20 ,40 and 80 minutes) after the addtion of lincomycin and GGPP to the translation assay. The addition of lincomycin to the translation assay allowed us to follow polysomal "run off" of the nascent chains associated with polysomes by preventing the reinitiation of ribosomes to mRNAs whereas chlorophyll-synthesis was not affected. The amount of chlorophyll available for the Chl *a*-P was limited due to the endogenous protochlorophyllide (~300pmol/4,2×10^7 plastids) that was converted to the chlorophyll-precursor chlorophyllide enzymatically by light and endogenous NADPH/H$^+$. The quantification of the polysomal fraction at all points of time shows the polysomal "run off" within 20 minutes. The quantification of the Chl *a* - binding proteins (P700, CP47, CP43, D2 and D1) shows the stabilization and assembly of all Chl *a* -P of PSII within 20 minutes chase in a sucrosegradient-fraction of about 100-150 kDa. In contrast to this, the Chl *a*- P of P700 can first be detected after 20 minutes and start to assemble at this point of time up to 80 minutes into a sucrosegradient-fraction of about 300 kDa. In parallel to the assembly of P700 the Chl *a*-P of PSII start to be degraded in a high amount up to 80 minutes. (Fig 2). This fact indicates a relation between the degradation of the Chl *a* -P of PSII and the stabilization and assembly of the Chl *a* -P of P700. The reason of this phenomenon might be a Chl *a* transfer between the Chl *a* -P of PSII to the Chl *a* -P of P700 because of the fact of the instability of the Chl *a* -P by lacking Chl (Fig. 1). This fact suggests a posttranslational binding of Chl *a* to P700 at the expense of the Chl *a* -P of PSII reaction center particles.

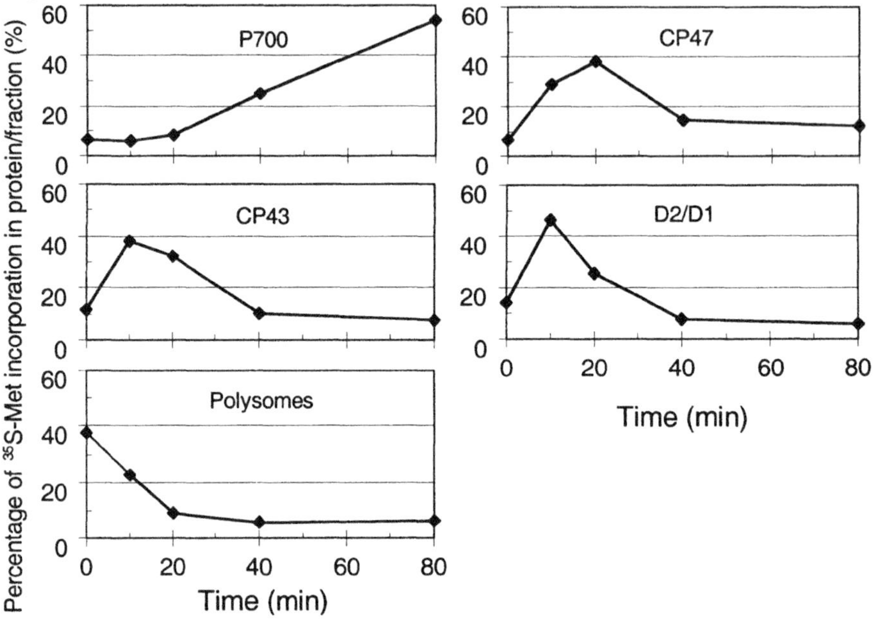

Figure 2: Quantification of radiolabeled Chl *a* binding proteins and polysomes
Chl *a* -binding proteins P700; CP47, CP43, D2/D1 and nascent chains associated with polysomes were quantified. Radiolabel incorporation per point of time was calculated as ratio between the total amount of radiolabel incorporated at one point of time and at sum of all time points.

References:

1. Ohashi, K., Tanaka, A., Tsuji, H. (1989) Formation of the Photosynthetic Electron Transport System during the Early Phase of Greening in Barley Leaves, Plant Physiol. **91**, 409-414
2. Egnéus, H., Reftel, S. and Selldén, G. (1972) The Appearence and Development of Photosynthetic Activity in Etiolated Barley Leaves and Isolated Etio-chloroplasts, Physiol. Plant. **27**, 48-55
3. Burkey K.O. (1986) Chlorophyll-protein complex composition and photochemical activity in developing chloroplasts from greening barley seedlings, Photosynthesis Research **10**: 37-49
4. Wellburn A.R. and Hampp R. (1978) Appearance of Photochemical Function in Prothylakoids during Plastid Development, Biochimika et Biophysika Acta, **547**: 380-397
5. Fradkin, L.I., Domanskii, V.P., Samoilenko, A.G., Shlyk, A.A. (1982) Doklad. Biophys. **264**, 93-97
6. Franck, F. (1993) Photosynthetic Activities during Early Assembly of Thylakoid Membranes, In: Ryberg M and Sunqvist C (eds) Pigment-Protein Complexes in Plastids: pp 365-381. Academic Press, San Diego, CA
7. Tanaka, A., Yoshiharu, Y., and Tsuji, H. (1991) Formation of Chlorophyll-Protein Complexes during Greening. 2. Redistribution of Chlorophyll among Apoproteins, Plant Cell Physiol. **32(2)**: 195-204
8. Eichacker, L. A., Helfrich, M., Rüdiger, W. and Müller, B. (1996) Stabilization of chlorophyll a-binding apoproteins P700, CP47, CP43, D2, and D1 by chlorophyll a or Zn-pheophytin a, J. Biol. Chem. **271**, 32174-32179

RECONSTITUTION OF CO-TRANSLATIONAL TARGETING OF POLYTOPIC MEMBRANE PROTEINS TO THE THYLAKOIDS IN A HOMOLOGOUS CHLOROPLAST TRANSLATION SYSTEM

E. HOUBEN[1], R. NILSSON[1], J.W DE GIER[1], J. BRUNNER[2], N. E. HOFFMAN[3] and K. J. VAN WIJK[1#]
[1]Dept. of Biochemistry, Stockholm University, S-10691 Stockholm, Sweden; [2]Lab. fur Biochemie, ETHZ, CH-8092 Zurich, Switzerland;[3]Dept. of Plant Biology, Carnegie Inst. of Washington, Stanford, CA, USA. # for correspondence: klaas@biokemi.su.se

1. Introduction

Chloroplasts in higher plants posses circular DNA containing 35 genes encoding for essential proteins located in the thylakoid membranes. The mechanisms of targeting and insertion of these proteins are unknown, despite their importance. Based on a number of observations in chloroplasts, it can be postulated that targeting and insertion of the polytopic chloroplast-encoded membrane proteins occurs co-translationally (1-3). Thus in order to reconstitute this targeting and insertion process, a homologous chloroplast *in vitro* initiation/translation system is required in which plasmid derived transcripts can be faithfully translated. The recent discovery of a translation system isolated from tobacco chloroplasts has opened up novel possibilities to address these important processes at a molecular level (4). In this paper, we have set out to evaluate the interaction of soluble stromal components cpSRP54 (5), cpSRP43 (6) and SecA (7) with the chloroplast encoded D1 protein, using this translation system. We show that ribosome D1 nascent chain complexes (D1 rncs) can be targeted very efficiently to the membrane and make functional interactions. In addition we show that cpRP54 interacts specifically with D1 rncs of defined length, implying a role for cpSRP54 in D1 biogenesis. To study the extent of conservation and mechanisms of the chloroplast targeting mechanisms with prokaryotes, we translated *E.coli* Leader peptidase (Lep) in the chloroplast system and attempted targeting Lep to the thylakoid membrane. Lep has two membrane spans and has served as a model protein to study targeting and insertion to the *E. coli* inner membrane (8-10) as well as to ER membranes (e.g.11). Lep is made without a cleavable signal sequence and its N- and C-terminal domains face the periplasmic side of the inner membrane (Fig. 4). In *E.coli* , insertion of Lep is SecA, SecY and SRP dependent (8-10).

J. H. Argyroudi-Akoyunoglou and H. Senger (eds.),
The Chloroplast: From Molecular Biology to Biotechnology, 237–242.

2. Material and Methods

2.1 DNA CONSTRUCTS AND PREPARATION OF MRNA TEMPLATES

A plasmid containing the full length 5'-UTR and 5'-coding region of *psbA* cloned behind the T7 promotor was a kind gift from Drs. Hirose and Sugiura. For incorporation of UV-activatible crosslinkers using amber supressors (12,13), stop condons were introduced by PCR into the N-terminus of the D1 protein at position F48, F53, I60 and F93. The Lep coding region was cloned behind the 5´UTR of D1 in the same plasmid. DNA templates for transcription of mRNAs were prepared by using appropriate restriction sites within the coding region or by PCR.

2.2 IN VITRO TRANSLATIONS AND ISOLATION OF RNC COMPLEXES

Chloroplast translation extracts were isolated from tobacco leaves and *in vitro* translations were carried out, principally according to (4) but with some modifications for co-translational targeting. For isolation of rncs, CAP was added to the translation reactions and translations were overlayed on 0.5 M sucrose cushions in 50 mM Hepes-KOH (7.7), 50 mM KOAC, 5 mM MgOAC, 2 mM DTT, protease inhibitors (medium A) and CAP. Sucrose cushions were spun for 45 min at 65.000 rpm in a TLA100 rotor.

2.3 UV-INDUCIBLE AND SULHYDRYL CROSSLINKING

To obtain site-specific crosslinking of the D1 nascent chain with interacting components, the photo-activated amino acid L-4'-(3-(trifuoromethyl)-3H-diazirin-3-yl)phenylalanine (TmdPheo) was incorporated at different amber positions within the nascent chain. The incorporation of this non-natural amino acid was mediated by an amber suppressor tRNA chemically aminoacylated with (Tmd)Pheo (12,13). Crosslinking was performed by irradation of the samples for 90" on ice at 385 nm using a long wave UV lamp. Alternatively, stable rnc's were incubated with sulfhydryll crosslinkers 0.5 mM BMH or 0.5 mM S-MBS. After incubation, the crosslinkers were quenched by addition β-mercaptoethanol (BMH) and glycine (S-MBS).

2.4. TARGETING ASSAYS AND POST-TARGETING TREATMENTS

Thylakoids were isolated from young pea seedlings. Endogenous DNA and RNA were removed from the thylakoids by nuclease treatments.For co-translational targeting, translations were carried out in the light in the presence of thylakoids. To test the affinity and insertion of nascent chain complexes, membranes were washed with medium A with 500 mM KOAc or with 4.5 M urea or in 0.1 M NaCO3/NaOH (12.5). or membranes were treated with proteinase K.

3. Results and Discussion

3.1 PRODUCTION OF D1 RNCS.

To trap the elongation process at different stages, truncated *psbA* mRNA's without a stop codon were generated which yielded peptides of 36, 87, 109 189 amino acid residues. Upon translation, the ribosome halts at the end of the transcript and a stable rnc should be generated, similarly as has been observed in eukaryotic and prokaryotic translation systems. The topology of the such rnc complexes with D1 nascent chains of different length is shown in Fig. 1, assuming that 35 aa are buried in the ribosome tunnel (14). Comparing these constructs could reveal sequence requirements and interacting domains as well as optimal nascent chain length for interaction with soluble stromal components (Fig 1). Upto 80% of the peptides were isolated as rncs, as deduced by sedimentation through sucrose cushions and gradients.

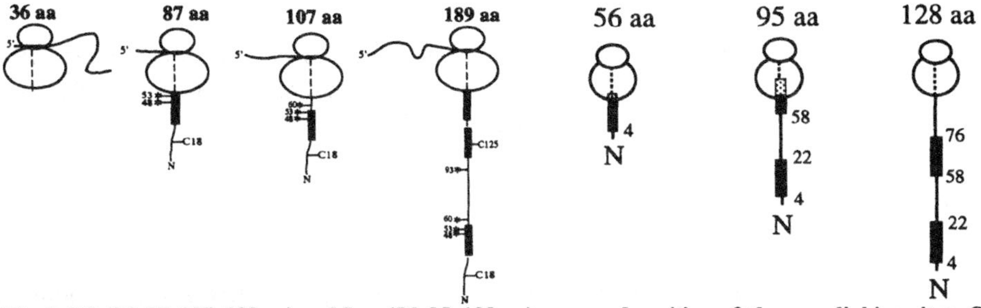

Fig. 1. D1 (36, 87, 107, 189 aa) and Lep (56, 95, 128 aa) rncs and position of the crosslinking sites. C= cysteine residue, * = photo-activatable crosslinker.

3.2 INTERACTION OF THE D1 NASCENT CHAIN WITH STROMA

To identify interacting soluble components with the D1 rnc's, we used a photocrosslinking approach, followed by immunoprecipitation with specific antibodies. UV inducable crosslinkers were incorporated at specific sites by introducing amber codons into the *psbA* gene in the first TM at position 48 and 53 and in the lumenal loop at position 60 and 93 (Fig 1), followed by translation of the transcripts in presence of supressor tRNA. Amber supression efficiencies upto 50% were achieved (not shown). By altering the placement of the crosslinker and the length of the nascent chain we hoped to gain insight in the interacting domains of the D1 protein with the soluble stromal components.

Fig 2 shows crosslinking of D1 rnc's with nascent chains of 87, 107 and 189 aa containing the amber codon at position 53 in the first TM, followed by immuoprecipitation with antiserum directed against cpSRP54 and SecA. A specific interaction was observed between cpSRP54 with the nascent chain of 87 and 107 aa, but not with the 189 aa chain. Photocrosslinking with the constructs containing stop codons at the other postitions did reveal very weak or no interaction between cpSRP54 and the nascent chain, indicating that the interaction between D1 and cpSRP54 depends on the length of the nascent chain and crosslink position. No significant immunoprecipitation of nascent chains using cpSRP54 antiserum was found with the supernatants, indicating that cpSRP54 did not interact with D1 nascent chains released from the ribosome (not shown). No interactions with SecA or cpSRP43 were detected.

The D1 protein contains 2 cysteines (C18 and C126) (Fig 1), which were used to further investigate interacting stromal components. After translation of the truncated proteins (87, 107 and 189 aa), stable rnc's were purified and interacting proteins were crosslinked with BMH and S-MBS. With all three constructs and with both crosslinkers, a clear and distinct crosslinking pattern was observed (Fig 3).

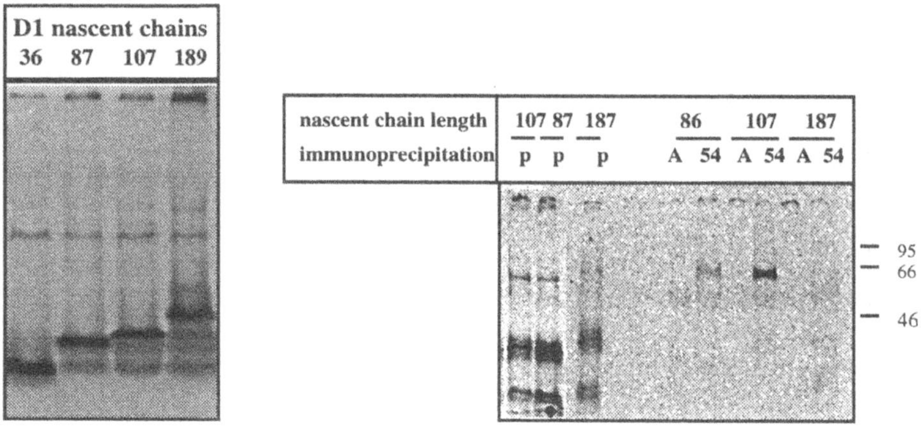

Fig. 2. Probing the interaction between cpSRP54 ("54") or SecA ("A") and D1 rnc's by photoactivated crosslinkers (at position 53) and immunoprecipitation. "P" = 10 % of the rnc pellet.

Immunoprecipitation with cpSRP54 antibody, immunoprecipitated a 50-60 kDa product with the 87 aa nascent chain after crosslinking with BMH and product of 60-70 kDa with the 107 aa nacsent chain after crosslinking with S-MBS. Very weak interaction was observed for the 189 aa nascent chain. Thus clearly the shorter nascent chains could interact with SRP54, but probably not with SecA or with cpSRP43. Immunoprecipitation with the antisera on the supernatants of the sucrose cushions did not yield any radioactive products.

3.3. TARGETING AND INSERTION OF D1 RNC'S

Truncated mRNA's were translated in presence of nuclease treated membranes and the interaction of the nascent chains with the membranes were probed by alkali, urea or high salt incubations. About 30-40% of the nascent chains were resistent to the washings, indicating at least partial integration (not shown).

3.4 TARGETING AND TRANSLATION OF LEP

To test whether Lep could be targeted to the thylakoid, Lep was translated in the presence of thylakoid membranes. 50% of translated Lep was found to be associated with the thylakoids, of which 20% was resistent to NaOH washing (Fig. 4), indicating that at least one TM was inserted into the lipid bilayer in unkown orientation. Protease incubations revealed a protected band of approximately 30 kDa (Fig 4), indicating trans-orientation of the C-terminal loop (see Fig. 4). Thus Lep was targeted and inserted into the thylakoid membranes in the same orientation as into the *E.coli* innermembrane. An overall insertion efficiency of around 10% was routinely observed, comparable to targeting to the E. coli inner membrane(15).

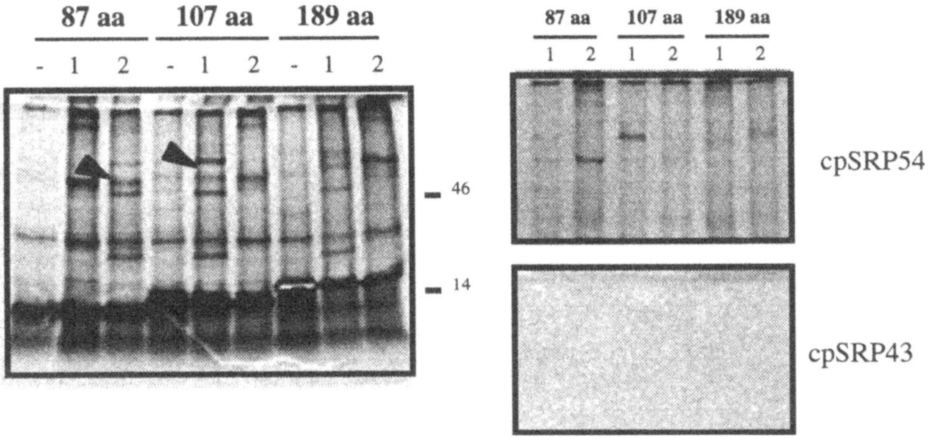

Fig 3. Interaction with D1 rnc's with cpSRP54 after crosslinking to cysteine residues (- no crosslinking; 1, 2 = crosslinking with S-MBS or BMH).

To investigate whether targeting and insertion of Lep occurred co- or post-translationally, translation of Lep was carried out in the absence of membranes. Subsequently, puromycin and lincomycin were added and after 5 minutes incubation, thylakoid membranes were added. After 15 and 30 minutes less than 1-2% of the total amount of translation product was associated with the membranes, of which none was resistant against proteinase K (not shown). Apparently, Lep is able to anker itself into the membrane post-translationally, although very inefficiently. However, translocation of the C-terminal tail (P2) did not occur significantly in a post-translational manner.

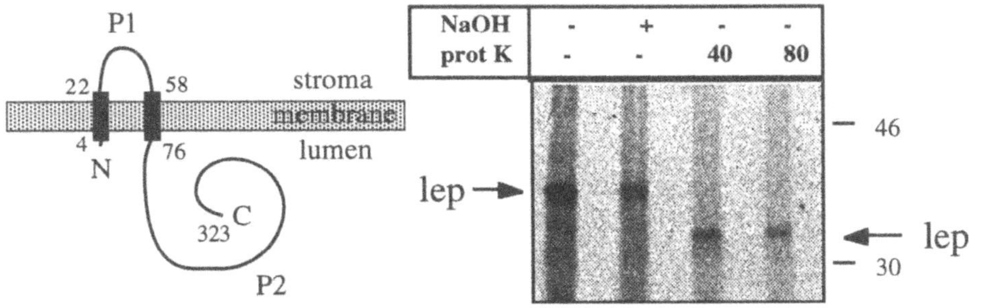

Fig. 4. Predicted topology of Lep and results of targeting and insertion of full length leader peptidase into the thylakoid membranes.

3.5 TARGETING AND INSERTION OF TRUNCATED VERSIONS OF LEP

To further verify the co-translational targeting process of Lep, elongation was trapped at different stages by translation of truncated transcripts of different length, resulting in truncated Lep of 56, 95 or 128 aa (Fig. 1). About 70-85% of the nascent chains could be purified as stable rnc complexes on sucrose cushions (not shown). About 60-80% of the nascent chains were found associated with the membrane in a salt, alkli and urea resistent manner (not shown). This implies that the rnc's were targeted with high efficiency to the membranes and that ribosomes were docked via the nascent chains (16). Comparing the 50-70% efficiency for targeting and insertion of the TM's, contrasts to the translocation of the C-terminal loop (P2 loop, Fig 4) indicating that the requirement for translocation of the long C-terminus was more complex than insertion of the TM's.

3.6 EFFECT OF SECA AND ΔpH ON INSERTION OF FULL LENGTH LEP

To investigate the dependence of the translocation of the P2 domain on SecA and the ΔpH, azide and nigericin respectively, were added prior to translation of full length Lep. The amount of NaOH resistant Lep associated with the membrane was not significantly affected by either azide or nigericine (Fig. 5). Nigericine did also not significantly effect the accumulation of protease protected Lep, whereas azide decreased the amount of protease protected Lep. Combining both nigericine and azide led to a virtually absence of protease protected Lep (Fig. 5). Thus we can conclude that the translocation of the P2 loop of Lep was sensitive to azide, implying that this process could depend on the ATP driven translocation by SecA.

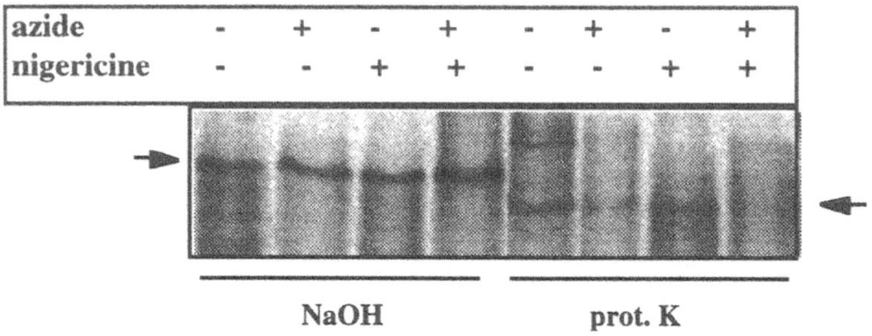

Fig. 5 Effect of azide (5 mM) and nigericin (1.0 μM) on insertion of full length Lep (indicated by arrows).

Acknowledgements

Antisera directed agianst SecA and cpSRP43 were kindly provided by Prof. T. Endo and Dr. L. Nussaume. This study was supported by NFR and CTS.

References

1. Margulies et al. (1975) BBA 402, 297-
2. Herrin, D & Michaels, A (1985) FEBS Lett. 184, 90-
3. Klein, R.R. et al. (1988) J. Cell Biology 106, 289-.
4. Hirose, T & Sugiura, M (1996) EMBO J. 15, 1678-
5. Franklin, AE. & Hoffman, NE (1993) J. Biol. Chem. 268, 22175-
6. Schuenneman, D. et al.(1998) PNAS (in press)
7. Yuan, J. et al. (1994) Science 266, 796-
8. Wolfe, P.B., Rice M., Wickner W. (1985) JBC 260, 1836-
9. Valent , Q.A., et al.. (1998) EMBO J. 17, 2540-
10. de Gier et al. (1996) FEBS Lett. 399, 307-
11. Mothes, W., et al. (1997) Cell 89, 523-
12. Brunner, J (1996) Trends in Cell Biol. 6, 154-
13. Martoglio, B & Dobberstein, B (1996) Trends in Cell Biol. 6, 142-
14. Blobel, G. & Sabatini, D.D. (1970) J. Cell. Biol. 45, 130-
15. van Klompenburg, W., et al.. (1997) FEBS Lett. 413, 109-
16. Borel, A.C. and Simon,S.M. (1996) Cell 85, 379-389

AGGREGATIONAL STATES OF *IN VITRO* SYNTHESIZED D1 AND D2 PROTEINS

STEFAN K. MUEHLBAUER AND LUTZ A. EICHACKER
Botanisches Institut der LMU München, Menzingerstr. 67,
80638 München, Germany

1. Introduction

Studies of gene expression in mature chloroplasts suggested that a light regulation of translation elongation is required for assembly of reaction center proteins [1-3]. In the light, analysis of ribosome distribution on psbA mRNA supported an accumulation of ribosomes at the site of initiation, but also revealed that ribosomes were pausing during translation elongation [4-6]. Translation elongation between pausing sites was shown to be regulated by the photosynthetic pH gradient [3], which was also shown to be necessary for membrane translocation of proteins into the thylakoid lumen [7]. To investigate the influence of regulatory events during translation on assembly of reaction center proteins, we separated translation from membrane insertion by studying the capability of *in vitro* expressed pD1 and D2-protein for insertion in the thylakoid membrane.

2. Material and Methods

The psbA gene from barley (Hordeum vulgare L.) was cloned into the expression region of vector pET21a+ (Novagen) after PCR-amplification from plastid DNA. A fusion protein with the presequence of the small subunit of ribulose-1,5-bisphosphate carboxylase was constructed using the tobacco rbcS gene [8]. mRNA was produced using the MEGAscript kit (Ambion) with T7 polymerase, and expressed in rabbit reticulocyte lysate (Promega) or E. coli S30 extract (Serva) according to the manufacturers' description. Proteins were separated on 12.5% polyacrylamide SDS-gels containing 4 M urea. Barley chloroplasts and etioplasts were isolated as described [9]. Plastid membranes were prepared by lysis in 10 mM Hepes/KOH pH8.0 and centrifugation. For separation of protein complexes, samples were treated with 1% dodecyl maltoside and centrifuged for 16 h at 160000 x g through a 0.1-1M sucrose gradient in 10 mM Tris-HCl, pH 6.8, 20 mm KCl, 10 mM $MgCl_2$.

J. H. Argyroudi-Akoyunoglou and H. Senger (eds.),
The Chloroplast: From Molecular Biology to Biotechnology, 243–246.

3. Results and Discussion

3.1. D1 AGGREGATES IN *IN VITRO* TRANSLATION SYSTEMS

Insertion of *in vitro* synthesized D1-protein into thylakoid membranes readily resulted in proteins that cofractionate with membranes. In order to distinguish between a membrane inserted and an aggregated state of the newly expressed D1-protein, we first investigated the aggregation properties of in-vitro-synthesized protein. One surprising result was that most of the D1 protein, when synthesized in an E. coli S30 translation system, could be pelleted at the low centrifugal forces of a microfuge at 20.000xg, even when no membranes were present in the reaction (Fig. 1A). This indicated that the protein is found in a very large complex, probably an aggregate, when it is not integrated into thylakoid membranes, and that it is not kept in a soluble form by nascent chain binding factors of the E. coli lysate. When the protein was translated in a rabbit reticulocyte system, only a small portion could be pelleted, indicating that here more - or more effective - protein factors were present that kept the protein soluble. However, the low amount of protein found in the membrane fraction after insertion assays indicated that the protein translated in reticulocyte lysate was not insertion competent.
In order to improve the insertion competence of D1 translated in E.coli S30 lysate, we first investigated whether its solubility could be increased by an N-terminal sequence extension. We constructed a fusion protein of the presequence of the soluble stromal enzyme subunit "small subunit of ribulose 1,5 bisphosphate carboxylase" (pSSU) and the D1 protein. However, neither the aggregation in the E.coli lysate nor the solubility in the reticulocyte lysate was affected.

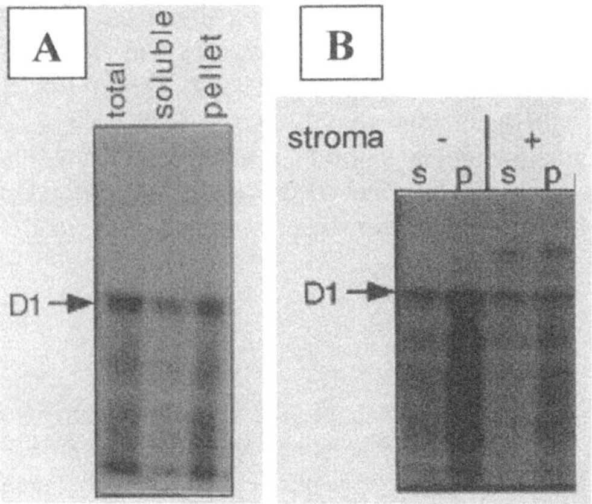

Figure 1: Aggregation of D1-proteins during translation
A) In-vitro transcribed psbA-mRNA was translated for 30 minutes in E. coli S30 extract in the presence of ^{35}S-methionine. After translation, half of the reaction was centrifuged in a microfuge at 20.000xg at 4 °C for 5 min. The supernatant (soluble), the non-visible pellet (pellet) and the uncentrifuged reaction (total) were separated on a 12,5%SDS-polyacrylamide gel.
B) Chloroplast stroma does not prevent D1 aggregation
psbA-mRNA was translated in E. coli S30 extract in the absence (-) or presence (+) of barley chloroplast stroma. After translation the reactions were separated in a soluble (s) and a non-soluble (p) phase as described in the legend to figure 1A.

3.2. CHLOROPLAST STROMA DOES NOT PREVENT AGGREGATION

We therefore tested an antiaggregating influence of chloroplast stroma on the translation products. However, the solubility of D1-proteins was not affected by stroma preparations (Fig. 1B). This supported our hypothesis that D1 is released directly into the membrane, when translated in the chloroplast. We therefore propose that no stromal factors may be required in chloroplasts during translation and membrane insertion.

Also, no antiaggregating influence could be detected when translation was carried out in the presence of detergents (PTE, octyl glucoside, dodecyl maltoside). Furthermore, the pelleted protein could not be completely dissolved by incubation in mild detergents or denaturing agents (urea, Na_2CO_3), indicating that resuspension of membranes in these reagents is not an appropriate means to test for membrane insertion, as the method does not allow to distinguish insertion from strong aggregation phenomena.

3.3. AGGREGATION OCCURS BY PROTEIN-PROTEIN INTERACTIONS

To investigate the formation of the D1-aggregates we tested for a participation of polyribosomes. Addition of RNase or puromycin, which releases nascent chains from ribosomes during translation, did not prevent the formation of aggregates (Fig. 2), indicating that protein-protein interactions between the D1 translation intermediates lead to the observed aggregation phenomenon. This aggregation was effective within 30 seconds after the onset of the translation reaction, when no full-length protein is present yet.

In addition, the amount of aggregated protein increased when translation reactions were further incubated after blocking of translation. Therefore, also the non-aggregated protein is not useful to study membrane insertion, as aggregation did still occur. We therefore propose that aggregation of the D1 protein can only be prevented by translation and membrane insertion on thylakoid bound polysomes.

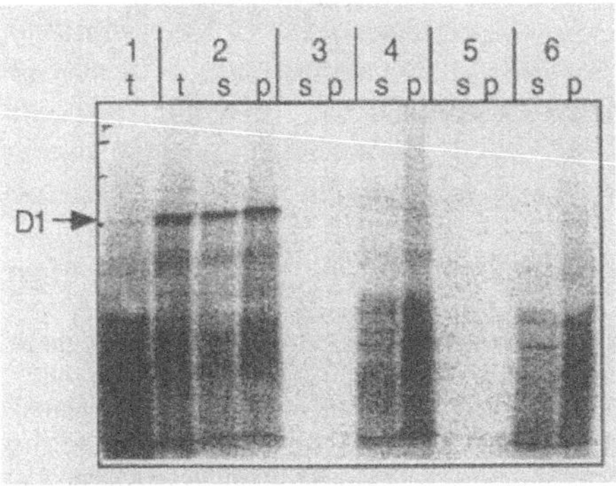

Figure 2: Aggregation of D1 is caused by translation intermediates, but not by polyribosomes.
In vitro transcribed psbA-mRNA was translated in E. coli S30 extract. After translation, half of the reaction was centrifuged at 20.000x g for 5 minutes. The supernatant (s), the non-visible pellet (p) and the uncentrifuged reaction (t) were separated on a 12,5%SDS-PAA gel. After 2 minutes the reaction was either stopped (1), or a chase with unlabeled methionine was performed for 10 more minutes (2). Puromycin was added to the reactions either from the beginning of a 12 minute translation (3) or after 2 minutes for 10 more minutes (4). Alternatively, RNase A was added either from the beginning (5) or after 2 minutes of labeling for 10 more minutes (6).

3.4. D1 AND D2 AGGREGATES BIND MEMBRANE LIPIDS

To test whether the size or composition of the D1 or D2 aggregates was changed in the presence of thylakoid membranes, samples from translation and insertion reactions were treated with dodecyl maltoside (DM) and centrifuged through sucrose density gradients. Most of the protein pelleted through 60% sucrose, as the aggregation was not dissolved by DM; a small part, however, was found in a distinct fraction near the top of the gradient at about 20% sucrose (Fig. 3,□). If translation reactions were performed in the presence of thylakoid membranes, the protein occured also in a second fraction with a slightly higher molecular weight (Fig. 3,■). However, this fraction did not correspond to a fraction in which native D1and D2 proteins are found in chloroplasts, when assembled into photosystem II reaction centres. Furthermore, the extent or position of the second peak did not change, if insertion reactions were performed i) cotranslationally or posttranslationally, ii) in darkness or light or iii) with chlorophyll-free etioplast membranes, or iv) when extracted lipids of etioplast membranes were used. This indicated that the increase in molecular weight is an effect of lipid binding to the protein rather than assembly into membrane complexes.

These experiments were also performed with in-vitro-translated D2 protein, which reacted identically. Also coexpression of the two proteins, which in photosystem II reaction centers are bound to each other, did not affect their solubility or behaviour in density gradients.

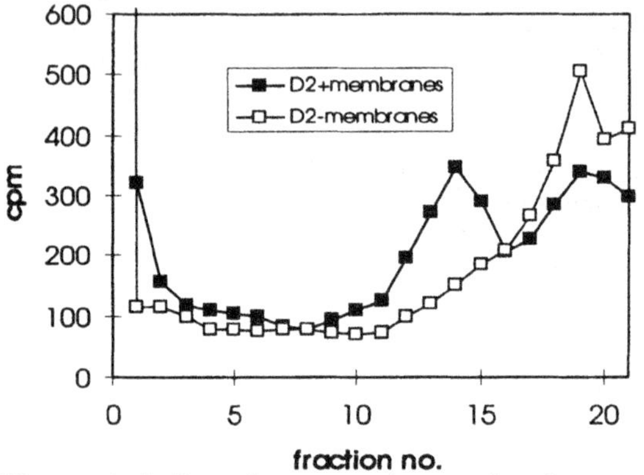

Figure 3: D2 forms higher molecular weight aggregates in the presence of membrane lipids.
psbD-mRNA was translated in E. coli S30 extract in the absence (□) or presence (■) of barley thylakoid membranes. After translation, the reactions were centrifuged, and the pellets were solubilized in 1% dodecyl maltoside and centrifuged through a 15 - 60% sucrose density gradient for 16 hours at 160.000 g. Gradients were fractionated from the bottom into 21 equal fractions, and radioactivity in the fractions was counted in a scintillation counter.

We conclude from these experiments that the exposure of nascent D1 or D2 apoprotein chains to a hydrophilic solution results in aggregation of the apoproteins, whereby an investigation of co- and posttranslational membrane insertion is counteracted. From our data, membrane insertion can only be studied faithfully, if nascent apoprotein chains interact cotranslationally with the membrane insertion apparatus on the thylakoid membrane.

References:

1. v.Wijk, K.-J. and Eichacker, LA (1996) Light is required for efficient translation elongation and subsequent integration of the D1-protein into photosystem II, FEBS Lett. **388**, 89-93
2. Edhofer, I., Mühlbauer, S. K., and Eichacker, L. A. (1998) Light regulates the rate of translation elongation of chloroplast reaction center protein D1, Eur. J. Biochem, in press
3. Mühlbauer, S. K., and Eichacker, L. A. (1998) Light dependent formation of the photosynthetic proton gradient regulates translation elongation in chloroplasts, J. Biol. Chem., in press
4. Kim, J-M., and Mullet, J.E. (1994) Ribosome-binding sites on chloroplast rbcL and psbA mRNAs and light-induced initiation of D1 translation, Plant Mol. Biol. **25**, 437-448
5. Kim, J-M., Gamble-Klein, P., and Mullet, J.E. (1991) Ribosomes pause at specific sites during synthesis of membrane-bound chloroplast reaction center protein D1, J. Biol. Chem. **266**, 14931-14938
6. Kim, J-M., Gamble-Klein, P., and Mullet, J.E. (1994) Vir-115 gene product is required to stabilize D1 translation intermediates in chloroplasts, J. Biol. Chem. **269**, 17918-17923
7. Robinson, C., and Klösgen, R. B. (1994) Targeting of proteins into and across the thylakoid membrane - a multitude of mechanisms, Plant Mol. Biol. **26**, 15-24
8. Klein, R. R., and Salvucci, M. E. (1992) Photoaffinity labeling of mature and precursor forms of the small subunit of ribulose-1,5-bisphosphate carboxylase/oxygenase after expression in Escherichia coli, Plant Physiol. **98**, 546-553
9. Eichacker, L. A., Helfrich, M., Rüdiger, W., and Müller, B. (1996) Stabilization of chlorophyll a-binding apoproteins P700, CP47, CP43, D2, and D1 by chlorophyll a or Zn-pheophytin a, J. Biol. Chem. **271**, 32174-32179

THE ABUNDANCE OF Cab and psbA TRANSCRIPTS AND OF THEIR PRODUCTS IN GREENING ETIOLATED LEAVES TRANSFERRED TO DARKNESS

J.H. GEORGAKOPOULOS, A. PROMBONA and
J.H. ARGYROUDI-AKOYUNOGLOU
Institute of Biology, NCSR "Demokritos"
Aghia Paraskevi, Attiki, Greece

1. Introduction

The growth of the photosynthetic membrane depends on coordinated synthesis and assembly of nuclear-coded and plastid-encoded proteins. An example of such a control of coordinated expression is that of Photosystem II; the reaction center (RC) proteins (D1/D2) are synthesized in the chloroplast and are cotranslationally inserted in the membrane, whereas the surrounding Light-harvesting (LHC) proteins are synthesized in the cytoplasm and after processing of the transit sequence they are imported in the organelle and anchored in the membrane as a lipophilic complex *via* Chl binding. On the other hand, the photosynthetic membrane is a dynamic system which responds to the changing environmetal conditions with a number of control mechanisms directed not only to the transcriptional and translational level, but most important to the posttranslational level. Such a post translational control is evident at the early stages of greening where new PS units are still beeing formed [1-3]. At these early stages a developmentally regulated thylakoid-bound protease degrades LHCs when not rescued by Chl binding [4]. This can be observed in etiolated plants exposed to intermittent light-dark cycles [1, 2], but also upon transfer to the dark of plants briefly preexposed to continuous light [2, 3]; in this case the preaccumulated LHCs in thylakoids are degraded, and their Chl a is used for stabilization of the new RCs. Thus the photosynthetic membrane is a dynamic system, under complex control mechanisms of synthesis, coordinated assembly and turnover of components, which the plants have developed to maximize photosynthetic efficiency under the changing environmental conditions. To understand how the coordination in PSII unit assembly and turnover may be controlled during chloroplast development, we studied how D1 and LHCII levels (immunodetection in total LDS extracted leaf proteins) are correlated with the levels of their transcripts (Northern analysis) in etiolated plants exposed to continuous light, but also following their transfer to the dark.

J. H. Argyroudi-Akoyunoglou and H. Senger (eds.),
The Chloroplast: From Molecular Biology to Biotechnology, 247–252.

2. Materials and Methods

Etiolated *Phaseolus vulgaris* plants grown on perlite at 22°C and 80% humidity were used. 6 days from sowing, cotyledons were harvested, one cotyledon was removed, and the leaves attached to the remaining one, were placed in covered petri dishes on wet filter paper, and exposed to continuous light (CL, 2,000 lux, fluorescent and incadescent lamps); transfer of samples from CL to the dark was at 0, 6, 12, 24 or 48 h CL exposure. Samples of leaves were frozen in liquid nitrogen until use.
For total leaf protein extraction, 1 g leaf sample was solubilized in 10 ml 50 mM Na_2CO_3-2% LDS-12% sucrose-0.05 M DTT, boiled for 5 min in a water bath and centrifuged in an Eppendorf centrifuge for 5 min; the supernatant was frozen until use. SDS-PAGE on 15% gels and Western blot analysis were as before [5]. 5 µg or 200 µg total protein was loaded per slot for LHC-II or D1 detection, respectively.
Anti-D1, raised against the D-E loop, was a kind offer of Drs I. Ohad (Israel) and E.-M. Aro (Finland). Anti LHC-II was a kind offer of Dr. K. Kloppstech (Germany).
Northern analyses was according to Sambrook *et al.* [6] and Sofianopoulou and Scazzocchio [7]. Leaves ground with sand in a mortar under liquid nitrogen were extracted in a mixture of 2 ml phenol- 3 ml buffer (200 mM Tris-HCl, pH 8.5-250 mM NaCl-50 mM EGTA containing 4.8% p-aminosalicylic acid and 0.8% triisopropylnaphthalene sulphonic acid) (10 min vigorous shaking-30 sec bursts). 2 ml $CHCl_3$: isoamyl alcohol (24:1) were added and after phase separation, the large RNA molecules were precipitated from the supernatant by 2 M LiCl. The RNA, washed at RT with 3 M Na acetate, pH 5.2, and cold 70% ethanol was dissolved in diethylpyrocarbonate-treated water, denatured in glyoxal/DMSO and analyzed by 1.2% agarose electrophoresis. Northern hybridization was at 42°C with a-^{32}P-d-CTP-nick translated Cab cDNA (690 bp PstI fragment) or psbA cDNA, in the presence of 50% formamide. For Cab mRNA analysis, 30 µg RNA was loaded per slot; for D1 mRNA, 10 µg per slot. Estimations of ribosomal RNA bands stained by methylene blue were used as a loading control of total RNA applied on slots.

3. Results and discussion

Figure 1 A,B shows the accumulation of Cab mRNA in the etiolated leaves exposed to CL for 6, 12, 24, 36 and 48 h (o), and in these leaves after their transfer to the dark for 24 and 48 h (•). As shown, Cab mRNA levels are very low in the dark, increasing drastically following exposure to CL. The accumulation of Cab mRNA in CL followed circadian oscillations, the amplitude of which was gradually decreased; the oscillations persisted after transfer of light-pretreated plants to the dark. The pattern of oscillations was similar to that earlier observed in etiolated leaves exposed to a single light flash and kept in the dark thereafter [8]. In Figure 1A the Northern blot analysis is shown, in Figure 1B the quantitative presentation of the densitometric scans obtained from the autoradiograms are shown. Figure 1A also shows the immunoblots obtained from the respective samples. As shown, the LHCII level is gradually

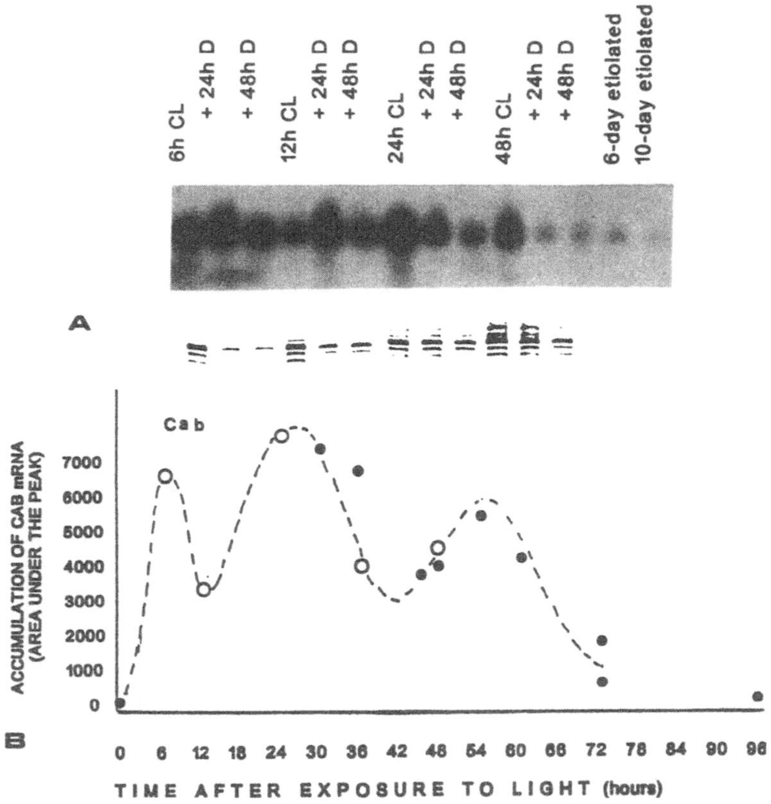

Figure 1. **Accumulation of Cab transcripts as monitored by Northern blot analysis A. Upper part: Autoradiogram of a Northern blot obtained from 6 day etiolated leaves exposed to CL, or transferred to the dark at various times thereafter. Lower part: Immunoblots against anti LHCII of total leaf proteins of the respective leaf samples. B. Quantitation of Cab mRNA accumulation (densitographic analysis of the area under the Cab mRNA band on autoradiogram in A). (o): leaves exposed to continuous light; (•): leaves after their transfer to the dark for 24 and 48 h.**

increased in the plants exposed to CL (see lanes 1, 4, 7, 10 from the left, corresponding to 6, 12, 24 and 48 h of CL respectively). A drastic degradation of LHCII is observed in the dark following 6 h of light exposure (see lanes 2 and 3, corresponding to 24 and 48 h of dark incubation, respectively), but this degradation is greatly reduced in the sample transferred to the dark after exposure to CL for 24 or 48h (lanes 11, 12 corresponding to 24 and 48 h dark incubation, respectively). These results suggest that although at the early stages of exposure to CL Cab transcription is high, the limited amout of Chl present is not enough to rescue the preexisting LHCII from protease action, since this Chl is required for stabilization of the newly formed reaction center proteins and for the assembly of the new PSU cores.

Figure 2 shows the results obtained for psbA mRNA accumulation under the conditions described above. Fig. 2A shows the autoradiograms of Northern blots and Fig. 2B the quantitative representation of these results. As shown, the psbA mRNA level was very low in 6 day old etiolated leaves, but it increased considerably as etiolation was prolonged (see psbA mRNA level of 10-d etiolated leaves -lane 5 from the left). The psbA accumulation, however, in CL was greatly enhanced. Transfer of plants from light to dark resulted in increased psbA mRNA level only when the transfer was at the early stages of greening; the increase continued in the dark at least for 48 h. This was not observed when transfer to the dark was at later stages of greening (e.g. after 48 h of CL); in this case the transcript level remained

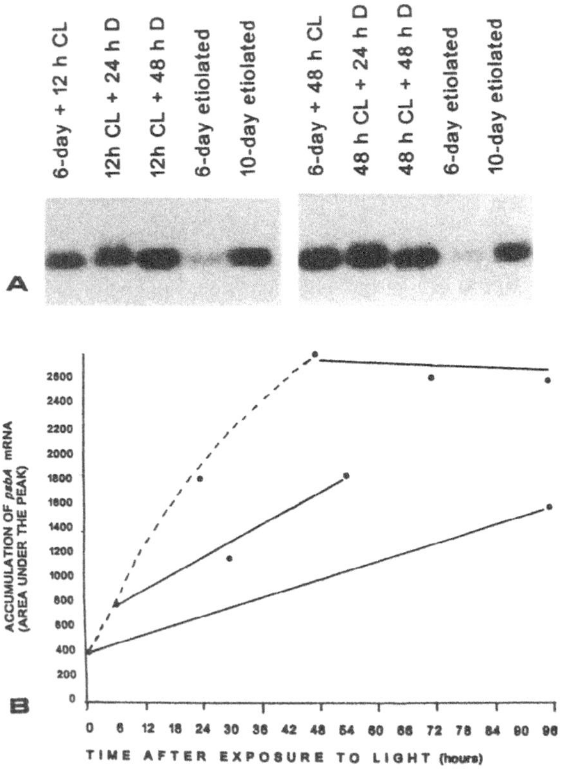

Figure 2. **Accumulation of psbA transcripts as monitored by Northern blot analysis**
A. Autoradiogram of a northern blot of psbA mRNA in leaves of 6 day old etiolated bean plants, first exposed to CL for 12 or 48 h and then transferred to the dark for an additional 24 and 48 h.
B. Quantitaion of the results in A, as monitored by densitographic analysis of autoradiograms . Dotted line : exposure to CL; solid line transfer to the dark.

more or less unaffected by the transfer to the dark. So the shorter was the preillumination period, the larger was the increase in psbA transcript level.More or less similar results were obtained with the immunodetectable D1 level in total leaf protein extracts of leaves exposed to CL and then transferred to the dark, with the exception that in etiolated tissue, despite the presence of psbA transcript, D1 was not detectable no matter how prolonged the etiolation was. In this case. no mature 32 kDa D1 form was detectable, but only polypeptides of lower molecular weight (23 and 17 kDa, which may be either translation-arrest or degradation products) (see Fig. 4a,b in 6 or 10 day-old plants) [9]. D1 accumulation was greatly enhanced during exposure to CL up to at least 48 h. The lower molecular weight D1 products

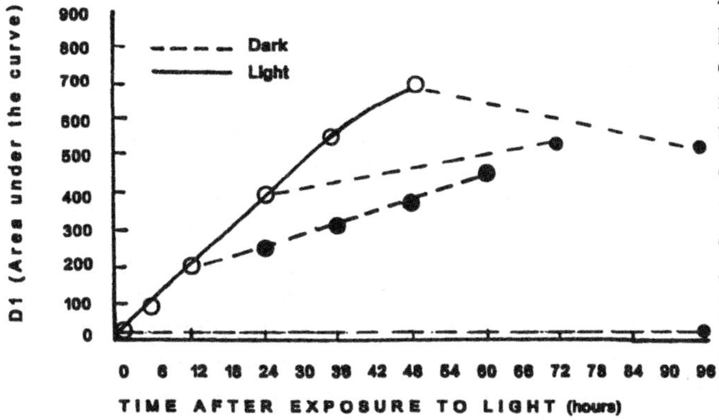

Figure 3. Accumulation of D1 protein in total leaf protein extracts, as monitored by immunodetection on Western blots (densitographic estimation of the immunostain under the 32 kDa D1 band). 6 day etiolated plants were either exposed to CL (solid line) or transferred to the dark following preillumination for various time periods (dotted line).

disappeared during exposure to CL in parallel to the appearance of the 32 kDa mature D1 form. Transfer of plants to the dark following light exposure resulted in increased D1 levels; the increase continued for at least 48 h, but again the increase was dependent on preillumination period, the shorter the preillumination the larger the increase.

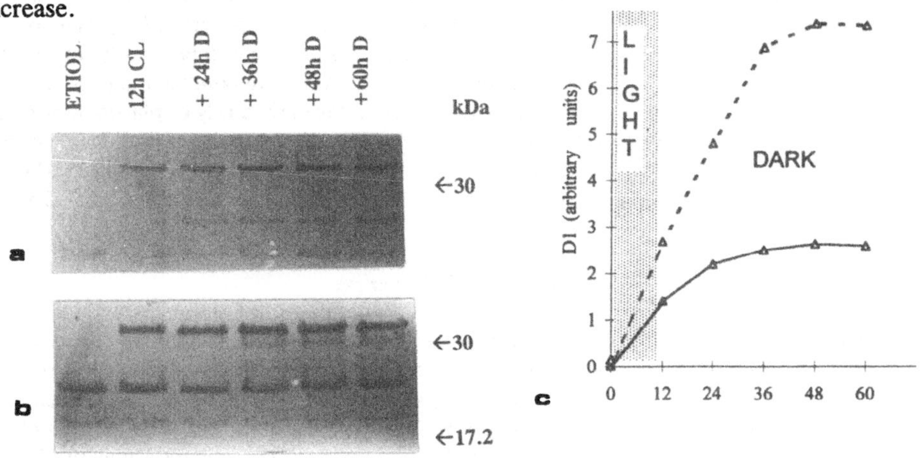

Figure 4 . Immunodetection of D1 on Western blots of total leaf protein extracts obtained from 5 (a) or 10 (b)-day old etiolated (ETIOL.) bean leaves, prior to or after their exposure to CL for 12 h, and subsequent transfer to the dark (D). c: densitographic estimation of the 32 kDa band in (a) (solid line) or in (b) (dotted line) .

4. Conclusions

In young etiolated tissue a low amount of psbA mRNA as well as of CAB mRNA is present, which increase as etiolation is prolonged. Both transcript levels increase during exposure of the etiolated tissue to CL; the rate of transcription or stabilization of transcripts is greatly increased in the light as compared to that in the dark. Although the CAB transcript level is under control of the circadian oscillator, that of the psbA seems to be unaffected by the clock. This suggests that the control of the oscillator is mainly exerted at the nucleus. Despite the presence of both psbA and LHCII transcripts in the etiolated tissue very little to none of their protein products can be detected. In the case of D1 this possibly reflects a translational arrest in the dark, due either to the absence of ATP, or of redox control, or of transmembrane pH gradient, etc; this is suggested from the detection of lower MW immunodetectable D1 polypeptides, which disappear upon transfer to CL. In the case of LHCII most important may be the absence of stabilizing Chl, required to rescue the protein from the proteolytic attack. In plants transferred to the dark following brief preillumination, the preexisting Chl associated with LHCII becomes available for binding to the newly synthesized D1 upon degradation of preaccumulated LHCII. Thus, newly synthesized D1 may use the Chl released from LHCII and may be stabilized in the dark. This is suggested from the immunodetection of the mature D1 protein in plants transferred to the dark following brief preillumination. This supports the earlier finding that under these conditions new PS unit cores, as monitored by photosynthetic activity measurements and P700/Chl ratios are being synthesized [3, 10].

6. References

1. Argyroudi-Akoyunoglou, J.H. and Akoyunoglou, G. (1979) The Chlorophyll-protein complexes of thylakoids in greening plastids of *Phaseolus vulgaris*, *FEBS Lett.* **104**,78-84.
2. Akoyunoglou, G. and Argyroudi-Akoyunoglou, J. H. (1986) Post-translational regulation of chloroplast differentiation, in G. Akoyunoglou and H. Senger (eds.), *Regulation of Chloroplast Differentiation,* Alan R Liss, New York, pp.571-582.
3. Argyroudi-Akoyunoglou, J.H., Akoyunoglou, A., Kalosakas, K. and Akoyunoglou, G. (1982) Reorganiation of the PSII unit in developing thylakoids of higher plants after transfer to darkness, *Plant Physiol* **70**, 1242-1248.
4. Tziveleka, L.A. and Argyroudi-Akoyunoglou, J.H. (1998) Implications of a developmental stage-dependent thylakoid-bound protease in the stabilization of the light-harvesting pigment-protein complex serving PS II dyring thylakoid biogenesis in red kidney bean, *Plant Physiol* **117**, 961-970.
5. Georgakopoulos J.H. and Argyroudi-Akoyunoglou J.H. (1997) Implication of D1 degradation in phosphorylated induced state transitions, *Photosynth Res* **53**, 185-195, 1997.
6. Sambrook, J., Fritsch, E.F. and Maniatis, T. (1989) *Molecular Cloning*, Cold Spring Harbor Laboratory Press, New York.
7. Sofianopoulou, V. and Scazzocchio, C. (1989) The proline transport protein of *Aspergillus nidulans* is very similar to amino acid transporters of *Saccharomyces cerevisiae*, *Mol Microbiol* **3**, 705-714.
8. Tavladoraki, P., Kloppstech, K. and Argyroudi-Akoyunoglou, J.H. (1989) Circadian rhythm in the expression of the mRNA coding for the apoprotein of the light-harvesting complex of photosystem II, *Plant Physiol* **90**, 665-672.
9. Mullet, L. E., Klein, P.G. and Klein R.R. (1990) Chlorophyll regulates accumulation of the plastid-encoded chlorophyll apoproteins CP43 and D1 by increasing apoprotein stability, *Proc Natl Acad Sci USA* **87**, 4038-4042.
10. Akoyunoglou A, and Akoyunoglou G. (1985) Reorganization of thylakoid components during chloroplast development in higher plants after transfer to darkness. Changes in PS I unit and cytochromes, *Plant Physiol* **79**,425-431.

CHLOROPHYLL-BINDING PROTEINS IN CYANOBACTERIA

WIM F.J. VERMAAS
Department of Plant Biology, and Center for the Study of Early Events in Photosynthesis, Arizona State University, Box 871601, Tempe, AZ 85287-1601, USA

1. Introduction

A main difference in the photosynthetic apparatus in thylakoids of cyanobacteria and plants is the light-harvesting antenna. In plants, light-harvesting chlorophyll *a/b*-binding (LHC) proteins are the main component of the antenna (reviewed in [1]), whereas in cyanobacteria phycobilisomes constitute the peripheral antenna system (see [2] for a review). In cyanobacteria, all chlorophyll was thought to be associated with reaction center complexes. Indeed, most of the chlorophyll in cyanobacteria is associated with the photosystem I (PS I) reaction center, as upon deletion of the PS I complex in *Synechocystis* sp. PCC 6803 the amount of chlorophyll per cell is reduced to about 15% of the amount in wild type [3]. However, the concept of all chlorophyll to be associated with reaction centers was proven to be an oversimplification as upon genetic deletion of both PS I and photosystem II (PS II) in *Synechocystis* sp. PCC 6803 chlorophyll still remained in the organism in apparently protein-bound form [4]. The existence of additional chlorophyll-binding proteins was also suggested from the ratio of chlorophyll and PS II observed in PS I-less *Synechocystis* sp. PCC 6803 strains: this ratio is 100-110 [3], which is at least twice the number of chlorophylls present in isolated oxygen-evolving PS II core complexes [5,6].

With the completion of the entire genome sequence of *Synechocystis* sp. PCC 6803 [7,8] and the accessibility of this sequence through CyanoBase [9], it has now become possible to search for potential chlorophyll-binding proteins in cyanobacteria in two ways. (1) Using chlorophyll-binding proteins from plants and other systems, open reading frames that may code for chlorophyll-binding proteins can be identified in *Synechocystis* sp. PCC 6803 and the function of corresponding proteins may be probed by reverse genetics approaches. One of the outcomes (corroborating research of the past decade) is that there are few obvious evolutionary relationships between most types of light-harvesting complexes, suggesting either that different classes of organisms may have developed their own light-harvesting system, or that the evolutionary rate of these complexes is relatively rapid. (2) N-terminal amino acid sequences of proteins associated with chlorophyll can be determined, and the corresponding gene can be identified. We have found a protein resembling a Type IV pilin that may be involved with transfer or storage of newly synthesized chlorophyll.

J. H. Argyroudi-Akoyunoglou and H. Senger (eds.),
The Chloroplast: From Molecular Biology to Biotechnology, 253–258.

2. Materials and Methods

Sequence comparisons of longer proteins (>70 residues) were done using the BLAST algorithm [10] available at the NCBI and CyanoBase web sites. For small proteins (LH1 and LH2, the light-harvesting proteins of purple bacteria) and for single membrane spans of LHC, the query sequence was matched against the translated *Synechocystis* sp. PCC 6803 genome sequence in six frames using the TFASTA program.

Methods used for characterization of *Synechocystis* sp. PCC 6803 mutants were identical to those described in recent publications from our group [11,12].

3. Results and Discussion

3.1. SIMILARITY SEARCHES

The known light-harvesting chlorophyll proteins found in membranes of photosynthetic systems can be divided into at least four groups: (1) reaction center proteins; (2) LHC-type proteins from plants and algae; (3) small antenna proteins (LH1 and LH2) from purple bacteria; and (4) antenna proteins from prochlorophytes. Moreover, well-characterized soluble or membrane-associated (but not integral) chlorophyll-binding proteins include the Fenna-Matthews-Olson (FMO) protein (connecting the chlorosome with the membrane) in green bacteria [13], and a peridinin/chlorophyll-binding protein (PCP) from the dinoflagellate *Amphidinium carterae* [14].

Reaction center proteins in plants are highly conserved, and core reaction center proteins have a clear homolog in the material encoded by the cyanobacterial genome. However, when searching the cyanobacterial genome for open reading frames predicted to code for polypeptides that resemble LHC from plants, the closest matches are provided by five open reading frames that code mostly for small, 6-8 kDa proteins (Table 1). These open reading frames are similar to *hliA*, a gene from *Synechococcus* sp. PCC 7942 that was induced upon high-light treatment and that coded for a protein predicted to have a single membrane span and with a sequence similar to those of the first and third transmembrane helices of LHC [15]. Four of the five open reading frames in *Synechocystis* sp. PCC 6803 that are similar to *hliA* are small, and the fifth one is a 3' extension of the ferrochelatase gene. As we have not observed a specific increase in expression of these genes in high light in *Synechocystis* sp. PCC 6803, these genes have been named *scpA-E*, with scp denoting small Cab-like protein [16]. Most of the residues known to interact with pigments in LHCII [17] have been conserved in the cyanobacterial SCPs, suggestive of a role of these SCPs in chlorophyll binding.

One hypothesis regarding the evolution of LHC involves a fusion of genes of SCP-type proteins with those for small proteins similar to the second helix of LHC, followed by gene duplication [15]. Even though searching for genes in the *Synechocystis* sp. PCC 6803 genome that are similar to DNA sequences coding for the second transmembrane helix of LHC has been unsuccessful, this does not necessarily imply the absence of such a gene, as there is little conservation of the second helix of LHCs between for example LHCI and LHCII species.

TABLE 1. Genes for known chlorophyll-binding proteins in a range of photosynthetic organisms in comparison to those found in the cyanobacterium *Synechocystis* sp. PCC 6803. Genes for reaction center components have been omitted as these are conserved in all oxygenic photosynthetic systems. For protein class abbreviations, see text.

protein class	occurrence	presence of similar genes in *Synechocystis* sp. PCC 6803
LHC	major antenna system in plants	small *scp* genes coding for LHC-type proteins with single membrane spans
PCB	chlorophyll *a/b*-binding protein in prochlorophytes	*isiA*, a gene that resembles *psbC* and that is induced upon iron starvation
LH1/LH2 proteins	antenna proteins in purple bacteria	none; no genes for clearly similar have been detected in other classes either
FMO	green bacteria, connecting chlorosome and RC	none; not detected yet in any photosynthetic organisms except green bacteria
PCP	soluble peridinin/chlorophyll protein in dinoflagellates	none; no closely related proteins in other photosynthetic organisms either

Small antenna proteins from purple bacteria include the α and β subunits of LH1 and LH2. Although there is variability among the primary sequences of these bacterial antenna proteins, an invariant feature of these proteins is a pigment-binding His residue (occasionally an Asn) toward the periplasmic side of the membrane. No homologs of genes for the purple bacterial LH1 and LH2 proteins are found in the *Synechocystis* sp. PCC 6803 genome (Table 1) and are also absent in other groups, so the evolutionary ancestry of the antenna complex in purple bacteria remains unclear.

Chlorophyll *a/b*-binding antenna proteins from prochlorophytes (PCB) are unrelated to LHC and are similar to IsiA (a CP43-like protein expressed predominantly under conditions of Fe depletion) in cyanobacteria [18]. This similarity is very convincing, and fits well with the concept that prochlorophytes and cyanobacteria are closely related and perhaps are even evolutionarily intertwined [19].

Interestingly, for the two soluble chlorophyll-binding proteins for which a crystal structure is available, the FMO protein from green bacteria [13] and the peridinin/chlorophyll-binding protein from dinoflagellates [14], there are no open reading frames in other photosynthetic systems that are convincingly similar (Table 1). This suggests that these proteins have evolved along pathways that are indepedent from those of other chlorophyll-binding proteins.

3.2. TRANSFER AND STORAGE OF CHLOROPHYLL

Chlorophyll is made by phytylation of chlorophyllide, and incorporated into chlorophyll-binding antenna or reaction center proteins. No detectable free chlorophyll accumulates, and levels of pigmented precursors are very low as well. Only in etiolated systems a significant amount of protochlorophyllide accumulates. The reason for the tight control of chlorophyll synthesis is that free chlorophyll is toxic in the light, as it can give rise to triplet states that can react with triplet (ground-state) oxygen to yield the highly oxidizing and reactive singlet oxygen molecule. Chlorophyll binding to carotenoid-containing proteins provides an effective mechanism for chlorophyll triplet quenching by carotenoids. How regulation of chlorophyll biosynthesis exactly occurs, whether chlorophyll can be "stored" temporarily bound to carotenoid-containing proteins, and what the steps are by which newly synthesized chlorophyll is transferred to newly made chlorophyll-binding proteins are largely unexplored issues.

One approach we used in trying to elucidate at least some of the proteins involved in the process of chlorophyll storage and/or transfer was to analyze the proteins associated with chlorophyll under conditions where the chlorophyll biosynthesis capacity far exceeded the demand for the pigment. In such a state proteins involved with chlorophyll storage and/or transfer are expected to be fully loaded with the pigment. Chlorophyll-binding proteins present in PS I-less / PS II-less *Synechocystis* sp. PCC 6803 strains were analyzed using a procedure involving non-denaturing gel electrophoresis, cutting out green bands, denaturing gel electrophoresis of the green bands, and N-terminal sequencing of prominent polypeptides. From the N-terminal sequence, the corresponding gene sequence can be determined directly by analysis of the genomic sequence. A striking result of this experiment was that a ~17 kDa polypeptide encoded by *sll1694* (open reading frame nomenclature according to CyanoBase) and resembling a Type IV pilin was identified in the chlorophyll-containing fraction. Upon deletion of the *sll1694* gene in a PS I-less/*chlL*$^-$ background (i.e., a strain lacking chlorophyll synthesis in darkness), chlorophyll synthesis was delayed upon exposure to light, and the 77K fluorescence emission spectrum remained blue-shifted by several nm for hours after the start of illumination [11]. Deletion of *sll1694* did not confer a phenotype in wild type under our laboratory conditions [11]. We interpret these data to indicate that *sll1694* codes for a protein that is involved in, but not absolutely required for, transfer of chlorophyll from the site of synthesis to where it can be functionally bound in the reaction center and antenna complexes. This is the first example of a protein involved in this transfer / storage process.

3.3. ORIGIN OF THE 77K FLUORESCENCE EMISSION PEAK AT 685 NM

From the two previous sections, the existence of cyanobacterial chlorophyll-binding proteins that are not associated with reaction centers has become apparent. As these proteins may not have a regular antenna function, energy transfer to reaction centers may be inefficient and therefore the fluorescence yield of these chlorophylls will be high as the lifetime of the excited state is expected to be long. The question is whether this fluorescence can be detected and if so, where. In plants and cyanobacteria, three 77K fluorescence emission maxima are found upon excitation with light that is absorbed by chlorophyll. A 725-735 nm peak originates from PS I. A 695 nm peak is associated specifically with one or more chlorophylls in PS II, most prominently the chlorophyll associated with His114 of the CP47 protein [20]. The amplitude of the 695 nm peak is

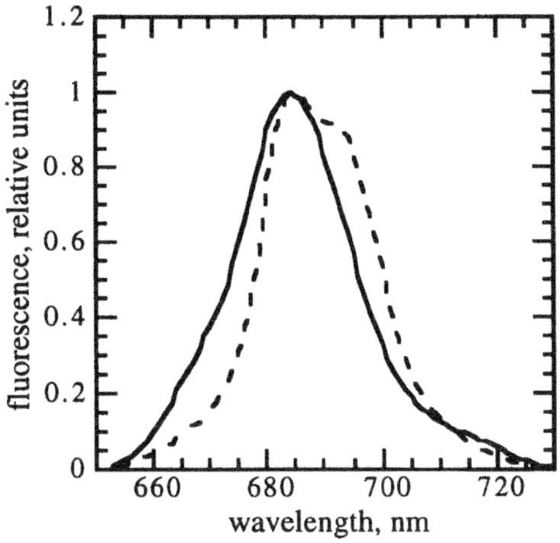

Figure 1. 77K Fluorescence emission spectrum of intact cells of the *psbDIC⁻/psbDII⁻* /PS I-less (broken lines) and PS I-less mutants (solid lines). Excitation was at 435 nm. Bandwidths for excitation and emission were 10 and 2 nm, respectively.

also sensitive to site-directed changes in particular residues of the D2 protein of PS II [12]. The third peak that is apparent in 77K fluorescence emission spectra is at 685 nm. In isolated complexes this peak unequivocally reflects PS II, but in intact cyanobacteria at least a major part of this fluorescence is likely to originate from other chlorophyll-binding proteins instead. The reasons for this statement are as follows: (1) in the *psbDIC⁻/psbDII⁻*/PS I-less mutant that lacks both PS II and PS I a 683 nm emission peak remains (Figure 1); (2) in mutants with a very much decreased PS II content the 685 nm peak remains essentially the same (for example, see [20]); and (3) upon chlorophyll synthesis, newly synthesized chlorophyll associates with protein to yield a 685 nm peak [21]. The 695 nm peak develops only later. Therefore, the 77K fluorescence emission peak at 685 nm in cyanobacterial systems originates to a significant degree from chlorophyll that is associated with complexes other than PS II. This may include the SCPs (the small Cab-like proteins) as well as proteins involved in chlorophyll transfer/storage. The identity of the major contributors to 683-685 nm fluorescence emission in PS I-less/PS II-less systems remains to be determined.

4. Acknowledgement

Financial support for this research is provided by a grant from the US Department of Energy (DE-FG03-95ER20180).

5. References

1. Bassi, R., Sandonà, D., and Croce, R. (1997) Novel aspects of chlorophyll *a/b*-binding proteins, *Physiol. Plant.* **100**, 769-779.
2. Sidler, W. (1994) Phycobilisome and phycobiliprotein structures, in D.A. Bryant (ed.), *The Molecular Biology of Cyanobacteria*, Kluwer Academic Publishers, Dordrecht, pp. 139-216.
3. Shen, G., Boussiba, S., and Vermaas, W.F.J. (1993) *Synechocystis* sp. PCC 6803 strains lacking photosystem I and phycobilisome function, *Plant Cell* **5**, 1853-1863.
4. Shen, G. and Vermaas, W.F.J. (1994) Chlorophyll in a *Synechocystis* sp. PCC 6803 mutant without photosystem I and photosystem II core complexes: Evidence for peripheral antenna chlorophylls in cyanobacteria, *J. Biol. Chem.* **269**, 13904-13910.
5. Tang, X.-S., and Diner, B.A. (1994) Biochemical and spectroscopic characterization of a new oxygen-evolving photosystem II core complex from the cyanobacterium *Synechocystis* sp. PCC 6803, *Biochemistry* **33**, 4594-4603.
6. Patzlaff, J.S., and Barry, B.A. (1996) Pigment quantitation and analysis by HPLC reverse phase chromatography: a characterization of antenna size in oxygen-evolving photosystem II preparations from cyanobacteria and plants, *Biochemistry* **35**, 7802-7811.
7. Kaneko, T., Sato, S., Kotani, H., Tanaka, A., Asamizu, E., Nakamura, Y., Miyajima, N., Hirosawa, M., Sugiura, M., Sasamoto, S., Kimura, T., Hosochi, T., Matsuno, A., Muraki, A., Nakazaki, N., Naruo, K., Okumura, S., Shimpo, S., Takeuchi, C., Wada, T., Watanabe, A., Yamada, M., Yasuda, M., and Tabata, S. (1996) Sequence analysis of the genome of the unicellular cyanobacterium *Synechocystis* sp. strain PCC 6803. II. Sequence determination of the entire genome and assignment of potential protein-coding regions, *DNA Research* **3**, 109-136.
8. Kaneko, T., Sato, S., Kotani, H., Tanaka, A., Asamizu, E., Nakamura, Y., Miyajima, N., Hirosawa, M., Sugiura, M., Sasamoto, S., Kimura, T., Hosochi, T., Matsuno, A., Muraki, A., Nakazaki, N., Naruo, K., Okumura, S., Shimpo, S., Takeuchi, C., Wada, T., Watanabe, A., Yamada, M., Yasuda, M., and Tabata, S. (1996) Sequence analysis of the genome of the unicellular cyanobacterium *Synechocystis* sp. strain PCC 6803. II. Sequence determination of the entire genome and assignment of potential protein-coding regions (supplement), *DNA Research* **3**, 185-209.
9. Nakamura, Y., Kaneko, T., Hirosawa, M., Miyajima, N., and Tabata, S. (1998) CyanoBase, a WWW database containing the complete nucleotide sequence of the genome of *Synechocystis* sp. PCC 6803, *Nucleic Acids Res.* **26**, 63-67.
10. Altschul, S.F., Madden, T.L., Schaffer, A.A., Zhang, J., Zhang, Z., Miller, W., and Lipman, D.J. (1997) Gapped BLAST and PSI-BLAST: a new generation of protein database search programs, *Nucleic Acids Res.* **25**, 3389-3402.
11. He, Q., and Vermaas, W. (1998) Genetic deletion of proteins resembling type IV pilins in *Synechocystis* sp. PCC 6803: Their role in binding or transfer of newly synthesized chlorophyll, *Plant Mol. Biol.*, submitted.
12. Lince, M.T., and Vermaas, W. (1998) Association of His117 in the D2 protein of photosystem II with a chlorophyll that serves as a conduit of excitation energy between the antenna and the reaction center, *Eur. J. Biochem.*, in press.
13. Tronrud, D.E., Schmid, M.F., and Matthews, B.W. (1986) Structure and X-ray amino acid sequence of a bacteriochlorophyll *a*-protein from *Prosthecochloris aestuarii* refined at 1.9 Å resolution, *J. Mol. Biol.* **188**, 443-454.
14. Hofmann, E., Wrench, P.M., Sharples, F.P., Hiller, R.G., Welte, W., and Diederichs, K. (1996) Structural basis of light harvesting by carotenoids: Peridinin-chlorophyll-protein from *Amphidinium carterae*, *Science* **272**, 1788-1791.
15. Dolganov, N.A.M., Bhaya, D., and Grossman, A.R. (1995) Cyanobacterial protein with similarity to the chlorophyll *a/b*-binding proteins of higher plants: evolution and regulation, *Proc. Natl. Acad. Sci. USA* **92**, 636-640.
16. Funk, C., and Vermaas, W.F.J. (1998) A cyanobacterial gene family coding for single-helix proteins resembling part of the light-harvesting proteins from higher plants, submitted.
17. Kühlbrandt, W., Wang, D.N., and Fujiyoshi, Y. (1994) Atomic model of plant light-harvesting complex by electron crystallography, *Nature* **367**, 614-621.
18. La Roche, J., van der Staay, G.W., Partensky, F., Ducret, A., Aebersold, R., Li, R., Golden, S.S., Hiller, R.G., Wrench, P.M., Larkum, A.W., and Green, B.R. (1996) Independent evolution of the prochlorophyte and green plant chlorophyll *a/b* light-harvesting proteins, *Proc. Natl. Acad. Sci. USA* **93**, 15244-15248.
19. Urbach, E., Robertson, D.L., and Chisholm, S.W. (1992) Multiple evolutionary origins of prochlorophytes within the cyanobacterial radiation, *Nature* **355**, 267-270.
20. Shen, G., and Vermaas, W.F.J. (1994) Mutation of chlorophyll ligands in the chlorophyll-binding CP47 protein as studied in a *Synechocystis* sp. PCC 6803 photosystem I-less background, *Biochemistry* **33**, 7379-7388.
21. Wu, Q., and Vermaas, W.F.J. (1995) Light-dependent chlorophyll *a* biosynthesis upon *chlL* deletion in wild-type and photosystem I-less strains of the cyanobacterium *Synechocystis* sp. PCC 6803, *Plant Mol. Biol.* **29**, 933-945.

MOLECULAR CLONING AND CHARACTERIZATION OF ClpX, A POTENTIAL REGULATOR OF CHLOROPLASTIC Clp PROTEASE

TAMAR HALPERIN, HANAN ITZHAKI and ZACH ADAM
Dept. of Agricultural Botany, The Hebrew University
Rehovot 76100, Israel

1. Introduction

Bacterial Clp protease is an ATP-dependent serine protease. It has a barrel-like structure, composed of two rings of 7 subunits of the proteolytic subunit ClpP, nested between two rings of hexamers of regulatory subunits, either ClpA or ClpX. The association between the proteolytic and the regulatory ATPase subunits requires binding of ATP, whereas degradation of proteins by this complex is dependent on ATP hydrolysis. The chloroplastic proteolytic subunit ClpP is encoded in the chloroplast genome (1, 2), but apparently it has also a nuclear-encoded, chloroplast-targeted homologue (3). The chloroplastic structural and functional homologue of the regulatory subunit ClpA, designated ClpC, is encoded in the nucleus and imported into the chloroplast (4, 5). ClpP and ClpC are localized to the stroma and expressed constitutively (6-8). Similar to the bacterial protease, the association between the two is dependent on ATP. When the regulatory subunit ClpC is immuno-precipitated with a specific antibody, the proteolytic subunit ClpP is also precipitated, but only in the presence of Mg·ATP or its non-hydrolyzable analog ATP-γ-S (9). Co-immunoprecipitated ClpC-P is proteolytically active, and can degrade mature OE33 (9), which we previously demonstrated to be degraded when mistargeted to the stroma, by a soluble ATP-dependent serine protease (7).

The regulatory subunits ClpA and ClpX in *E. coli* were shown to have chaperone activity independent of the proteolytic subunit (10). Thus, the chloroplast homologues are likely to play roles in processes other than proteolysis as well. In fact, it has recently been observed that ClpC is associated with the chloroplast import apparatus, on its stromal face (11). It is possible that ClpC functions analogously to Hsp70 in mitochondrial import, by interacting with partially translocated proteins and provides the driving force for the import process. However, it is not clear yet whether this species is indeed ClpC that has dual roles in proteolysis and import, or else, the related protein ClpD that cannot interact with ClpP (A. Clarke, personal communication).

Since in *E. coli* the proteolytic subunit ClpP can interact with either ClpA or with the smaller ATPase ClpX, we searched for a plant homologue of ClpX. We report here on the isolation of a plant cDNA encoding ClpX, characterize its structure and expression, and discuss its possible roles.

2. Methods

2.1. Cloning and sequencing of AtClpX cDNA - *Arabidopsis* expressed sequence tag 207D12T7 showing homology to *E. coli* ClpX and the *Arabidopsis* cDNA library λ-Zip-Lox were obtained from the *Arabidopsis* Resource Center at Ohio State University. A

J. H. Argyroudi-Akoyunoglou and H. Senger (eds.),
The Chloroplast: From Molecular Biology to Biotechnology, 259–264.

1500-base pair *Eco*RI fragment was isolated from the expressed sequence tag clone, radioactively labeled, and used as a DNA probe to screen the cDNA library. Isolated phages were excised *in vivo*, plasmid DNA was prepared, and the longest clones were further analyzed. Automated sequencing of both strands was performed in an Applied Biosystems Model 373 DNA Sequencer.

2.2. Expression of AtClpX - Arabidopsis seedlings were grown at 25°C, in light intensity of 100 $\mu E \cdot m^{-2} \cdot s^{-1}$ for 4 weeks. To test the effect of temperature on expression of ClpX, plants were transferred to either 40 or 15°C and leaf samples were harvested after 2 or 4 h. To assess the effect of light intensity, plants were transferred to either 40 or 700 $\mu E \cdot m^{-2} \cdot s^{-1}$ for 3 days before leaves were harvested and frozen. Samples were homogenized in 0.1 M Tris-HCl, pH 7.5, and 0.5% sarcosyl. Total RNA was extracted with phenol-chloroform, followed by precipitations with 2 M LiCl and ethanol. RT-PCR was performed with a commercially available kit (Promega), using 0.25 μg RNA. Preliminary experiments confirmed that reactions carried out with lower or higher amounts of RNA gave proportional signals.

3. Results and Discussion

3.1. An Arabidopsis cDNA encodes a ClpX homologue

A search of *Arabidopsis* data base revealed an expressed sequence tag encoding a protein with a considerable homology to bacterial ClpX. Since the cDNA did not contain the initiating Met, we used it as a probe to screen an *Arabidopsis* cDNA library and isolate the full-length corresponding clone. The 1956 base-pair-long cDNA (GenBank accession number AF036328), designated AtClpX, encodes a protein of 579 amino acids, with a calculated molecular mass of 66.9 kD and a pI of 9.57. A comparison to *E. coli* ClpX (EcClpX), presented in Fig. 1, shows that AtClpX has a 94 residues-long extension in its amino-terminus. This region of the protein is rich in hydroxylated and basic residues (22 and 19, respectively), contains only five acidic residues, and devoid of Gly and Pro in its first 37 residues (not shown). These characteristics are common in chloroplast-targeting sequences. The Walker A and B motives of the ATP-binding site are 100% conserved in these two proteins, and high degree of conservation is observed between and around these regions. Homology between the two proteins extends also to the carboxy-terminus, where two PDZ-like domains are found (see section 3.2.).

Comparison of AtClpX to related proteins shows that it resembles most ClpX from *Synechococcus*, and also those from other bacteria (Table I), with a similarity index higher than 50. Lower degree of homology is found between AtClpX and mitochondrial ClpX-like proteins deduced from yeast (Table I), human and mice sequences (not shown). It should be noted that EcClpX shares ca. 50% homology with another ATPase, HslU, which is found associated with the proteolytic subunit HslV in *E. coli*. However, the lower degree of homology between AtClpX and bacterial HslU (Table I) suggests that the *Arabidopsis* sequence is indeed a homologue of ClpX and not of HslU. The higher similarity to bacterial sequences, including *Synechococcus*, and the lower similarity to mitochondrial-targeted ClpX sequences, together with the characteristics of the putative transit peptide, support localization of AtClpX to the chloroplasts. Results of import experiments *in vitro* (not shown) also support this suggestion. However, immuno-localization is still needed for conclusive localization of AtClpX. These experiments are currently underway.

```
  95  SSQGDPPDLWQPPGDGVSVRVNGSSVNLGRGGGGGGSNPGGPGNGTGSNS
       :                 : :     |  |
   1  MTDKRKDGSGKLLYCSFCGKSQHEVRKLIAGPSVYICDECVDLCNDIIRE

 145  KEDCWGGSNLGSDFPTPKEICKGLNKFVIGQERAKKVLSVAVYNHYKRIY
                 |  |||:||   |: :||||| ||||| |||||||||:
  51  EIKEVAPHRERSALPTPHEIRNHLDDYVIGQEQAKKVLAVAVYNHYKRL-

 195  HESSQKRSAGETDSTAAKPADDDMVELEKSNILLMGPTGSGKTLLAKTLA
                              ||| |||||| ||||||||||| |||
 100  ----------------RNGDTSNGVELGKSNILLIGPTGSGKTLLAETLA
                                              A

 245  RFVNVPFVIADATTLTQAGYVGEDVESILYKLLTVADYNVAAAQQGIVYI
      | : ||| :|||||||:||||||||| |: |||   ||:|  || |||||
 134  RLLDVPFTMADATTLTEAGYVGEDVENIIQKLLQKCDYDVQKAQRGIVYI
                                                     B

 295  DEVDKITKKAESLYISRDVSGEGVQQALLKMLEGTIVNVPEKGARKHPRG
      ||:|||::| :   |:||||||||||||||::|||:  ||  | ||||
 184  DEIDKISRKSDNPSITRDVSGEGVQQALLKLIEGTVAAVPPQGGRKHPQQ

 345  DNIQIDTKDILFICGGAFVDIEKTISERRH-DSSIGFGAPVRANMRAGGV
      : :|:||  |||||||||  ::| || |    | |||||  |
 234  EFLQVDTSKILFICGGAFAGLDKVISHRVETGSGIGFGATV------KAK
                                 *     ****** *

 394  TTAAVASNLMETVETSDLIAYGLIPEFVGRFPVLVSLSALTENQLMQVLT
      :  |    |:  ||  ||| :||||||:|| ||: :|  |:|  |:|:|
 278  SDKASEGELLAQVEPEDLIKFGLIPEFIGRLPVVATLNELSEEALIQILK
                *                              *      *

 444  EPKNALGKQYKKMYQMNSVKLHFTESALRLIARKAITKNTGARGLRALLE
      |||||| |||  ::::: | | | : ||  ||:||: : |||||||  ::|
 328  EPKNALTKQYQALFNLEGVDLEFRDEALDAIAKKAMARKTGARGLRSIVE
                     * ****** *       *

 494  SILMDSMYEIPDEGTGSDMIEAVVVDEEAVEGEGR
        |:|:||::|      : :| ||:||  ::|: :
 378  AALLDTMYDLP----SMEDVEKVVIDESVIDGQSK
                          *      *
```

Fig. 1. Alignment of AtClpX and EcClpX: Amino acid sequences of ClpX from *Arabidopsis* and *E. coli* were aligned. Identical (|) and similar (:) residues are indicated. The Walker A and B motives of the ATP-binding sites are shaded. Single and double underlines indicate the first and second PDZ domains. Residues in PDZ-domains implicated in interaction with bound peptides are denoted (*).

***Table I.* Comparison between AtClpX and related proteins**

	Similarity index
Synechococcus ClpX	54.0
E. coli ClpX	52.1
Bacillus ClpX	51.6
Yeast ClpX	35.1
E. coli HslU	33.6

A pair-wise comparison between the mature portion of AtClpX and the indicated proteins was conducted using the Lipman-Pearson Protein Alignment module of the DNASTAR program. (Fully identical sequences have a similarity index of 100.0). The accession numbers of the sequences used are: AtClpX, AF036328; *Synechococcus* ClpX, U92039; *E. coli* ClpX, L18867; *Bacillus* ClpX, X95306; Yeast ClpX, Z36096; *E. coli* HslU, AU000467.

3.2. PDZ domains in AtClpX

The carboxy-terminus of AtClpX contains an interesting structural feature, two PDZ-like domains in tandem. PDZ domains occur in numerous eukaryotic signaling proteins, where they mediate protein-protein interactions (12). Recently it was reported that PDZ domains occur also in prokaryotic and plant proteases (13). Such domains in bacterial ClpX were shown to mediate binding specificity (14). Comparison of *Arabidopsis* and EcClpX shows similarities between the two proteins in these regions as well, including residues suggested to be involved in peptide binding (Fig. 1). Thus, the carboxy-terminus of AtClpX might be involved in determining substrate specificity and substrate binding.

The elucidation of the three-dimensional structure of a human PDZ domain revealed that the bound peptide is nested in a groove between the β–sheet B and the α-helix B stretches of the protein (15). When we compared the corresponding regions of ClpX sequences to the binding site of this PDZ domains we observed high homology, especially at the end of the pocket, where the carboxy-terminus of the bound peptide interacts with the PDZ domain (Fig. 2). The high similarity observed in totally different proteins suggests that the carboxy-terminus of ClpX is involved in substrate binding, through its two PDZ-like domains. Although it appears that the overall structure of such domains has been conserved, slight differences between these domains in different proteins may account for differential affinity to specific substrates. It should be noted that bacterial ClpA is expected to contain a single PDZ-like domain in its carboxy-terminus, and so is chloroplastic ClpC. It will be interesting to explore whether these domains in ClpC and ClpX display different substrate specificities.

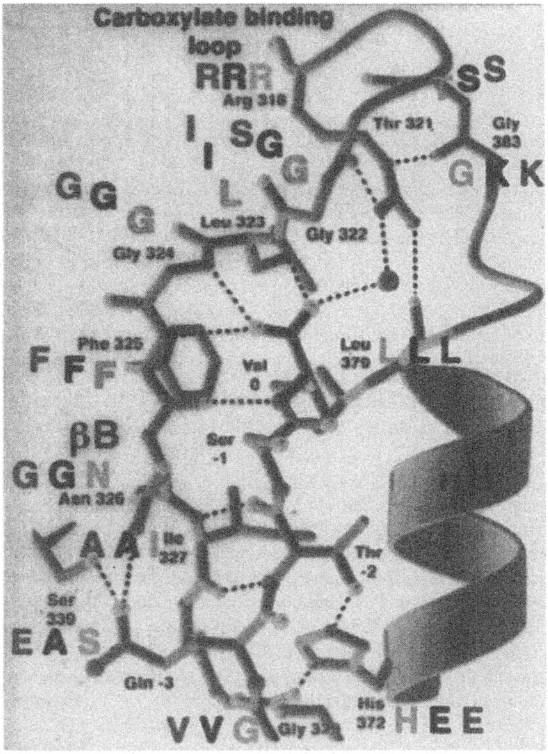

Fig. 2. Structure of the binding loop of a human PDZ domain and comparison to ClpX sequences. The structure of the binding loop of PDZ-3, the third domain of the brain synaptic PSD-95 protein, was adapted from Doyle *et al.* (15). Residues in PDZ-3 comprising the binding loop and interacting with the bound peptide are indicated in light letters, closer to the polypeptide skeleton. Residues from the corresponding regions in the first PDZ domain of EcClpX and AtClpX are indicated by dark and medium intensity letters, respectively.

3.3. Expression of AtClpC

To study the expression of AtClpX, we performed a RT-PCR analysis of RNA from *Arabidopsis* plants. ClpX transcript was found in leaves shoots ant roots, and smaller amounts were found also in flowers (Fig. 3). Changes in environmental conditions did not affect the expression: following exposure to higher (40°C) and lower (15°C) temperatures transcript levels remained unchanged. Exposure to higher or lower light intensities did not affect the transcript level either (Fig. 3). These findings are not surprising in light of the constitutive expression observed also for ClpC and ClpP (6, 8).

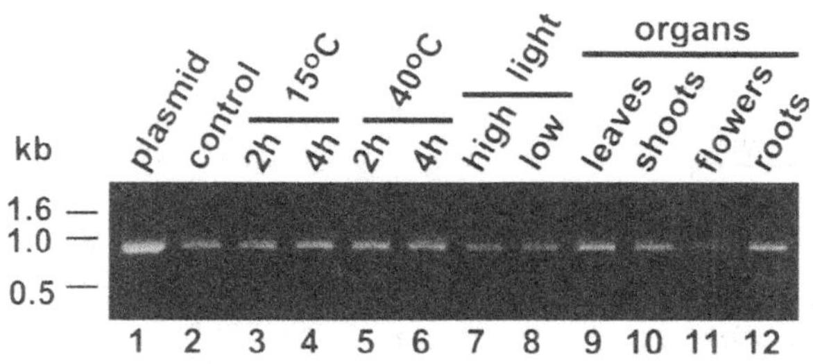

Fig. 3. Expression of AtClpX. Different organs from *Arabidopsis* plants, or samples from plants exposed to different temperatures or light intensities were harvested. RNA was extracted from the samples and subjected to RT-PCR analysis.

4. Conclusion

The isolation of a cDNA clone encoding a homologue of ClpX in *Arabidopsis* and its characteristics suggest the presence of a second regulatory subunit of Clp protease in chloroplasts, in addition to ClpC. However, the localization of ClpX to chloroplasts still needs to be confirmed. If it is indeed localized to the chloroplast, interaction with ClpP needs to be demonstrated. When these issues are settled, it will be important to explore substrate specificities of ClpX and ClpC, and the mechanisms underlying substrate recognition. Experiments dealing with these questions are currently underway in our lab.

5. References

1. Gray, J. C., Hird, S. M., and Dyer, T. A. (1990) *Plant Mol. Biol.* **15,** 947-950
2. Maurizi, M. R., Clark, W. P., Kim, S. H., and Gottesman, S. (1990) *J. Biol. Chem.* **265,** 12546-12552
3. Schaller, A., and Ryan, C. A. (1995) *Plant Physiol.* **108,** 1341
4. Gottesman, S., *et al.* (1990) *Proc. Natl. Acad. Sci. USA* **87,** 3513-3517
5. Moore, T., and Keegstra, K. (1993) *Plant Mol. Biol.* **21,** 525-537
6. Shanklin, J., Dewitt, N. D., and Flanagan, J. M. (1995) *Plant Cell* **7,** 1713-1722
7. Halperin, T., and Adam, Z. (1996) *Plant Mol. Biol.* **30,** 925-933
8. Ostersetzer, O., and Adam, Z. (1996) *Plant Mol. Biol.* **31,** 673-676
9. Adam, Z., Halperin, T., Itzhaki, H., Lindahl, M., and Ostersetzer, O. *The proteolytic machinery of chloroplasts: homologues of bacterial proteases.* in *XIth International Congress on Photosynthesis.* 1998. Budapest: Kluwer.
10. Gottesman, S., Wickner, S., and Maurizi, M. R. (1997) *Genes Dev.* **11,** 815-23
11. Nielsen, E., Akita, M., Davilaaponte, J., and Keegstra, K. (1997) *EMBO J.* **16,** 935-946
12. Saras, J., and Heldin, C. H. (1996) *Trends Biochem. Sci.* **21,** 455-458
13. Ponting, C. P. (1997) *Protein Sci.* **6,** 464-8
14. Levchenko, I., Smith, C. K., Walsh, N. P., Sauer, R. T., and Baker, T. A. (1997) *Cell* **91,** 939-47
15. Doyle, D. A., Lee, A., Lewis, J., Kim, E., Sheng, M., and MacKinnon, R. (1996) *Cell* **85,** 1067-76

LIGHT STRESS-ACTIVATED PROTEASES IN THE THYLAKOID LUMEN

IWONA ADAMSKA
Department of Biochemistry, Arrhenius Laboratories for Natural Sciences, StockholmUniversity, S-10691 Stockholm, Sweden

Introduction

Generally two types of proteolysis can be distinguished in the chloroplast: (i) protein processing which involves cleavage of signal sequences during protein import into organelles, and (ii) destructive proteolysis which leads to loss of particular proteins from various chloroplast compartments [1, 2, 3]. While the physiological substrates for the first type of proteolysis can be easily defined, the substrates degraded by the latter one are more complex and can be divided into three major classes.

One general class are non-functional proteins. Here belongs misfolded, truncated or mutant proteins which often contain abnormal amino acid composition. The accumulation of such non-functional proteins is especially enhanced under various stress conditions such as light or heat shock [4, 5]. It was shown by many groups that the D1-protein of photosystem II (PSII) reaction center is rapidly degraded under photoinhibitory conditions [6, 7]. Similar light stress-related turnover was observed for the D2- and CP43-proteins of PSII [8]. Also mistarget proteins undergo rapid degradation [5].

The second class of proteins directed for degradation are specific "timing proteins" which function is required for limited period of time. The good example for this type of proteolysis is the degradation of light stress proteins, ELIPs, after lowering of the light intensity [9, 10] or degradation of heat shock proteins after lowering of temperature [4]. Also long-term acclimation of plants from low to high intensity light involves the specific degradation of light-harvesting subunits in order to reduce the functional antenna size [11, 12].

A third target for destructive proteolysis are "proteins without partners" e.g. members of multi-protein complexes, where one subunit of the complex is missing or proteins without cofactors. It was demonstrated [5] showed that the absence of chloroplast-encoded large subunit of the ribulose-1,5-bisphosphate carboxylase the nuclear-encoded small subunit is normally imported into the plastid, but is rapidly degraded. The availability of prosthetic groups can also affect the stability of chloroplast proteins. When the supply of chlorophyll is limited, newly synthesized chlorophyll-binding proteins are targeted to the thylakoid membranes but fail to accumulate [13, 14].

Till the beginning of eightees nothing was known about proteolytic events in the chloroplast. At this time it was postulated that the chloroplast proteins, in order to be degraded, have to be exported out of the organelle. The vacuole was considered to be the place for the degradation of the chloroplast proteins. In 1984 it was shown for the first

J. H. Argyroudi-Akoyunoglou and H. Senger (eds.),
The Chloroplast: From Molecular Biology to Biotechnology, 265–270.

time that the isolated chloroplasts are able to degraded their own proteins [5]. During the following fourtheen years the attempt were made to identify and isolate some of the chloroplast proteases but till today only limited informations are available concerning proteolytic processes in the chloroplasts [5].
Here I present three proteolytic activities which are localized in the thylakoid lumen and activated under light stress conditions.

Identification and characterization of the thylakoid lumen proteases

Thylakoid lumen preparations were performed using mature pea leaves exposed to light stress for 3 hours (Fig. 1 A). The immunoblotting with antibodies directed against various proteins localized in the thylakoid membranes, thylakoid lumen or stroma

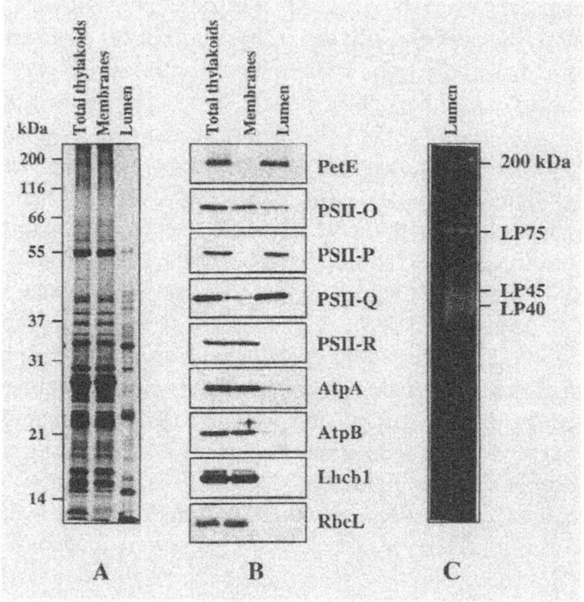

Figure 1. Protein composition of the thylakoid lumen. Mature pea leaves were exposed to light stress (2.000 $\mu mol\,m^{-2}s^{-1}$) for 3 hours and the thylakoid lumen was isolated by sonication of the purified thylakoid membranes. (A) Silver staining of the total thylakoids and thylakoids separated into membranes and lumenal fractions subjected to the SDS-PAGE; (B) Western blotting with antibodies directed against lumenal protein plastocyanin (petE), peripheral thylakoid membrane proteins, such as α- (AtpA) and β- (AtpB) subunits of ATP-synthase, 33 kDa (PSII-O), 23 kDa (PSII-P), 16 kDa (PSII-Q) and 10 kDa (PSII-R) proteins of oxygen-evolving complex, integral thylakoid light-harvesting chlorophyll a/b-binding protein of PSII (Lhcb1) and stroma-localized large subunit of ribulose-1,5-bisphosphate carboxylase (RbcL); (C) Assay of proteolytic activities using substrate-SDS-PAGE. In this assay the unspecific protease substrate gelatine (0.2%) was included into the gel. After separation of the proteins on the substrate-SDS-PAGE, the gel was incubated for 12-24 hours in the assay buffer to allow degradation of the substrate. After staining of the gel with the Coomassie blue white zones on the blue background indicate the position of the proteases. LP, luminal protease

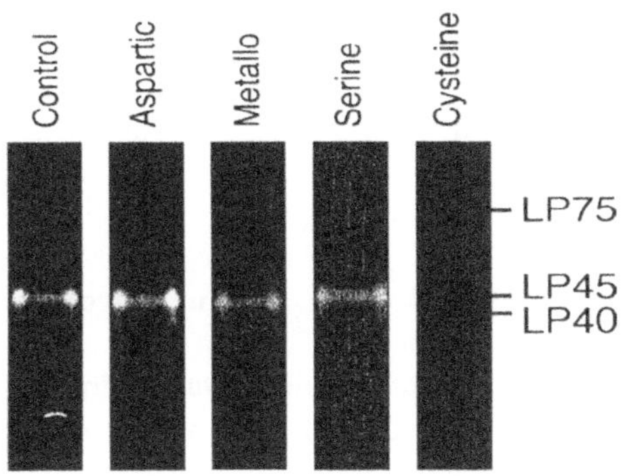

Figure 2. Classification of the lumenal proteases. Thylakoid lumen was isolated from pea leaves exposed to light stress and the proteolysis of chromogenic substrate azocoll was assayed in the absence (control) or in the presence of various protease inhibitors directed against aspartic-, metallo-, serine- or cysteine-classes of endopeptidases.

fractions confirmed that the lumenal preparations contained mainly soluble lumenal proteins. Plastocyanin as well as three major polypeptides of the oxygen-evolving complex of PSII, which are known to be peripherally attached to the thylakoid membranes at the lumenal side were found in these preparations (Fig. 1 B). Only traces of other extrinsic thylakoid proteins, such as α- and β-subunits of the ATP-synthase complex were assayed. The lumenal fractions contain neither stromal nor intrinsic thylakoid membrane proteins.

Using thylakoid lumen fractions for protease activity assays I could demonstrate that at least three proteolytic activities are present and active in this compartment under light stress conditions. These activities with the apparent molecular masses of 75, 45 and 40 kDa (designed here as lumen proteases: LP75, LP45 and LP40, respectively) recognized and degraded unspecific substrates, such as gelatine (Fig. 1 C), acid-pretreated hemoglobine or collagen-derivates commertially named azocoll and hide powder azure (not shown).

According to their catalytic centers all endopeptidases of animal and plant cells are divided into 4 types: serine- (containing Ser, His and Asp as catalytic triad residues), cysteine- (containing Cys and His), aspartic- (containing Asp, Ser/Thr, Glu) and metallo-endopeptidases (containing metal binding motif His-Glu-X-X-His). Recently, other groups of proteases have been discovered which catalytic mechanism still remain to be elucidated. Using set of inhibitors directed to various group of proteases I could classified the LP45 and LP40 to be cysteine-type of endopeptidases (Fig. 2). The activities of these proteases were completely abolished in the presence of E64, a cysteine protease inhibitor of microbial origin, and leupeptin which inhibits both, serine and cysteine-types of endopeptidases. The activity of LP75 was not influenced by any of the inhibitors tested suggesting that this proteolytic activity has an unknown catalytic mechanism.

Kinetics of activation of lumenal proteases by light stress were measured (not shown). The data revealed that the LP45 and LP40 were constitutively active under light stress conditions and their activity was reduced only after lowering of the light intensity. The LP75 was active only transiently after 3 hours exposure of leaves to light stress and its activity decreased thereafter.

Working modell for regulation of lumenal protease activity

Studies with inhibitors of cytoplasmic (cycloheximide) or chloroplast protein biosynthesis (chloramphenicol) indicated that all three investigated proteases were present in the thylakoid lumen under low light conditions in their non-active form (not shown). Illumination of the plants or isolated intact chloroplasts activated these enzymes via still unknown mechanism. Considering a very narrow pH range between pH 5-6 where the LP75, LP45 and LP40 reached the maximal activity and the fact that an increase of the transthylakoidal delta-pH was crutial for the activation of two of these proteases, the LP45 and LP40 (not shown) at least two step activation mechanism of these proteases is considered.

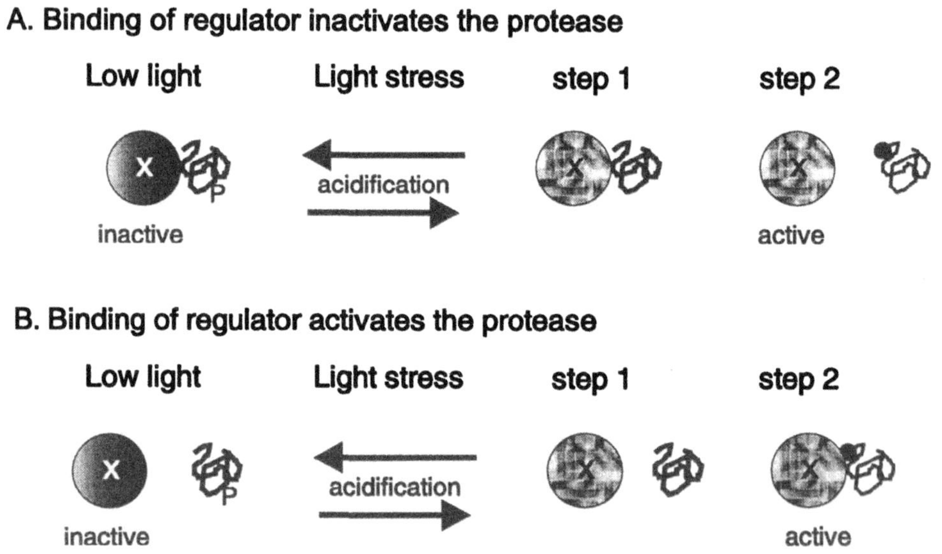

Figure 3. Model for activation mechanism of the lumenal proteases.In the first step the acidification of the thylakoid lumen under light stress conditions leads to the modification of putative regulatory component which might bind to the inactive protease or alternatively might be released from the inactive protease/regulator complex. In the second step the changes in the interaction between protease and its regulator influence the activity of the protease.

The nature of proposed protease regulator is still not clear. Such a regulator can be represented by a protease inhibitor which under moderate light conditions binds reversibly to the protease and inhibits its activity. Another possibility is that the protease regulator causes covalent modification of the protease which changes its activity.

This working model for regulation of proteolytic activity in the thylakoid lumen under light stress conditions is supported by the *in vitro* studies. Lumenal proteases which were activated by illumination of isolated intact chloroplast with high intensity light can be kept in their active state *in vitro* at low pH. Increasing the pH value to pH 7.0 reversibly inactivated the LP45 and LP40. In contrast the inactive form of LP45 and LP40 present in the lumen isolated from control chloroplasts kept at ambient light regime or from chloroplasts transferred after illumination to recovery condition could not be activated by the pH shift and needed a short exposure to light stress for their activation.

Acknowledgement

This work was supported by the Deutsche Forschungsgemeinschaft (Ad 94/4-1; Ad 94/4-2), Bonn, Germany

References

1. Cline, K., Henry, R., Li, C. and Yuan, J. (1993) Multiple pathways for protein transport into or across the thylakoid membranes, *EMBO J* **12**, 4105-4114
2. Rechsteiner, M.,and Rogers, S.W. (1996) PEST sequences and regulation by proteolysis , *TIBS* **21**, 267-271.
3. Pohlschroder, M., Prinz, W.A., Hartmann, E., Beckwith, J. (1997) Protein translocation in the three domains of life:variations on a theme, *Cell* **91**, 563-566
4. Gottesman, S. (1996) Proteases and their targets in *Escherichia coli*, Annu. *Rev. Genet.* **30**, 465-506.
5. Adam, Z. (1996) Protein stability and degradation in chloroplast, *Plant Mol. Biol.* **32**, 773-783.
6. Prasil, O., Adir, N., and Ohad, I. (1992) Dynamics of photosystem II: Mechanism of photoinhibition and recovery process, *Top. Photosynth.* **11**, 295-348.
7. Aro, E.M., Virgin, I., and Andersson, B. (1993) Photoinhibition of photosystem II. Inactivation, protein damage and turnover, *Biochim. Biophys. Acta* **1143**, 113-134.
8. Christopher, D.A., and Mullet, J.E. (1994) Separate photosensory pathways co-regulate blue light/ultraviolet-A-activated psbD-psbC transcription and

light-induced D2 and CP43 degradation in barley (*Hordeum vulgare*), *Plant Physiol.* **104**, 1119-1129.

9. Adamska, I., Kloppstech, K., and Ohad, I. (1993) Early light-inducible protein in pea is stable during light stress but is degraded during recovery at low light intensity, *J. Biol. Chem.* **268**, 5438-5444.
10. Adamska, I., Lindahl, M., Roobol-Boza, M., and Andersson, B. (1996) Degradation of the light stress protein is mediated by an ATP-independent serine-type protease under low light conditions, *Eur. J. Biochem.* **236**, 591-599.
11. Lindahl, M., Yang, D.H., and Andersson, B. (1995) Regulatory proteolysis of the major light-harvesting chlorophyll a/b protein of photosystem II by a light-induced membrane-associated enzymic system, *Eur. J. Biochem.* **231**, 503-509.
12. Walters, R.G., and Horton, P. (1995) Acclimation of *Arabidopsis thaliana* to the light environment:regulation of chloroplast composition, *Planta* **197**, 475-481.
13. Bennett, J. (1981) Biosynthesis of the light-harvesting chlorophyll *a/b*-protein. Polypeptide turnover in darkness, *Eur. J. Biochem.* **118**, 61-70.
14. Kuttkat, A., Edhofer, I., Eichacker, L., and Paulsen, H. (1997) Light-harvesting chlorophyll *a/b*-binding protein stably inserts into etioplast membranes supplemented with Zn-pheophytin a/b, *J. Biol. Chem.* **272**, 20451-20455.

THE CARBOXYL-TERMINAL PROCESSING OF PRECURSOR D1 PROTEIN OF PHOTOSYSTEM II REACTION CENTER BY A NUCLEAR-ENCODED PROTEASE (CtpA)

YUMIKO YAMAMOTO AND KIMIYUKI SATOH
Department of Biology, Faculty of Science, Okayama University
3-1-1 Tsushima-naka, Okayama 700-8530, Japan

1. Introduction

The major proteins involved in the primary processes of photosynthesis are encoded by genes located in the chloroplast genome, and the expression of this gene information is regulated by nuclear-encoded factors often at the later stages, *e.g.*, the translational and post-translational levels, in response to the development and the varying environmental conditions. The carboxyl-terminal (C-terminal) processing of the precursor D1 protein (pD1) is a typical example of post-translational regulation.

The D1 subunit (D1) constitutes the core part of the photosystem II reaction center (PSII RC), and provides the binding sites for the pigments and redox co-factors engaged in the primary photochemistry [1]. Despite its importance in the structure and function of PSII, the D1 has an extraordinarily high rate of turnover. In this dynamic process, D1 is translated by thylakoid-bound ribosomes on the stromal surface as a precursor with C-terminal extension, which is co-translationally translocated into the thylakoidal lumen and cleaved off by a nuclear-encoded processing protease (CtpA) (Fig. 1). The failure in the removal of this extension by the action of protease blocks the assembly of the Mn-cluster in the PSII, allowing the primary photochemistry to proceed normally, as shown by studies on LF-1 mutant of *Scenedesmus obliquus* [2, 3] and site-directed mutants of *Synechocystis* sp. PCC 6803 [4, 5]. However, the role of the C-terminal extension, as well as the functional meaning of the processing, if any, has not been elucidated [4].

In this article, we deal with the C-terminal processing of pD1 as an example of nuclear-regulation of the chloroplast gene expression at the post-translational level, and briefly summarize our recent results, together with those of other relevant works.

2. Substrate — Precursor D1 Protein

2.1. C-TERMINAL EXTENSION AND CLEAVAGE SITE

To date, more than 50 *psbA* genes coding for pD1 have been sequenced from a wide variety of organisms, ranging from cyanobacteria to higher plants [6]. The proteolytic cleavage of pD1 occurs on the C-side of Ala-344, resulting in the removal of 8-16 amino acid residues [4]. Interestingly, the sequence upstream of the cleavage site is highly conserved throughout all putative D1, whereas the C-terminal extension is variable in terms of both chain length and sequence. Ala-345 after the cleavage site (+1 position), however, is highly conserved among different species, with exceptions of Ser in some cyanobacteria and green algae, and Val in *Chlorella vulgaris* C-27. The C-terminal extension is absent in the deduced sequence of D1 in *Euglena gracilis*.

J. H. Argyroudi-Akoyunoglou and H. Senger (eds.),
The Chloroplast: From Molecular Biology to Biotechnology, 271–276.

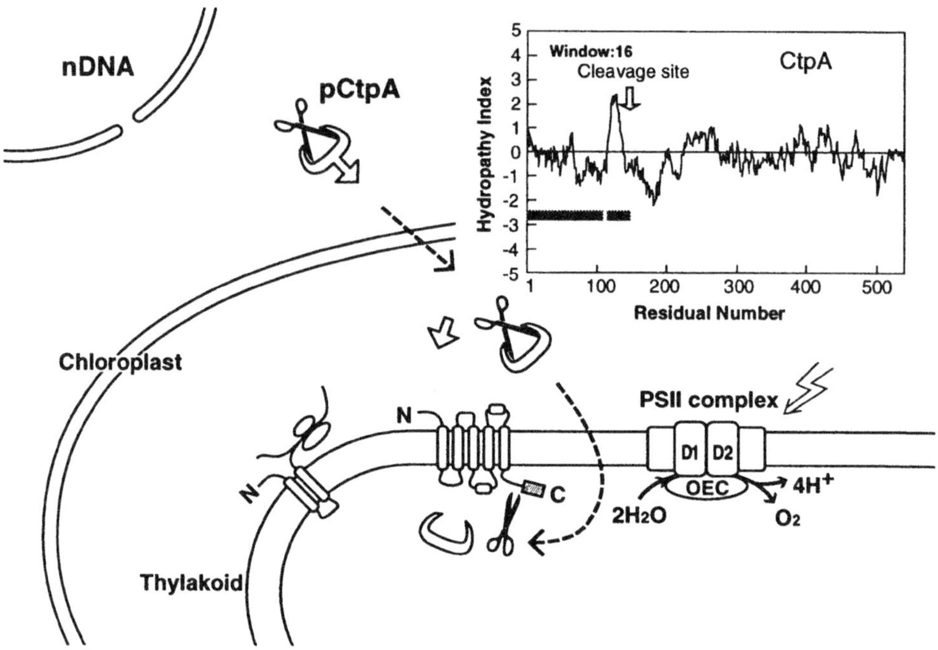

Figure 1. A model for the C-terminal processing of pD1 by CtpA

2.2. *IN VITRO* SUBSTRATE FOR ANALYSIS

In order to analyze the molecular interaction in the C-terminal processing of pD1, we utilized the following three classes of substrate: (i) the oligopeptides corresponding to the C-terminal sequence of pD1 [7]; (ii) the *in vitro* translated full-length pD1 [8]; and (iii) the pD1 in the PSII complex on the thylakoid membrane from the LF-1 mutant of *Scenedesmus*, which is deficient in the processing activity [9].

2.3. RECOGNITION SIGNAL ON SUBSTRATE

The recognition signal around the cleavage site was investigated by analyzing the effectiveness of C-terminal synthetic oligopeptides with amino acid substitution(s), both as substrate and as competitive inhibitor [7, 9]. This *in vitro* analysis demonstrated the following: (i) the negative charge of Asp-342 at the -3 position plays a crucial role in the cleavage; (ii) the Leu-343 at the -2 position is essential for the enzyme-substrate interaction; and (iii) the amino residue at the +1 position considerably influences the rate of cleavage (substitution of Val or Gly at this position served to decrease oligopeptide effectiveness as both substrate and competitive inhibitor, and substitution of Pro resulted in tolerance of the oligopeptide to proteolysis). On the other hand, the mutation at the 345th position was introduced into *Chlamydomonas reinhardtii* [10] and *Synechocystis* sp. PCC 6803 [4]. However, the influence of substitutions on the processing rate *in vivo* has not been established [10], except in the case of the Ser345Pro mutant in which unprocessed D1 was accumulated [4]. A mixed culture growth experiment, on the other hand, demonstrated the disadvantages of Val- and Gly-substitutions in the viability under high light conditions [submitted].

3. Enzyme — C-Terminal Processing Protease for pD1 (CtpA)

3.1. PURIFICATION

The protease involved in the C-terminal processing of pD1 has been extracted from thylakoids of spinach, *S. obliquus* and pea by Triton X-100 at low concentration or by sonication, together with thylakoid-lumenal extrinsic proteins, such as PsbO, PsbP and PsbQ [8, 11]. The purification has been achieved for spinach [8] and *Scenedesmus* [12], by a series of chromatographies. The purified enzyme is active in the monomeric state without added co-factors. The molecular mass of the protease was estimated from gel-filtration chromatography to be 40-45 kDa [8, 12].

3.2. PROTEIN/GENE SEQUENCING AND HOMOLOGY

Shestakov *et al.* [5] identified a *ctpA* gene by means of a genetic complementation analysis for a PSII-deficient mutant of *Synechocystis*. The homologous gene was sequenced based on the partial amino acid sequence of the purified C-terminal processing protease for pD1 in spinach [13] and *Scenedesmus* [12].

The deduced amino acid sequence of CtpA indicated that the protease is synthesized as a precursor with N-terminal transit peptide, as will be discussed in section 3.4.1. The mature CtpA exhibits significant homology with both the C-terminal processing protease of *Bartonella bacilliformis* and the tail-specific protease (Tsp; known to be identical with the C-terminal processing protease for penicillin-binding protein 3 (Prc)) of *E. coli* [14]. As shown in Fig. 2, these proteases share an extremely high similarity in two domains, although the over-all homology is very low.

3.3. OVER-EXPRESSION IN *E. coli*

We have succeeded in over-expressing the mature CtpA of spinach in the BL21(DE3) strain of *E. coli* by using the expression vector of pET-15b(+), which carries an N-terminal His-tag, six consecutive His residues, and a thrombin-sensitive cleavage site to permit easier detection and purification of the expressed protein [in preparation]. The induction by IPTG at 37°C resulted in the expression of proteins within insoluble inclusion bodies. However, lowering the temperature of induction to 20°C led some portions of the protein to be soluble and active for specific C-terminal processing of pD1, suggesting that these portions are precisely folded. It is intriguing that CtpA in the functionally active state could be expressed in *E. coli* without affecting the growth of *E. coli*, although there is a significant sequence homology between Tsp and CtpA. The CtpA can be rapidly purified in homogeneity from a soluble fraction of *E. coli* grown at 20°C by Ni-chelating chromatography, utilizing the His-tag at the N-terminus.

3.4. CHARACTERIZATION OF THE ENZYME

3.4.1. *Localization in Chloroplast*

In order to investigate the localization of CtpA in the chloroplast, spinach thylakoids were subjected to trypsin-treatment before or after extraction of CtpA with Triton X-100, and the processing activity in the extract was assayed, as shown in Fig. 3. As can be seen, trypsin-treatment after Triton X-100 extraction completely destroyed the activity for C-terminal processing of pD1 (lane 4), whereas that before extraction had no effect on the activity (lane 3), confirming that the CtpA is located in the thylakoidal lumen, as suggested by Oelmüller *et al.* [15]. On the other hand, the analysis of the putative amino acid sequence of eukaryotic CtpA revealed that the precursor CtpA possesses a bipartite N-terminal pre-sequence characteristic for proteins transported from the cytosol to the thylakoidal lumen [12, 13]. The former portion of the spinach pre-

```
Spinacia oleracea      LSEENRIFLE AWRTIDRAYV DKTFN--GQS WFRYREN--- -----ALRNE  40
Scenedesmus obliquus   VTSEQLLFLE AWRAVDRAYV DKSFN--GQS WFKLRET--- -----YLKKE
Synechocystis 6803     FTEEQKLLLQ SWRLVNQSYL DETFN--HQN WWLLREK--- -----YVKR-
E. coli (Tsp)          PWPKNEAELN ALWDSKVKFD ELSLKLTGKT DKEIRETLTR RYKFAIRRL-

                       PMNSREETYT AIRKMVATLN DPFTRFLEPE KLKSLRSGTQ SSLTGVGISI  90
                       PMDRRAQTYD AIRKLLAVLD DPFTRFLEPS RLAALRRGTA GSVTGVGLEI
                       PLRNREETYT AIEEMLATLD EPFTRLLRPE QYGNLQVTTT GELSGVGLQI
                       AQTNSEDVFS LAMTAFAREI DPHTNYLSPR NTEQFNTEMS LSLEGIGAVL

                       GPTAVDQSST GLVVISATPG APASRAGIL- PGDVILAIDD ASTDKMGIYE 139
                       --TYDGGSGK DVVVLTPAPG GPAEKAGAR- AGDVIVTVDG TAVKGLSLYD
                       N---INPETN QLEIMAPLAG SPAEEAGLQ- PHDQILAIDG VDTQTLSLDE
                       QMDDDYTVIN SMVAGGPAAK SKAISVGDKI VGVGQTGKPM VDVIGWRLDD

                       AANILQGPDG SSVDLTI--- CSRDEIKHVV LKRERITLSP VKSRLCEMPG 186
                       VSDLLQGEAD SQVEVVLHAP GAPSNTRTLQ LTRQKVTINP VTFTTCSNVA
                       AAARMRGPKN TKVSLEILSA GT-EVPQEFT LTRQLISLSP V---------
                       VVALIKGPKG SKVRQEILPA GKGTKTRTVT LTRERIRL-- ------EDRA

                       SAKDAP---- PKVGYIKLTS FTENASDAVK EAIETLRSNN VNAFVLDLRD 232
                       AAALPPGAAK QQLGYVRLAT FNSNTTAAAQ QAFTELSKQG VAGLVLDIRN
                       AAQLDDSRPG QSVGYIRLSQ FSANAYKEVA HALHQLEEQG ADGYILDLRN
                       VKMSVKTVGK EKVGVLDIPG FYVGLTDDVK VQLQKLEKQN VSSVIIDLRS
                                                                    A

                       NSGGLFPEGI EIAKIWLNKG VIVYICDSRG -VRDIYDVEG SSAVAGSEPL 281
                       NGGGLFPAGV NVARMLVDRG DLVLIADSQG -IRDIYSADG NS-IDSATPL
                       NPGGLLQAGI DIARLWLPES TIVYTVNRQG -TQESFTANG EAATD--RPL
                       NGGALTEAV SLSGLFIPAG PIVQVRDNNG KVREDSDTDG QVFYKG--PL
                                                                    B

                       VVLVNKGTAS ASEILAGALK DNKRAVVFGE PTYGKGKIQS VFELSDGSGL 331
                       VVLVNRGTAS ASEVLAGALK DSKRGLIAGE RTFGKGLIQT VVDLSDGSGV
                       VVLVNQGTAS ASEILAGALQ DNQRATLVGE KTFGKGLIQS LFELSDGAGI
                       VVLVDRFSAS ASEIFAAAMQ DYGRALVVGE PTFGKGTVQQ YRSLN---RI

                       AVTVARYETP AHTDIDKVGI KPDHPL-PAS FPKDENDFCT CVQDPSSTCY 381
                       AVTVARYQTP AGVDINKIGV SPDVQLDPEV LPTDLEGVCR VLGSDAAPRL
                       AVTVAKYETP QHHDIHKLGI MPDEVVEQPL ISFAEITSPA DVQYQAALDL
                       YDQMLRPEWP A---LGSVQY TIQKFYRVNG GSTQRKGVTP DIIMPTGNEE

                       LNGVQLFSR*                                             389
                       FG*
                       LTGGVAIAHK SSSIPAMATA HKPN*
                       TETGEKFEDN ALPWDSIDAA TYVKSGDLTA FEPELLKEHN ARIAKDPEFQ
```

Figure 2. Alignment of the amino acid sequence of mature CtpA from different organisms and homologous Tsp from *E. coli*. Asterisks on the upper part show amino acids which are conserved in CtpA from different origin. Completely conserved residues are indicated by boxes. Underlines, A and B, indicate highly conserved regions which are thought to be involved in the catalytic mechanism.

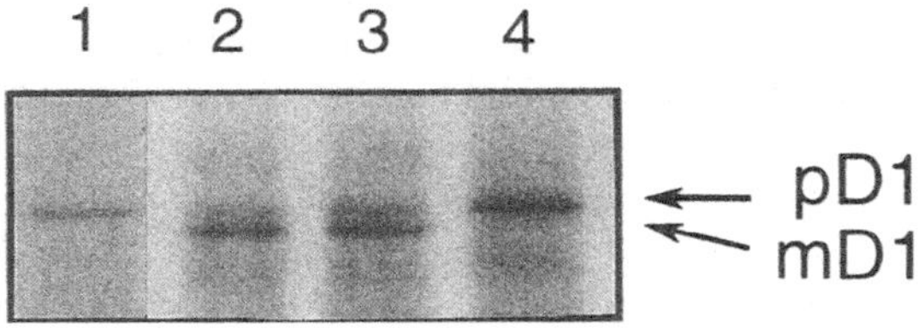

Figure 3. Trypsin-treatment of thylakoid membranes. Spinach thylakoids (1 mg chl/ml) were subjected to trypsin-treatment (10 μg/ml, at 27°C for 30 min) before (lane 3) or after (lane 4) Triton X-100 extraction (at 0.25 % (w/v), at 4°C for 30 min). Trypsin-treatment was terminated by incubation with 50 μg/ml trypsin inhibitor at 27°C for 2 min. In lane 2, thylakoids were extracted by Triton X-100 without trypsin treatment. Neither Triton X-100 extraction nor trypsin-treatment was carried out in lane 1. After these treatments, thylakoids were centrifuged at 188,000 xg and the processing activity in the resultant supernatant was assayed using *in vitro* translated pD1 as the substrate.

sequence which contains 115 amino acids (39 amino acids in the case of *S. obliquus*) is overall basic and enriched in hydroxyl residues, which are typically observed in well known stromal targeting signals. The latter portion, consisted of approximately 35 amino acids, exhibits features common to many known lumenal targeting signals, *i.e.*, a positively charged N-terminal region, a significantly hydrophobic stretch, and a relatively conserved C-terminal motif of AXA (SXS in the case of spinach). In prokaryotic cyanobacteria that are believed to be predecessors of chloroplasts, CtpA harbors an undivided pre-sequence for transport across the thylakoid membrane, reflecting the cyanobacterial intracellular compartment [5]. More recently, Karnauchov *et al.* [16] demonstrated that the CtpA is efficiently transported into the thylakoid membrane via both ΔpH-dependent and Sec-dependent pathways.

3.4.2. *Enzymatic Properties*

The proteolytic activity for the cleavage of synthetic oligopeptides, as well as that for cleavage of the *in vitro* translated full-length pD1, exhibits a drastic pH dependency with an optimum pH of 7.5-8.0. The activity is quite low in the acidic region of pH 5.0-6.0, which corresponds to the pH region of the lumenal space in illuminated thylakoids, where the processing protease is expected to be highly active. In contrast, the optimum pH for the processing of pD1 embedded in the thylakoid membrane was approximately 6.0. This discrepancy may suggest the involvement of some additional factors in the enzyme-substrate interaction for the substrate embedded in the thylakoid membrane. The estimated high K_m value of about 0.3 mM for the reaction using synthetic oligopeptides [7] may also be related to this phenomenon. CtpA is resistant to any of the known protease inhibitors tested, which include serine, cysteine, aspartic and metal protease inhibitors.

3.4.3. *Catalytic Center*

By comparison with Tsp, the active residues of which have been analyzed, it is proposed that CtpA could be classified as a novel type of serine protease with Ser/Lys (hydroxyl/amine) dyad as its catalytic center, which employs Ser and Lys residues as a nucleophile and a general base in its catalytic mechanism, respectively [14]. The thylakoid processing protease (TPP) which has been identified in *Arabidopsis thaliana*, seems to belong to this group [17]. The catalytic centers with hydroxyl/amine dyad are found predominantly in proteases from prokaryotes sources, such as β-lactamase, LexA repressor and leader peptidase. This fact convinces us that the *ctpA* gene has been transferred into the nucleus from chloroplast of endosymbiotic origin.

4. Acknowledgments

This work was supported by a Grant-in-Aid for Scientific Research (B) (No. 09440268) from the Ministry of Education, Science, Sports and Culture of Japan.

5. References

1. Nanba, O. and Satoh, K. (1987) Isolation of a photosystem II reaction center consisting of D-1 and D-2 polypeptides and cytochrome *b*-559, *Proc. Natl. Acad. Sci. USA* **84**, 109-112.
2. Metz, J. and Bishop, N.I. (1980) Identification of a chloroplast membrane polypeptide associated with the oxidizing side of photosystem II by the use of select low-fluorescent mutants of *Scenedesmus*, *Biochem. Biophys. Res. Commun.* **94**, 560-566.
3. Rutherford, A.W., Seibert, M. and Metz, J.G. (1988) Characterisation of the low-fluorescent (LF1) mutant of *Scenedesmus* by EPR, *Biochim. Biophys. Acta* **932**, 171-176.
4. Nixon, P.J., Trost, J.T. and Diner, B.A. (1992) Role of the carboxy terminus of polypeptide D1 in the assembly of a functional water-oxidizing manganese cluster in photosystem II of the cyanobacterium *Synechocystis* sp. PCC 6803: Assembly requires a free carboxyl group at C-terminal position 344, *Biochemistry* **31**, 10859-10871.
5. Shestakov, S.V., Anbudurai, P.R., Stanbekova, G.E., Gadzhiev, A., Lind, L.K. and Pakrasi, H.B. (1994) Molecular cloning and characterization of the *ctpA* gene encoding a carboxyl-terminal processing protease, *J. Biol. Chem.* **269**, 19354-19359.
6. Svensson, B., Vass, I. and Styring, S. (1991) Sequence analysis of the D1 and D2 reaction center proteins of photosystem II, *Z. Naturforsch* **46c**, 765-776.
7. Taguchi, F., Yamamoto, Y. and Satoh, K. (1995) Recognition of the structure around the site of cleavage by the carboxyl-terminal processing protease for D1 precursor protein of the photosystem II reaction center, *J. Biol. Chem.* **270**, 10711-10716.
8. Fujita, S., Inagaki, N., Yamamoto, Y., Taguchi, F., Matsumoto, A. and Satoh, K. (1995) Identification of the carboxyl-terminal processing protease for the D1 precursor protein of the photosystem II reaction center, *Plant Cell Physiol.* **36**, 1169-1177.
9. Yamamoto, Y. and Satoh, K. (1998) Competitive inhibition analysis of the enzyme-substrate interaction in the carboxy-terminal processing of the precursor D1 protein of photosystem II reaction center using substituted oligopeptides, *FEBS Lett.* **430**, 261-265.
10. Takahashi, Y., Utsumi, K., Yamamoto, Y., Hatano, A. and Satoh, K. (1996) Genetic engineering of the processing site of D1 precursor protein of photosystem II reaction center in *Chlamydomonas reinhardtii*, *Plant Cell Physiol.* **37**, 161-168.
11. Bowyer, J.R., Packer, J.C.L., McCormack, B.A., Whitelegge, J.P., Robinson, C. and Taylor, M.A. (1992) Carboxyl-terminal processing of the D1 protein and photoactivation of water-splitting in photosystem II, *J. Biol. Chem.* **267**, 5424-5433.
12. Trost, J.T., Chisholm, D.A., Jordan, D.B. and Diner, B.A. (1997) The D1 C-terminal processing protease of photosystem II from *Scenedesmus obliquus*, *J. Biol. Chem.* **272**, 20348-20356.
13. Inagaki, N., Yamamoto, Y., Mori, H. and Satoh, K. (1996) Carboxyl-terminal processing protease for the D1 precursor protein: cloning and sequencing of the spinach cDNA, *Plant Mol. Biol.* **30**, 39-50.
14. Paetzel, M. and Dalbey, R.E. (1997) Catalytic hydroxyl/amine dyads within serine proteases, *Trends biochem. Sci.* **22**, 28-31.
15. Oelmüller, R., Herrmann, R.G. and Pakrasi, H.B. (1996) Molecular studies of CtpA, the Carboxyl-terminal processing protease for the D1 protein of the photosystem II reaction center in higher plants, *J. Biol. Chem.* **271**, 21848-21852.
16. Karnauchov, I., Herrmann, R.G., Pakrasi, H.B. and Klösgen, R.B. (1997) Transport of CtpA protein from the cyanobacterium *Synechocystis* 6803 across the thylakoid membrane in chloroplasts, *Eur. J. Biochem.* **249**, 497-504.
17. Chaal, B.K., Mould, R.M., Barbrook, A.C., Gray, J.C. and Howe, C.J. (1998) Characterization of a cDNA encoding the thylakoidal processing peptidase from *Arabidopsis thaliana*, *J. Biol. Chem.* **273**, 689-692.

PROTEOLYTIC MECHANISM IN LHCII STABILIZATION

L.A. TZIVELEKA and J.H. ARGYROUDI-AKOYUNOGLOU
Institute of Biology, NCSR Demokritos
Aghia Paraskevi, Attiki, Greece

1. Introduction

It is now generally accepted that the complicated mechanism regulating Photosynthetic unit (PSU) synthesis/assembly/turnover, in addition to the well known control at the transcriptional and translational level, also involves stringent control at the post-translational level. This became evident to us [1] when working with etiolated plants exposed either to Intermittent light-dark cycles (ImL) (which form only 1/7th of the Chl in plants exposed to continuous light [2]) or first to continuous light and then transferred to the dark. It was then found that (a) the LHCII, which is transcibed and translated in ImL plants [3], can not be stabilized in their thylakoids [4], unless the dark phase in the light-dark cycles is shortened [5]; (b) the LHCII preaccumulated in plants exposed to continuous light (CL) is degraded upon transfer of plants to the dark and its Chla is used for the assembly and stabilization of new PSU cores only when preexposure to CL is brief [6]; (c) new LHCII is not stabilized in plants transferred to CL following prolonged preexposure to ImL and no new Chl is formed [7] although the plants are deficient in Chl and LHCII. These results pointed to the involvement on one hand of Chl in stabilizing the newly formed Reaction center and LHC proteins, but also to the involvement of a proteolytic system that could degrade LHC proteins when not rescued by Chl binding. Today an astonishingly large number of chloroplast proteases (stromal, lumenal, thylakoidal) are being characterized and their role in the regulation of PSU assembly and turnover is questioned in many labs. Here we describe our recent findings on the thylakoid-bound protease involved in LHCII stabilization, its developmental regulation and the attempts made for its isolation.

2. Methods

Etiolated *Phaseolus vulgaris* plants were grown on perlite in the dark (22°C, 80% humidity); leaves attached to one cotyledon were exposed to ImL or CL. Thylakoids were isolated from intact plastids, separated by percoll gradient centrifugation [8] or from 80% intact plastids, separated by differential centrifugation [6]; they were used prior to or after TX-100 solubilization [9] and their proteolytic activity was assessed by immunodetection on Western blots [10, 11] of the remaining endogenous, or

J. H. Argyroudi-Akoyunoglou and H. Senger (eds.),
The Chloroplast: From Molecular Biology to Biotechnology, 277–282.

exogenously added purified LHCII [12], following incubation at 37°C. Assay mixtures had 1 μg/μl thylakoid protein (sample protein + LHCII) in 40 mM Tris-HCl, pH 8.6 and a sample protein to LHCII ratio of 5 [13]. Whenever TX-100 was present in assays, it was at 0.5%. NaBr washing was as in [13]. Proteolytic activity against exogenously added azocoll was assessed as in [14], Chl was measured as in [15], protein as in [16] and SDS-PAGE was as in [17]. NaBr-wash proteins were analyzed on 7% SDS gel [18], containing exogenous LHCII (0.2 mg/ml) and 10 μg BSA /ml.

3. Results

3.1. DEVELOPMENTAL CONTROL OF THE THYLAKOID-BOUND PROTEASE

Chloroplasts possess proteolytic activity against LHCII which is developmentaly regulated. It is low in 6 day etiolated plants, but increases with prolonged etiolation up to about 10 days from sowing, declining thereafter. The activity is specifically directed against POR and LHCII, and localized in the membrane fraction. This activity is low in the stroma, where it remains constant during development; in contrast in the mebrane fraction the activity increases gradually during chloroplast development. The protease is light activated: pretreatment of etiolated plants with a 2-min light pulse or exposure to ImL enhances drastically the activity (see Fig. 1A). In contrast, exposure of plants to CL results in a drastic decline of activity, following an initial rise for a few hours (see Fig. 1B). The decline is coincident with Chl accumulation. The latter can be alleviated in samples solubilized with TX-100, a finding which points to the shielding effect of Chl on the LHCII substrate. Thus, to detect proteolytic activity against LHCII in isolated thylakoids from mature green plants solubilization in TX-100 is required. This is not necessary in the early stages of greening, where similar activity is detected before or after TX-100 solubilization [13].

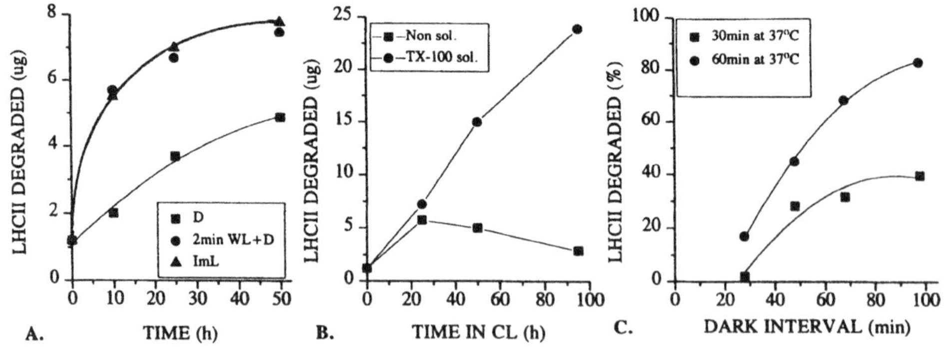

Figure 1. Developmental regulation of the thylakoid-bound proteolytic activity in 6 day etiolated plants (A) prior to (■) or after exposure to a 2 min white light flash (●), to ImL (2 min Light-98 min Dark in cycles) (▲); (B) to CL prior to (■) or after solubilization in TX-100 (●); and (C) to 20 Light-Dark cycles with varying dark phase duration.

These results, therefore, agree with those concerning LHCII stabilization during development. High proteolytic activity in ImL plants, which actually is increased as ImL exposure is prolonged, explains the insufficient LHCII stabilization in thylakoids of these plants as well as their inability to stabilize new LHCII or Chl when transferred to CL following prolonged exposure to ImL. Rescuing of LHCII by the Chl accumulated in CL plants on the other hand explains how plants cope with this high proteolytic activity. The question that remained unanswered, therefore, was why ImL plants with long dark phase duration in the light-dark cycles do not accumulate LHCII, in contrast to those exposed to cycles with short dark phase duration [5]. We thus assessed the proteolytic activity in primary thylakoids isolated from etiolated plants exposed to cycles with varying dark phase duration (2 min light every 28, 48, 68 or 98 min in cycles). We found that in plants exposed to the same number of such cycles (i.e. having equal Chl accumulation), the activity was greatly enhanced as the dark phase duration was prolonged (see Fig. 1C). At equal Chl accumulation, therefore, the proteolysis of LHCII is greater in the long dark-phase ImL cycles than in the short dark-phase ImL cycles. However, proteolysis was found to be similar in thylakoids of plants exposed to the two ImL regimes for equal time period (i.e. at equal developmental stage).

3.2. ATTEMPTS TO ISOLATE THE PROTEASE

The protease is SDS-stable up to 0.1% SDS and heat stable (almost no reduction of activity when thylakoids are heated for 5 min up to 60°C). On the other hand, the thylakoid-bound proteolytic activity can be washed off by NaBr-treatment. As shown in Figure 2 thylakoids are inactive following washing, but the NaBr-wash gains a high proteolytic activity. The activity in the NaBr wash as affected by SDS concentration and pretreatment at increasing temperatures is shown in Table I.

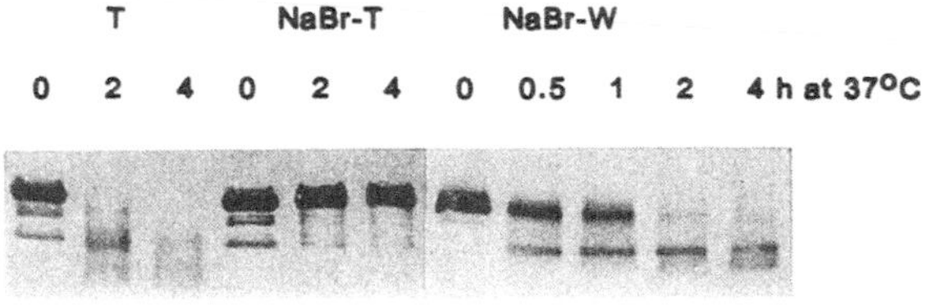

Figure 2. Removal of proteolytic activity from thylakoids by NaBr treatment

Using these findings we have further attempted to separate the protease. Thylakoids solubilized in SDS and analyzed by SDS-PAGE, followed by gel fractionation and incubation of gel slices with either exogenous azocoll, or purified LHCII, showed at least 5 peaks of proteolytic activity against azocoll, and one against LHCII. The latter was localized at the origin of the resolving gel (see Fig. 3A). A similar result was obtained with thylakoids heated in the dark at 56°C for 3 days. In this case analysis of thylakoids, showed no coomassie stainable protein bands following SDS-PAGE (see Fig. 3B). Nevertheless, these thylakoid extracts actively degraded exogenous LHCII.

Table I. Effect of SDS (A) or heat pretreatment (B) on the proteolytic activity against exogenous LHCII in the NaBr wash.

A.

SDS %	LHCII Degraded µg/15 min
0	6.4
0.1	4.6
0.2	2.3
0.4	1.0
0.8	0

B.

Pretreatment ^{o}C x 5 min	LHCII Degraded µg/15 min at 37°C
0	6.1
30	5.7
60	4.4
90	0.3

This activity was localized again at the origin of the resolving gel following SDS-PAGE of thylakoid samples prior to TX-100 solubilization. However, samples following TX-100 solubilization resolved an additional activity band at about 45 kDa.

Figure 4 shows resolution patterns of the activity found in the NaBr-wash on substrate containing gels (see Fig. 4A) and standard non substrate containing gels (see Fig. 4B). As shown, a band of reduced coomssie stain (suggesting proteolytic degradation of the substrate) is observed at the origin of the LHCII-containing gel (see Fig. 4A). On the other hand resolution of the NaBr-wash by SDS-PAGE on non-substrate-containing gels resulted in the separation of the high molecular weight activity band, but also of an additional band at 45 kDa (see Fig. 4B).

These results, therefore, suggest that the protease is of high molecular mass, probably a supramolecular complex, part of which is a 45 kDa component.

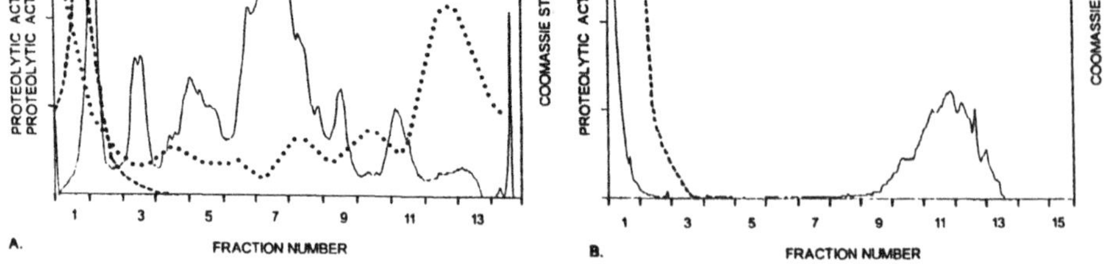

Figure 3. Resolution by SDS-PAGE of proteolytic activity against azocoll and LHCII. The gel fractions (0.4 cm/slice) were incubated overnight with the substrates (5mg/ml azocoll; 45µg/ml LHCII). The activity against LHCII is monitored by immunodetection of the remaining LHCII on Western blots of each fraction. A: Non-heated thylakoids. B: Heated thylakoids (non-solubilized in TX-100).

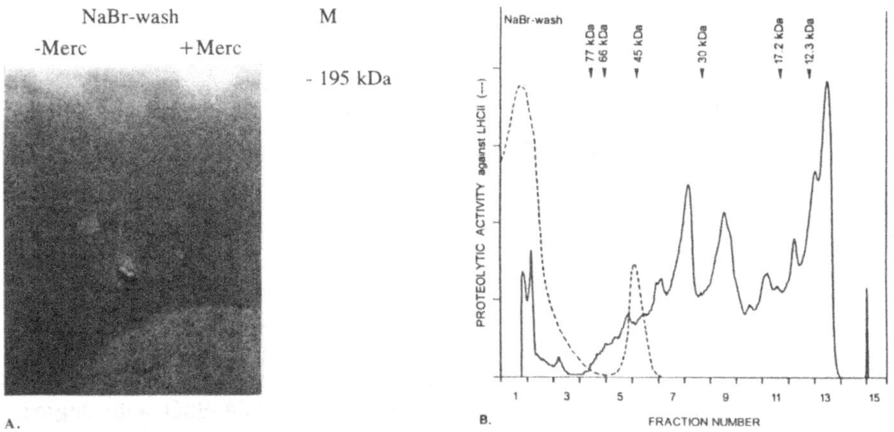

Figure 4. Resolution of proteolytic activity against LHCII in the NaBr-wash by A: SDS-PAGE on Coomassie stained substrate-containing gel. B: SDS-PAGE on non substrate-containing gels.

The inability to detect the 45 kDa activity peak in substrate containing gels may indicate that substrate to protease binding, which might be expected to occur in the substrate containing gels, does not allow dissociation of the supramolecular complex.

Estimates of the protease purification during these steps are given in Table II.

TABLE II. Purification procedure of the LHCII protease from *Phaseolus vulgaris* thylakoids

SAMPLE	ASSAY CONDTTIONS		DEGRADATION RATE	TOTAL	
	Protein/LHCII (mg/mg)	%SDS	µg LHCII/mg.min	Protein (mg)	Activity
A. Thylakoids	1.0	0	11.3	65.0	734
NaBr-wash	1.0	0	29.0	3.4	98
B. NaBr-wash	3.0	0	5.5	3.4	19
Protease peak	3.0	0.66*	6.4	0.24	1.5
Protease peak	3.0	0	64.0	0.24	15

* Calculated on the basis of 100% elution from the gel.

4. References

1. Akoyunoglou, G. and Argyroudi-Akoyunoglou, J.H. (1986) Post translational regulation of chloroplast differentiation, in: G. Akoyunoglou and H. Senger (eds.) *Regulation of Chloroplast Differentiation,* Alan R. Liss, NY, pp. 571-581.

2. Argyroudi-Akoyunoglou, J.H. and Akoyunoglou, G. (1970) Photoinduced changes in the Chl a to Chl b ratio in young bean plants, *Plant Physiol* **46,** 247-249.
3. Viro, M. and Kloppstech, K. (1982) Expression of genes for plastid membrane proteins in barley under intermittent light conditions, *Planta* **154**, 18-24.
4. Argyroudi-Akoyunoglou, J.H. and Akoyunoglou, G. (1979) The Chlorophyll-protein complexes of the thylakoid in greening plastids of *Phaseolus vulgaris*, *FEBS Lett.* **104,** 78-84.
5. Tzinas, G., Argyroudi-Akoyunoglou, J.H. and Akoyunoglou, G. (1987) The effect of dark interval in intermittent light on thylakoid development: photosynthetic unit formation and light-harvesting protein accumulation, *Photosyn Res* **14,** 241-258.
6. Argyroudi-Akoyunoglou, JH., Akoyunoglou, A., Kalosakas, K. and Akoyunoglou, G. (1982) Reorganization of Photosystem II unit in developing thylakoids of higher plants after transfer to darkness. Changes in Chlorophyll b, light harvesting content and grana stacking, *Plant Physiol* **70,** 1242-1248.
7. Akoyunoglou, G. and Argyroudi-Akoyunoglou, J.H. (1978) Control of thylakoid growth in *Phaseolus vulgaris*, *Plant Physiol* **61,** 834-837.
8. Schindler, C. and Soll, J. (1986) Protein transport in intact, purified pea etioplasts, *Arch Biochem Biophys* **247,** 211-220.
9. Anastassiou, R. and Argyroudi-Akoyunoglou, J.H. (1995) Thylakoid-bound proteolytic activity against LHC-II apoprotein in bean, *Photosyn Res* **43,** 241-250.
10. Towbin, H., Staehelin, T. and Gordon, J. (1979) Electrophoretic transfer of proteins from polyacrylamide gels to nitrocellulose sheets: procedure and some applications, *Proc Natl Acad Sci USA* **76,** 4350-4354.
11. Blake, M.S., Johnston, K.H., Russel, G.J. and Gotschlich, E.C. (1984) A rapid sensitive method for detection of alkaline phosphatase conjugated anti-antibody on western blots, *Anal Biochem* **136,** 175-179.
12. Burke, J.J., Ditto, C.L. and Arntzen, C.J. (1978) Involvement of the light-harvesting complex in cation-regulation of excitation energy distribution in chloroplasts, *Arch Biochem Biophys* **187,** 252-263
13. Tziveleka, L.A. and Argyroudi-Akoyunoglou, J.H. (1998) Implications of a developmental stage-dependent thylakoid-bound protease in the stabilization ofd the light-harvesting pigment-protein complex serving photosystem II during thylakoid biogenesis in red kidney bean, *Plant Physiol* **117**, 961-970.
14. Anastassiou, R. and Argyroudi-Akoyunoglou, J.H. (1995a) Thylakoid-bound proteolytic activity *vs.* exogenous azocoll during chloroplast development in bean, *J Plant Physiol* **145,** 24-30.
15. Mackinney, G. (1941) Absorption of light by chlorophyll solutions, *J Biol Chem* **140,** 315-322
16. Lowry, O.H., Rosenbrough, N.J, Farr, A.L. and Randall, R.J. (1951) Protein measurement with Folin phenol reagent, *J Biol Chem* **193,** 265-275.
17. Laemmli, U.K. (1970) Cleavage of structural proteins during assembly of the head of bacteriophage T4, *Nature* **277**, 680-685.
18. Lacks, S.A. and Springhorn, S.S. (1980) Renaturation of enzymes after polyacrilamide gel electrophoresis in the presence of sodium dodecyl sulfate, *J Biol Chem* **255,**7467-7473.

THE REGULATORY ROLE OF POLYAMINES ON THE STRUCTURAL AND FUNCTIONAL PHOTOADAPTATION OF THE PHOTOSYNTHETIC APPARATUS

K. KOTZABASIS[a], B. STRASSER[c], E. NAVAKOUDIS[a], H. SENGER[b] AND D. DÖRNEMANN[b]

[a] *Department of Biology, University of Crete, P.O. Box 1470, GR-71110 Heraklion/Crete, Greece.*

[b] *Fachbereich Biologie, Philipps Universität Marburg, Karl-von-Frisch-Str., D-35032 Marburg, Germany.*

[c] *Laboratory of Bioenergetics, University of Geneva, CH-1254 Jussy-Geneva, Switzerland*

1. Introduction

The adaptation of the photosynthetic apparatus to low and high light intensities is a well documented phenomenon, as well in higher plants [1], as in green algae [2]. The organisms respond to shade conditions by increased amounts of chlorophyll at a lowered photosynthetic capacity. Their photosynthetic compensation point is shifted to lower light intensities and the respiratory oxygen uptake decreased. Low light intensities can be mimiced by monochromatic red light in higher plants [3] and by blue light in algae [4]. The photoreceptors responsible for the adaptation to the different light qualities are phytochrome [5] and blue light photoreceptors [6], respectively. Changes in the molecular organization of the photosynthetic apparatus of green algae adapting to low intensities of white light result in an increase of the light harvesting complex, preferentially of photosystem II [7]. The ability of green algae to adapt unexpectedly fast (in about 8h, [8]) to new irradiation conditions demonstrates the high capability of the photosynthetic apparatus to respond to changes in the environmental conditions.
In this paper the influence of the directed inhibition or induction of specific steps of polyamine biosynthesis on the photoadaptation mechanism are studied.

2. Materials and Methods

<u>*2.1 Organism, growth and illumination*</u>: Cultures of the unicellular green alga *Scenedesmus obliquus*, wild type strain D3 were grown autotrophically in liquid culture medium [9] in a temperature-controlled water bath (30°C). For the photoadaptation experiments under various white light intensities, autotrophically pre-grown cultures were irradiated for 36 h with low (55 μmol m^{-2} s^{-1}) and high (235 μmol m^{-2} s^{-1}) light intensities. For the adaptation of the cells to monochromatic light conditions cultures were irradiated for 36 h with blue and red light (40 μmol m^{-2} s^{-1}).

J. H. Argyroudi-Akoyunoglou and H. Senger (eds.),
The Chloroplast: From Molecular Biology to Biotechnology, 283–286.

2.2 Polyamine analysis and determination: The polyamine analyses were performed according to the method of Kotzabasis *et al.* [10].
2.3 Polarography: Rates of oxygen evolution and uptake (maximum photosynthetic and respiration rates) of the adapted cultures were determined polarographically at 30°C with a Micro Clarc electrode system (Gilson Medical Electronics Middleton, WI, USA). The cell density was adjusted to an absorbance of 0.3 at 679nm.
2.4 Measurements of fluorescence induction: For the fluorescence induction measurements the portable "Plant Efficiency Analyser, PEA" (Hansatech Instruments Ltd) was used. The maximum yield of photochemistry (Fv/Fm), the absorbance per reaction center (antenna size), the rate of primary photochemistry per reaction center , the electron transport per reaction center and the number of photosynthetic reaction centers per unit area were measured according to the JIP-method of Strasser and Strasser [11].

3. Results and discussion

In the present work we describe investigations on the possible influence of polyamines on the development of structure and functioning of the photosynthetic apparatus during adaptation to different light intensities and qualities. Therefore their biosynthesis was specifically inhibited or enhanced at different steps of the biosynthetic pathway to reveal the influence of the intracellular levels of Put, Spd and Spm on the differentiation of an adapting photosynthetic apparatus.
Two sets of *Scenedesmus* cultures were pretreated with three different inhibitors of polyamine biosynthesis or exogeneous polyamines were added, both at a concentration of 1mM. The inhibitors were 1,4-diamino-2-butanone (1,4-DB), an inhibitor of Put biosynthesis, cyclohexylamine (CHA), an inhibitor of Spd biosynthesis and 1,3-diaminopropane (1,3-DP), which inhibits Spm biosynthesis. Added polyamines were Put, Spd and Spm. Furthermore, a control without the addition of polyamines or inhibitors was run. One set of cultures was then exposed to a low light intensitiy of 55 $\mu mol\ m^{-2}\ s^{-1}$, the second one to a high photon fluence rate of 235 $\mu mol\ m^{-2}\ s^{-1}$ for 36 h. The determination of the Chls and the calculation of the Chla/b ratios, as well as the rates of maximum photosynthesis and respiration (data not shown), suggest that the cultures generally show a stage of adaptation that correllates with the light intensity they have been exposed to before. The most significant difference, compared to the other cultures was observed with the 1,4-DB- and Spm-samples (data not shown).
Changes of the photosynthetic antenna size, in the number of reaction centers, in the photosynthetic rate and also in the Chl a/b ratio, were observed in the 1,4-DB-, Put-, 1,4-DB/Put- and Spm-treated cultures adapted under high light intensities in comparison to the control. This demonstrates that the inhibition of Put biosynthesis under high and low light conditions leads to changes in structure and functioning of the photosynthetic apparatus similar to those observed with low light adapted cultures, e. g. an enlargement of the LHC accompanied by a decrease of the number of reaction centers per unit area, and a decrease of the maximum photosynthetic rates, respiration and Chla/b ratios. This is in agreement with the results of Hoffmann and Senger [8] concerning the reorganization of the photosynthetic apparatus during the adaptation from high to low light intensities. In

our experiments the 1,4-DB-treated cultures showed changes similar to those described for the low light treatment (Tab. I).
The addition of Spm yielded, in comparison to the 1,4-DB-treated culture, similar results except that there was no decrease in the number of reaction centers per unit area (Tab. I). A possible explanation for this phenomenon could be the stabilizing effect of Spm on the PSII-RCs [12].

TABLE 1. Changes in %, in the photosynthetic antenna size [LHC], in the number of reaction centers [RCs], in the photosynthetic rate (Ph.R) and also in the Chl a/b ratio, as determined for the 1,4-DB-, Put-, 1,4-DB/Put- and Spm-treated cultures adapted under high and low light intensities, in comparison to the control.

	1,4DB/ HL	1,4DB/ LL	Put/ HL	Put/ LL	1,4DB +Put/HL	1,4DB +Put/LL	Spm/ HL	Spm/ LL
RCs	-47,1	-20,8	+5,7	-0,7	-1,4	-11,5	+15,7	-2,9
LHC	+56,0	+69,2	-9,1	+6,2	+12,7	+27,5	+42,6	+21,0
Ph. R.	-42,4	-27,7	-21,1	+1,3	-33,7	-0,2	-34,8	-18,6
Chl a/b	-27,5	-23,3	+5,4	-5,1	-13,8	-18,0	-20,7	-10,2

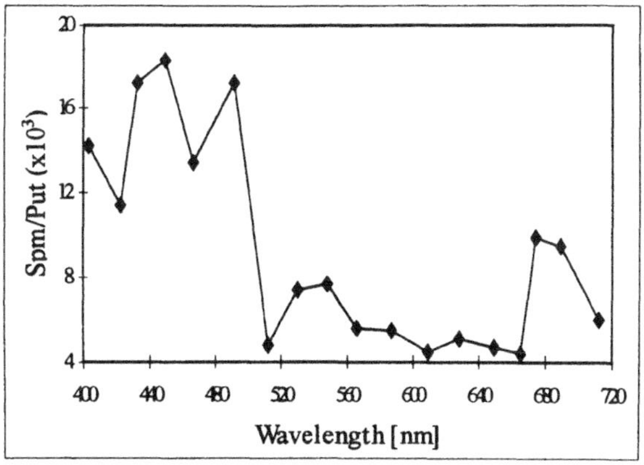

Figure 1. Action spectrum of the intracellular ratio of Spm/Put.

To examine the effect of blue and red light in the presence of 1,4-DB, Put, 1,4-DB in combination with Put and Spm, autotrophically grown cultures of *Scenedesmus* together with an untreated control were exposed to monochromatic blue (40 μmol m^{-2} s^{-1}) and red (40 μmol m^{-2} s^{-1}) light in the presence of the above compounds. The results were very similar to those obtained with low and high light irradiated cultures (data not shown).

The similar trend of the data obtained from the inhibition of Put biosynthesis and from the exogeneous supply of Spm are, in principal, in agreement with the results of Besford *et al.* [12], who demonstrated that both, a decrease of the Put level and also an increase of the cellular Spm content, retards effectively the degradation of the D1 and the D2 protein of PSII and additionally that of cytochromes in the thylakoid membranes by senescence. Our results suggest that the ratio of Spm/Put and not exclusively the levels of Put and Spm are the main regulatory factor for the development of structure and functioning, and therefore for the photoadaptation mechanism of the photosynthetic apparatus.
In order to ascertain the dependence of the intracellular Spm/Put contents from the applied light quality, and further, to elucidate the nature of the primary photoreceptor for the regulation of the Spm/Put ratio and for photoadaptation, action spectra were carried out. *Scenedesmus* cultures of a cell density of 1 μl PCV per ml cell suspension were irradiated

with 18 different light qualities ranging from 400 to 712 nm at a photon fluence rate of 2 $\mu mol\ m^{-2}\ s^{-1}$ for 24 h. According to our above results an increase of the Spm/Put ratio indicates an adaptation of the photosynthetic apparatus to the applied low light conditions. The action spectrum on the Spm/Put ratio in Fig. 1 thus gives evidence for the nature of the photoreceptors for the photoadaptation of the photosynthetic apparatus in low light conditions. It can be deduced that the changes in the Spm/Put ratio are mainly a blue light effect, although a minor participation of a red absorbing component at 680 nm, possibly the PSII-RC, can not be excluded. This is in agreement with data of Senger [6], who reported that in green algae the responsible receptor for the adaptation of the photosynthetic apparatus to low light conditions is a blue light photoreceptor.

Action spectra on the LHC size, on the primary photochemistry rate per reaction center on the electron transport per reaction center and on chlorophyll biosynthesis (characteristc parameters for the photoadapta-tion) are similar to the corresponding spectrum on the Spm/Put ratio (data not shown).

From the presented data it can be concluded that polyamines, especially the intracellular Spm/Put ratio participate in the regulation of photoadaptation of the photosynthetic apparatus. Changes in the Spm/Put ratio and in photoadaptation are regulated by the same photoreceptor(s). Furthermore, we speculate that the regulation of the Spm/Put ratio is part of the signal transduction chain from photoreception to the photoadaptation mechanism of the photosynthetic apparatus.

Aknowledgement

K. K. thanks the Alexander von Humboldt-Foundation for a fellowship which facilitated this research.

4. References

1. Björkman, O. (1972) in Photosynthetic adaptation to contrasting light climates, Carnegie Inst. Year Book 71, pp.82-85.
2. Senger, H., Bauer, B. (1987) The influence of light quality on adaptation and function of the photosynthetic apparatus, *Photochem. Photobiol.* **45**, 939-946.
3. Buschmann, C., Meier, D., Kleudgen, H.K. and Lichtenthaler, H.K. (1978) Regulation of chloroplast development by red and blue light, *Photochem. Photobiol.* **27**, 195-198.
4. Humbeck, K., Hoffmann, B. and Senger, H. (1988) Influence of energy flux and quality of light on the molecular organization of the photosynthetic apparatus in *Scenedesmus, Planta* **173**, 205-212.
5. Holmes, M.G. and Smith, H. (1977) The function of phytochrome in the natural environment-II. The influence of vegetation canopies on the spectrakl energy distribution of natural daylight, *Photochem. Photobiol.* **25**, 539-545.
6. Senger, H. (1987) Sun and shade effects of blue light in plants, in H. Senger (ed.), *Blue Light Responses: Phenomena and Occurence in Plants and Microorganisms,* CRC Press, Boca Raton, FL, pp.141-149.
7. Senge, M. and Senger, H. (1990) Functional changes in the photosynthetic apparatus during light adaptation of the green alga *Chlorella fusca, J. Photochem. Photobiol.* **B8**, 63-71.
8. Hoffmann, B. and Senger, H. (1988) Kinetics of photosynthetic apparatus adaptation in *Scenedesmus obliquus* to change in irradiance and light quality, *Photochem. Photobiol.* **47**, 737-739.
9. Bishop, N.I. and Senger, H. (1971) Preparation and photosynthetic properties of synchronous cultures of *Scenedesmus, Methods Enzymol.* **23**, 53-66.
10. Kotzabasis, K., Christakis-Hampsas, M.D. and Roubelakis-Angelakis, K.A. (1993) A narrow-bore HPLC method for the identification and quantitation of free conjugated and bound polyamines, *Anal. Biochem.* **214**, 484-489.
11. Strasser, B.J. and Strasser, R.J. (1995) Measuring fast fluorescence transients to adress environmental questions: The JIP-test, in P. Mathis (ed.), *Photosynthesis: from Light to Biosphere*, Kluwer Academic Publishers, Dordrecht, Boston, London, Vol. V, pp. 977-980.
12. Besford, R., Richardson, C., Capell, T. and Tiburcio, A.F. (1993) Effect of polyamines on stabilization of complexes in the thylakoid membranes of osmotically stressed oat leaves, *Planta* **189**, 201-206.

ALTERATIONS IN THE PLASTID MEMBRANE-ASSOCIATED POLYAMINES DURING CHLOROPLAST PHOTODEVELOPMENT

D. DÖRNEMANN[b], E. NAVAKOUDIS[a] AND K. KOTZABASIS[a]
[a] *Department of Biology, University of Crete, P.O. Box 1470, GR-71110 Heraklion/Crete, Greece.*
[b] *Fachbereich Biologie, Philipps Universität Marburg, Karl-von-Frisch-Str., D-35032 Marburg, Germany.*

1. Introduction

The effect of polyamines, based on their charge, has been observed as binding to macromolecules. The binding of polyamines to macromolecules may indicate their role in the regulation of basic processes in plants such as cell division, morphogenesis, senescence and response to stress [1]. The enzyme transglutaminase is capable to bind polyamines covalently between two carboxyl groups of two glutamic acid residues in the same protein [2]. The apoproteins of the chlorophyll-a/b antenna complex, LHCP II, (CP24, CP26, CP29) [3], the proteins D_1, D_2, Cyt f and the large subunit of ribulose-1,5-bisphosphate carboxylase/oxygenase [4] are some of the substrates of the plastidic transglutaminase. Kotzabasis and coworkers [5] reported that Put, Spd, and Spm are present in intact chloroplasts, particulary in thylakoid membranes, in PS II membranes, in the LHC-proteins and in the PS II-complex of spinach. However, when further fractionation of the photosynthetic membranes was performed, only Spm was detected in the photochemically active PS II-core complex and the PS II-reaction center. Furthermore, diamines and polyamines, exogenously supplied to osmotically stressed oat leaves, are reported to retard protein degradation, to inhibit chlorophyll (Chl) loss and to stabilize thylakoid membranes [4]. Isolated intact etioplasts from *Zea mays* contained great amounts of polyamines and exhibited high activities of the main polyamine biosynthetic enzymes, arginine decarboxylase (ADC) and ornithine decarboxylase (ODC), but conversly very low activity of the main catabolic enzyme, diamine oxidase (DAO), compared to the chloroplast enzyme levels [6].
In this contribution the levels of Put, Spd, Spm and norSpd are determined during darkness and chloroplast development in the light in isolated plastid membranes from the wild type (Wt) and the mutants C-2A' and C-6D of *Scenedesmus obliquus*.

2. Materials and Methods

2.1. Organisms and growth: In this study the wild type and the two pigment mutants C-2A' and C-6D of the unicellular green alga *Scenedesmus obliquus* were used. The cells

J. H. Argyroudi-Akoyunoglou and H. Senger (eds.),
The Chloroplast: From Molecular Biology to Biotechnology, 287–290.

were grown heterotrophically [7] in complete darkness at 32°C for 3 days. For the greening experiments 3-days dark-grown cultures were transferred to white fluorescent light of an intensity of 20 $W.m^{-2}$ and maintained at 30°C.

2.2. Preparation of membrane particles: For the preparation of thylakoid membranes the method of [8] was followed.

2.3. Analysis of polyamines: The polyamines were analysed acccording to [9].

3. Results and Discussion

Results from previous papers [5,6] gave rise for a study on possible changes in the level of the plastid membrane-bound polyamines in the wild type of *Scenedesmus* and its pigment mutants C-2A' and C-6D before and during their light-dependent development of photosynthetic capacity. In the Wt of *Scenedesmus*, Chl biosynthesis in darkness does not stop at the level of protochlorophyllide (Pchlide), but continues at lower rates compared to light conditions up to the formation of Chl a and b. Mutant C-6D synthesizes in darkness only Chl a and precursors of carotenoids, but no Chl b. It thus lacks an intact PSII and the LHC protein complexes [10]. Mutant C-2A', in contrast, behaves like an angiosperm being unable to convert Pchlide to chlorophyllide (Chlide) in darkness [11]. On the onset of light both mutants overcome the block and start, after a short lag-phase, to synthesize both chlorophylls in a regular manner. After 24h in light they exhibit a phenotype almost identical to the Wt.

For the experiments cultures were grown for three days in darkness and subsequently exposed to white fluorescent light of an intensity of 20 $W.m^{-2}$ for two days. Polyamines were extracted and analyzed from both, the separated plastidal membranes and from whole cells in defined intervalls over a total period of five days. The main polyamines that could be identified were Put, Spd, Spm and the more seldom norspermidine (norSpd). In all three cultures (Wt, C-6D and C-2A') the amount of the intracellular diamine Put decreased dramatically at the onset of illumination, whereas the levels of the other polyamines changed only slightly (data not shown).

Isolation of plastidal membranes from each sample and analysis of the polyamines showed that in thylakoids of the Wt the membrane-associated Put in both dark- and light-grown cells shows a gradual, slightly exponential decrease during the whole cultivation time in darkness and in the light. (Fig. 1). The corresponding level of Put in the plastid membranes of mutant C-6D is higher in darkness. In the light an immediate, drastical drop, comparable to that observed for whole cells, follows (Fig. 1). The shape of the Put content curve of pigment mutant C-2A' membranes is similar to the other curves in darkness. At the onset of light mutant C-2A' shows an immediate increase of the Put level with a maximum after five hours of illumination, followed by a rapid decrease (Fig. 1). The initial enhancement of the Put level in the plastid membranes of mutant C-2A' after its transfer to light could possibly be related to the conversion of Pchlide to Chlide. This hypothesis is supported by the fact that Wt and mutant C-6D do not accumulate Pchlide in darkness and, conclusively, should not demonstrate an increase in Put after the onset of illumination, which, in fact, can be seen in Fig. 1. All these data point out to a reduction of the thylakoid-bound Put level in strict correlation with

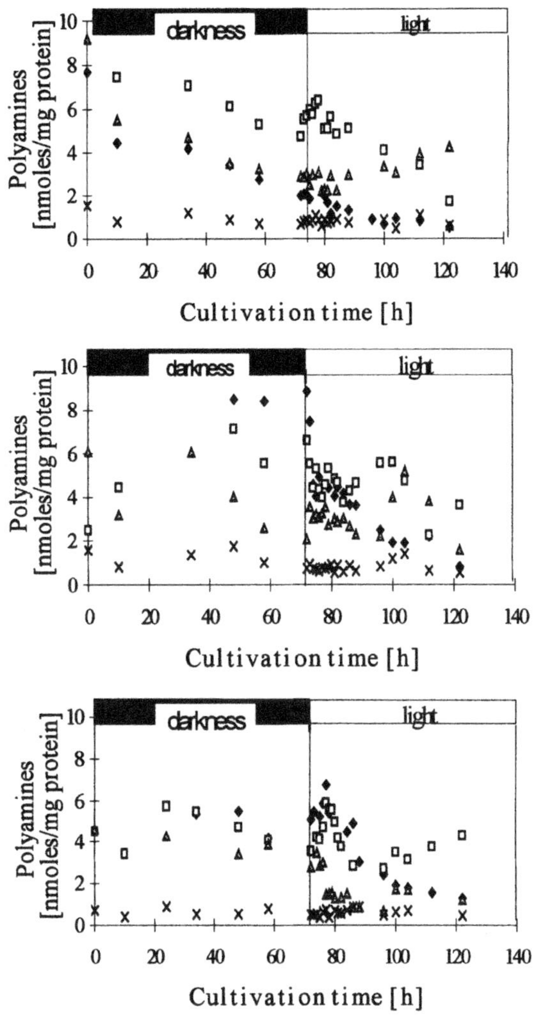

Figure 1. Changes in the amount of plastid membrane-bound putrescine (Put: ♦), spermidine (Spd: ▲) , spermine (Spm: x) and norspermidine (norSpd: □) in wildtype (top), mutant C-6D (middle) and mutant C-2A' (bottom). The cells were grown for 72 h in darkness, followed by 50 h in the light.

chloroplast development of the organism. Similar results from maize that mainly the content of these polyamines follows the formation of fully developed chloroplasts are described in [6]. In mutant C-6D, Spd reveals a maximum after about five hours in light (Fig. 1). Compared to the Wt and mutant C-2A' the other mutant, C-6D, represents again a destinct chloroplast development, however, with an intermediate starting point. As already mentioned, this mutant synthesizes only Chl a and precursors of carotenoids during heterotrophic growth in darkness. The cells demonstrate PS I activity, but lack an active PS II with the core complex CPa. Upon transfer to light the organism immediately forms the complete pattern of carotenoids and reveals an active PS II [10]. Only one hour after the onset of light a distinct peak of the photosystem II-complex CPa can be detected by gel electrophoresis. Humbeck *et al.* [12] reported in 1990 that mutant C-6D synthesizes the 45 and 48 kDa apoprotein of CPa in darkness, but they could also show that the apoproteins are only assembled to an intact complex in the light.

Regarding the rapid formation of CPa and the increase of the Spd level after the transfer to light, it is suggested that the incorporation of Spd into the complex is most probably an important premise for the assembly of CPa not only in this mutant.

During the light period Spd shows a second peak after 30 hours of illumination. Corresponding maxima are found for Spm and norSpd. These findings correlate to a certain extent with the formation of LHCP II. Its assembly starts roughly two hours after the onset of light and is completed by the formation of the oligomeric forms about 24 hours later [12]. This correlation between polyamine peaking and the formation of the fully developed LHCP II suggests again a role for the above mentioned polyamines in the assembly and stabilization of the oligomeric forms of LHCP II. This interpretation is supported by the fact that in LHCP II from spinach all of these polyamines could be detected [6]. Further support for our conclusions is given by Del Duca *et al.* [8] who showed that the apoproteins of the Chl a/b antenna complex, CP 24, CP 26 and CP29, are substrates for the plastidic transglutaminase, responsible for attaching polyamines to proteins. Furthermore, similar, but much lower increases in the amounts of thylakoid-bound Spd and Spm were found in the Wt and mutant C-2A' of *Scenedesmus* about 30 hours after the onset of light, which temporally correlates with the data from mutant C-6D (Fig. 1).

In conclusion, the present contribution provides corroboration of the existence of polyamines in the plastid membranes of the various chloroplast developmental stages. Furthermore, it is shown that chloroplast development occurs in three different ways (Wt, C-6D, C-2A') and is accompanied by specific fluctuations of thylakoid-bound polyamines. Finally, our data give evidence for a key role of polyamines in the assembly of the photosynthetic membrane complexes and thus the formation of the photosynthetic apparatus.

4. References

1. Slocum, R.D., Kaur-Sawhney, R. and Galston A.W (1984) The physiology and biochemistry of polyamines in plants, *Arch. Biochem. Biophys.* **235**, 283-303.
2. Isekson, I. and Apelbaum, A. (1987) Evidence of transglutaminase activity in plant tissue, *Plant Physiol.* **84**, 972-974.
3. Del. Duka, S., Tidu, V., Bassi, R., Esposito, C. and Srafini-Fracassini D. (1994) Identification of chlorophyll a/b proteins as substrates of transglutaminase activity in isolated chloroplasts of *Helianthus tuberosus L., Planta* **193**, 283-289.
4. Besford, R., Richardson, C., Capell, T. and Tiburcio, A.F. (1993) Effect of polyamines on stabilization of complexes in the thylakoid membranes of osmotically stressed oat leaves, *Planta* **189**, 201-206.
5. Kotzabasis, K. Fontinou, C. and Ghanotakis, D. (1993) Polyamines in the photosynthetic apparatus. Photosystem II highly resolved subcomplexes are enriched in spermine, *Photosynthesis Res.* **38**, 83-88.
6. Andreadakis, A. and Kotzabasis, K. (1996) Changes in the biosynthesis and catabolism of polyamines in isolated plastids during chloroplast photodevelopment, *J. Photochem. Photobiol.* **B33**, 163-170.
7. Bishop, N.I. and Wong, J. (1971) Observations of photosystem II mutants of *Scenedesmus*: Pigments and proteinaceous components of the chloroplast, *Biochim. Biophys. Acta* **234**, 433-445.
8. Senger, H. and Mell, V. (1977) Preparation of photosynthetically active particles from synchronized cultures of unicellular algae, in D.M. Prescott (ed.), *Methodes in Cell Physiology*, Balaban Int. Science Services, Philadelphia, Vol. XV pp. 433-455.
9. Kotzabasis, K., Christakis-Hampsas, M.D. and Roubelakis-Angelakis, K.A. (1993) A narrow-bore HPLC method for the identifaction and quantification of free, conjugated and bound polyamines, *Anal. Biochem,* **214**, 484-489.
10. Humbeck, H., Römer, S. and Senger, H. (1988) Changes in carotenoid compositon and function of the photosynthetic apparatus during light-dependent chloroplast differentiation in mutant C-6D of *Scenedesmus obliquus, Botanica Acta* **101**, 220-228.
11. Kotzabasis, K. and Senger, H. (1989) The influence of 5-aminolevulinic acid on protochlorophyllide and protochlorophyll accumulation in dark grown *Scenedesmus, Z. Naturforsch.* **45c**, 71-73.
12. Humbeck, K., Römer, S. and Senger, H. (1990) Light-dependent assembly of the components of photosystem II-core complex Cpa in mutant C-6D of *Scenedesmus obliquus, J. Plant Physiol.* **136**, 569-573.

CHLOROPHYLL FLUORESCENCE KINETIC DEPENDS ON AGE OF LEAVES AND PLANTS

Z. ŠESTÁK
Institute of Experimental Botany, Academy of Sciences of the Czech Republic, Na Karlovce 1a, CZ-160 00 Prague 6, Czech Republic

Abstract

Recently, parameters of chlorophyll (Chl) fluorescence induction kinetics have often been used as characteristics of stress effect on plant photosynthesis. In these determinations, age of tested leaves and plants is only exceptionally taken into account. Nevertheless, photosynthetic activities markedly change during life span of leaves and plants, they even vary at individual places of the leaf blade that differ in relative age of cells and chloroplasts. These changes are reflected in Chl fluorescence kinetic, and the ontogenic differences in its parameters may be larger than those induced by stress. Comparison of values from the literature has shown that relatively independent of leaf age is the often used ratio of variable and maximum fluorescence F_v/F_m: it rapidly declines only at leaf senescence, when its value is related to the Chl content in leaves.

General Information

During the development of a chloroplast, its size and ultrastructure, contents of pigments and pigment-protein complexes, composition and activities of photosystems and the respective enzymes of photosynthesis change. In a leaf, anatomical parameters, size and shapes of cells, the number and localisation of chloroplasts, and even the morphology undergo expressed developmental changes. As each leaf is formed in another phase of plant life and in different microclimatic conditions, also leaves of different position on stem or branches differ in their developmental course. All these differences are finally reflected in gas exchange, carbon and biomass accumulation in leaves and plants (for reviews see [1, 2]).

The ontogenic changes of all photosynthetic parameters are modified by environmental factors, most heavily by various stresses. These changes are reflected mainly in contents of Chls and their presence in pigment-protein complexes as components of various levels of photosystem organisation and function. As an easy way for detecting the stress effects, the well known Kautsky effect, *i.e.*, Chl fluorescence induction kinetics (for the history of this effect see [3]) has been chosen. At present, individual parameters of induction curves may be easily evaluated using various commercial devices and computer programmes (for reviews see [4 - 8]).

J. H. Argyroudi-Akoyunoglou and H. Senger (eds.),
The Chloroplast: From Molecular Biology to Biotechnology, 291–296.

The absorption and fluorescence spectra of monolayers of chloroplasts from leaves of the same age are almost uniform, but they diversify in whole leaves. Leaf spectra differ due to chloroplast localisation, radiation scattering and damping in the sample induced by variance in leaf anatomy. Leaf anatomy is modified by, *e.g.*, sun and shade position, water supply, elevation; different character of upper and lower sides is also reflected in Chl fluorescence parameters (for references see [9]). In leaves, a clear relation of the shape of fluorescence induction curve with Chl content, cell size, and leaf thickness exists (see, *e.g.*, [4, 8]). Chloroplast and cell age are also major determinants of differences in Chl fluorescence kinetics that exist within the area of one leaf (for references see [9]). Nevertheless, only a few users of the Chl fluorescence characteristics take into account that especially in sun leaves an expressed diurnal course exists [10]. Similarly, the differences appearing in the course of plant development and characterising leaves of various insertion (cf. [9]) are only rarely considered.

Many statements in recent papers on stress effects in plants are based on Chl fluorescence parameters. Unfortunately, only rarely the age and insertion of analysed leaves are given in the papers. The goal of this paper is to show, using examples from the literature, that leaf age should always be taken into account and which fluorescence characteristic is most proper for stress research.

General Course of Changes of the Chl Fluorescence Kinetics

The shape of the well known OIDPSMT curve changes during leaf development (for references see [9]). Both the rapid and slow phases are modulated: this holds both for the appearance of peaks, their heights, and time of their appearance. The height of the first peak (P) that characterises change from open to closed traps usually takes the course typical for all photosynthetic characteristics: it increases to leaf maturity and then declines again. The secondary peak (M), which characterises accumulation of $NADPH_2$ and ATP and is connected with pH rise in chloroplasts, may not appear in young leaves; in older leaves two or more such peaks may appear, but in senescent leaves a very simple kinetics may be found: a maximum followed by a slow continuous decline (cf. also [11]). The level of steady fluorescence (T) characterising the proceeding Calvin cycle usually declines from young to mature leaves, but increases again till leaf senescence. Times necessary to reach P or M mostly decline with leaf age.

Variance in leaf life span may explain differences found between leaves of different insertion. Thus, even if according to the rate of Chl accumulation per leaf area and changes in Chl *a/b* ratio both the primary and first trifoliate leaves of soybean developed at a similar speed, their life span and development of photosynthetic activities were different (Fig. 1): the formation of peak M was evident in the 11-d-old primary leaf and in the 12-d-old trifoliate leaf. In the 11-d-old primary leaf a decline in height of peak P was already evident, while this height increased till leaf age of 23 d in the first trifoliate leaf [12].

In the modulated fluorescence kinetic measurements (the method of Schreiber *et al.* [13]) saturating irradiation pulses are applied that suppress the photochemical part of fluorescence quenching: the resulting spikes on the curves are lower in young than mature leaves.

Values calculated from peaks and troughs of the kinetic curves (such as F_0, F_m) and their ratios (such as F_v, F_v/F_m, F_v/F_0, $\Delta F/F_m'$, Rfd, q_P, q_N, *etc.*) reflect these general developmental changes. During the whole leaf life span, the general course of a fluorescence parameter may be as follows: a fairly rapid increase to a maximum (or plateau) followed by a more or less linear decline (or, due to the character of a parameter, a mirror pattern of this course). If only one part of leaf development is studied, only an increase, decrease, or no changes at all are found.

Differences between leaves of different insertion or age measured at the same time may reflect the environmental effects in the period prior to measurements: this is expressed especially when needles of different years are compared.

F_v/F_m as an Example of Changes During Leaf Development

Various parameters of Chl fluorescence induction are differently affected by chloroplast and leaf age: degree of this is related to the site of photosynthetic mechanism they represent. Large differences were found, *e.g.*, in photochemical quenching q_P [14, 15] or in the coefficient Rfd (fluorescence decrease from F_m to steady state fluorescence) that indicates the potential photosynthetic activity of a leaf (see, *e.g.*, [16]).

The ratio F_v/F_m (calculated as F_m - F_0/F_m) is a most often used indicator of efficiency of photosystem 2. A comparison of some values for control leaves presented in recent literature (Table 1) shows that this indicator is fairly stable in some plant species in leaves that are near to maturity phase. Nevertheless, during leaf senescence the F_v/F_m value rapidly declines. These changes are related with changes in Chl content in leaf tissues. Therefore F_v/F_m is appropriate for testing the effect of stress factor, especially when it induces senescence-similar reactions. In some herbs and especially in conifers, differences in F_v/F_m during vegetation season and between needles formed in subsequent years are fairly large. Also the sensitivity of differently old leaves to the same agent may be different. No difference may result from opposite effects of leaf ageing and stressor action (cf., *e.g.*, [17]).

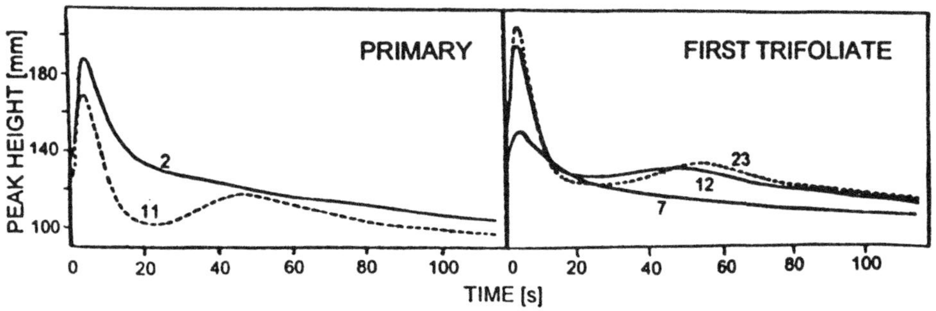

Figure 1. Fluorescence induction curves of the primary and first trifoliate leaves of *Glycine max* cv. Maple Arrow grown under fluorescent+incandescent lamps. Figures at curves show leaf age [d]. First leaf started its development in a 2-d-old plant, the first trifoliate one in 8-d-old plant. The leaves reached maximum chlorophyll contents per leaf area and maximum *a/b* ratio after 8-10 d of development. (Modified from [12].)

Table 1. Changes in F_v/F_m during leaf development or in leaves of various insertion on a plant found recently by various authors. Chl - chlorophyll; P_N - net photosynthetic rate.

Plant	Leaves and growth condition	F_v/F_m (max. Δ%)	Notice	Author
Hordeum vulgare	flag leaf; 15 June-7 July; field	0.78→0.32 (-59)	14 d of a plateau followed by 8 d of rapid decline; parallel to Chl/f.m.	[18, 19]
Nicotiana tabacum cv. Xanthi	expanding→developed→ mature leaf; growth chamber; low irradiance, low N	0.860→0.832 (-3) →0.801 (-7)	only increase to leaf maturity; correlation with quantum yield of oxygen evolution	[20]
Lycopersicon esculentum cv. Rehovot 13	third leaves 12 and 18 d-old	0.801→0.799 (0)	no difference upto 18 d; Chl/area, Chl/d.m., and Chl/f.m. decline	[21]
cv. Ailsa Craig	leaves 8+7 (green) to senescent 2+1 (yellow); 56 d-old plants; growth room	0.81→0.72 (-11)	F_v/F_m differences much smaller than in Chl/area (-79 %) or oxygen evolution (-70 %)	[22]
Phaseolus vulgaris cv. Kinghorn	primary leaf, 7 to 22 d-old; growth chamber	0.73→0.76 (+4) →0.73 (-4)	F_v/F_m does not follow the changes in Chl/area	[23]
Cucurbita pepo cv. Ambassador	young→mature leaves; greenhouse	0.807→0.784 (-3)	smaller difference than in q_P, q_N	[24]
Cucumis sativus	leaves 1 to 6; growth chamber	0.781→0.762 (-2) →0.783 (+3)	decline followed by increase; almost opposite to Chl/area	[25]
Spinacia oleracea cv. Savoy	younger third and fully expanded second leaves; 100 d-old plants; growth chamber	0.78→0.74 (-5)	(in second leaves larger decline induced by photoinhibition)	[26]
Gossypium hirsutum cv. T25 and HS26	leaves 1 (young, >7 cm^2) to 4 (on 4th node); 22 d-old plants; growth chamber	T25: 0.31→ 0.14 (-55) HS26: 0.37→ 0.18 (-51)	these differences found at growth temperature 28/20 °C (day/night) are larger than at 28/28 °C	[27]
Zea mays cv. Pride 5 and Pioneer 3902	leaves 8 (young), 6, and 4 (old); medium N nutrition; growth chamber	Pride: 0.80→ 0.64 (-20) Pioneer: 0.80 →0.67 (-16)	differences smaller than in Chl/area (-43 or -31 %) or P_N (-45 or -37 %)	[28]
Actinidia deliciosa	leaf expansion characterised by leaf diameter (30→170 or 13→180 mm); field	0.39→0.74 (+90) or 0.55→0.77 (+40)	increase with leaf size parallel to increase in Chl/area; strange sampling (means of leaves collected in six positions on the vine from October to May between 08:00 and 16:00 h)	[29]
Prunus persica cv. Dixired	young and mature leaves; 7-8 y-old trees; two successive years	0.76→0.78 (+3) or 0.78→0.79 (+1)	differences similar to that of q_P	[30]

Curatella americana	expanding (15-30 d-old) to yellowish (330-365 d-old); 5-6 m high trees	0.65→0.73 (+12)→0.68 (-7) or 0.66→ 0.75(+14)→ 0.66 (-12)	differences in Chl more expressed than in F_v/F_m, but both maxima in mature leaves	[31, 32]
Pinus sylvestris	seasonal changes in current year needles	0.81 (August) →0.10 (March)(-88) →0.84 (June) (+740)	differences similar to those in content of D1-protein and epoxidation state of carotenoids	[33]
Picea abies	seasonal changes in 1-3 y-old needles; 45-50 y-old trees	0.80→0.84 in July (+5)	values vary between 0.80 and 0.87 in individual months	[34]

Recommendation

Differences in Chl fluorescence parameters between leaves of different age or insertion are often larger than those induced by a stressor. Therefore it is always necessary to check, prior to starting experiments, the extent of differences in the used parameter that are induced by leaf and plant age, select leaves of the same or similar age, and in any case mention in the report leaves of which age and position on plant were tested. For studying the stress effects on plants, the ratio F_v/F_m is recommended.

References

1. Šesták, Z. (ed.) (1985) *Photosynthesis During Leaf Development*, Academia, Praha, Dr W. Junk Publishers, Dordrecht.
2. Čatský, J. and Šesták, Z. (1997) Photosynthesis during leaf aging, in M. Pessarakli (ed.), *Handbook of Photosynthesis*, Marcel Dekker, New York, pp. 633-660.
3. Lichtenthaler, H.K. (1992) The Kautsky effect: 60 years of chlorophyll fluorescence induction kinetics, *Photosynthetica* **27**, 45-55.
4. Lichtenthaler, H.K. and Rinderle, U. (1988) The role of chlorophyll fluorescence in the detection of stress conditions in plants, *CRC crit. Rev. anal. Chem.* **19**, S29-S85.
5. Holzwarth, A.R. (1991) Excited-state kinetics in chlorophyll systems and its relationship to the functional organization of the photosystems, in H. Scheer (ed.), *Chlorophylls*, CRC Press, Boca Raton, pp. 1125-1151.
6. Karukstis, K.K. (1991) Chlorophyll fluorescence as a physiological probe of the photosynthetic apparatus, in H. Scheer (ed.), *Chlorophylls*, CRC Press, Boca Raton, pp. 769-795.
7. Schreiber, U., Bilger, W. and Neubauer, C. (1994) Chlorophyll fluorescence as a nonintrusive indicator for rapid assessment of in vivo photosynthesis, in E.-D. Schulze and M.M. Caldwell (eds.), *Ecophysiology of Photosynthesis*, Springer-Verlag, Berlin, pp. 49-70.
8. Mohammed, G.H., Binder, W.D. and Gillies, S.L. (1995) Chlorophyll fluorescence: A review of its practical forestry applications and instrumentation, *Scand. J. Forest Res.* **10**, 383-410.
9. Šesták, Z. and Šiffel, P. (1997) Leaf-age related differences in chlorophyll fluorescence, *Photosynthetica* **33**, 347-369.
10. Demmig-Adams, B., Adams, W.W., III, Barker, D.H., Logan, B.A., Bowling, D.R. and Verhoeven, A.S. (1996) Using chlorophyll fluorescence to assess the fraction of absorbed light allocated to thermal dissipation of excess excitation, *Physiol. Plant.* **98**, 253-264.
11. Shavnin, S.A. and Fomin, A.S. (1993) [Seasonal changes in chlorophyll fluorescence of needles of *Pinus silvestris*], *Fiziol. Rast.* **40**, 209-213.
12. Dijak, M., Smith, D.L. and Ormrod, D.P. (1987) Effect of growth light quality on fluorescence characteristics of leaves of *Glycine max*, *Environ. exp. Bot.* **27**, 185-192.

13. Schreiber, U., Schliwa, U. and Bilger, W. (1986) Continuous recording of photochemical and non-photochemical chlorophyll fluorescence quenching with a new type of modulation fluorometer, *Photosynth. Res.* **10,** 51-62.
14. van Kooten, O., Meurs, C. and van Loon, L.C. (1990) Photosynthetic electron transport in tobacco leaves infected with tobacco mosaic virus, *Physiol. Plant.* **80,** 446-452.
15. Skórzyńska-Polit, E. and Baszyński, T. (1997) Differences in sensitivity of the photosynthetic apparatus in Cd-stressed runner bean plants in relation to their age, *Plant Sci.* **128,** 11-21.
16. Bach, T.J. and Lichtenthaler, H.K. (1983) Inhibition by mevinolin of plant growth, sterol formation and pigment accumulation, *Physiol. Plant.* **59,** 50-60.
17. Boese, S.R. and Huner, N.P.A. (1992) Developmental history affects the susceptibility of spinach leaves to *in vivo* low temperature photoinhibition, *Plant Physiol.* **99,** 1141-1145.
18. Humbeck, K., Kloppstech, K. and Krupinska, K. (1994) Expression of early light-inducible proteins in flag leaves of field-grown barley, *Plant Physiol.* **105,** 1217-1222.
19. Humbeck, K., Quast, S. and Krupinska, K. (1996) Functional and molecular changes in the photosynthetic apparatus during senescence of flag leaves from field-grown barley plants, *Plant Cell Environ.* **19,** 337-344.
20. Balachandran, S. and Osmond, C.B. (1994) Susceptibility of tobacco leaves to photoinhibition following infection with two strains of tobacco mosaic virus under different light and nitrogen nutrition regimes, *Plant Physiol.* **104,** 1051-1057.
21. Andre, M., Ducamp, M., Betsche, T.: Photoinhibition by high light in a shade plant. Role of O_2 concentration. - In: Baltscheffsky, M. (ed.): Current Research in Photosynthesis. Vol. II. Pp. 511-514. Kluwer Acad. Publ., Dordrecht - Boston - London 1990.
22. Gruenzweig, J.M., Rabinowitch, H.D. and Katan, J. (1993) Physiological and developmental aspects of increased plant growth in solarised soils, *Ann. appl. Biol.* **122,** 579-591.
22. John, I., Drake, R., Farrell, A., Cooper, W., Lee, P., Horton, P. and Grierson, D. (1995) Delayed leaf senescence in ethylene-deficient ACC-oxidase antisense tomato plants: molecular and physiological analysis, *Plant J.* **7,** 483-490.
23. Madey, E., Ghosh, S., Thompson, J.E. and Dumbroff, E.B. (1995) The effects of paclobutrazol on stabilization of photosynthetic components in senescing bean leaves, in P. Mathis (ed.), *Photosynthesis: from Light to Biosphere,* Vol. IV, Kluwer Academic Publishers, Dordrecht, pp. 541-544.
24. Mészáros, I., Tóth, V.R. and Váradi, G. (1995) Effect of deficiency on the net photosynthesis, chlorophyll fluorescence and pigment composition of cucumber, in P. Mathis (ed.), *Photosynthesis: from Light to Biosphere,* Vol. IV, Kluwer Academic Publishers, Dordrecht, pp. 717-719.
25. Ciompi, S., Castagna, A., Ranieri, A., Nali, C., Lorenzini, G. and Soldatini, G.F. (1997) CO_2 assimilation, xanthophyll cycle pigments and PSII efficiency in pumpkin plants as affected by ozone fumigation, *Physiol. Plant.* **101,** 881-889.
26. Gray, G.R., Boese, S.R. and Huner, N.P.A. (1994) A comparison of low temperature growth vs low temperature shifts to induce resistance to photoinhibition in spinach (*Spinacia oleracea*), *Physiol. Plant.* **90,** 560-566.
27. Warner, D.A. and Burke, J.J. (1993) Cool night temperatures alter leaf starch and photosystem II chlorophyll fluorescence in cotton, *Agron. J.* **85,** 836-840.
28. McCullough, D.E., Girardin, P., Mihajlovic, M., Aguilera, A. and Tollenaar, M. (1994) Influence of N supply on development and dry matter accumulation of an old and a new maize hybrid, *Can. J. Plant Sci.* **74,** 471-477.
29. Greer, D.H. (1995) Effect of canopy position on the susceptibility of kiwifruit (*Actinidia deliciosa*) leaves on vines in an orchard environment to photoinhibition throughout the growing season, *Aust. J. Plant Physiol.* **22,** 299-309.
30. Raggi, V. (1995) CO_2 assimilation, respiration and chlorophyll fluorescence in peach leaves infected by *Taphrina deformans, Physiol. Plant.* **93,** 540-544.
31. Sobrado, M.A. (1995) The influence of leaf age and drought on photosynthesis of the tropical evergreen tree *Curatella americana* L., in P. Mathis (ed.), *Photosynthesis: from Light to Biosphere,* Vol. IV, Kluwer Academic Publishers, Dordrecht, pp. 549-552.
32. Sobrado, M.A. (1996) Leaf photosynthesis and water loss as influenced by leaf age and seasonal drought in an evergreen tree, *Photosynthetica* **32,** 563-568.
33. Ottander, C. (1995) Seasonal changes in photosystem II organisation and pigment composition in *Pinus sylvestris, Planta* **197,** 176-183.
34. Godde, D. (1992) Photosynthesis in relation to modern forest decline; content and turnover of the D-1 reaction centre polypeptide of photosystem 2 in spruce trees (*Picea abies*), *Photosynthetica* **27,** 217-230.

SUCCESSIVE DEGRADATION OF THE LIGHT-HARVESTING SYSTEM OF THE PHOTOSYNTHETIC APPARATUS DURING SENESCENCE OF BARLEY FLAG LEAVES

K. HUMBECK[1] AND K. KRUPINSKA[2]
[1] *Institut für Pflanzen- und Zellphysiologie, Kröllwitzerstr. 44, D-06099 Halle*
[2] *Botanisches Institut III, Gyrhofstr. 15, D-50931 Köln*

1. Abstract

Levels of light-harvesting complexes were investigated during maturation and senescence of flag leaves derived from field-grown barley plants (*Hordeum vulgare* L.). In mature flag leaves possessing photosynthetically highly active chloroplasts levels of the major light-harvesting complex LHCII and also of the minor light-harvesting complexes CP29 of photosystem II and LHCI of photosystem I were high.
During growth of barley plants in the field the light conditions drastically changed. Variations in photosynthetically active radiation (PAR) in the range from 200 to 1500 $\mu mol{\cdot}m^{-2}{\cdot}s^{-1}$ did neither affect photosynthetic capacity nor photosystem II efficiency in the mature flag leaves, indicating that the photosynthetic apparatus was sufficiently protected against photodamage. The changes in PAR did also not affect levels of LHCII, LHCI and CP29. In contrast, levels of another member of the chlorophyll a/b-binding protein family, ELIP, drastically fluctuated showing high ELIP levels at days with high PAR.
During senescence of flag leaves of field-grown barley plants levels of minor light-harvesting complexes CP29 and LHCI decreased much faster than those of the bulk LHCII. The early decline in the minor complexes coincided with a decrease in photosystem II efficiency. The decrease in the levels of minor complexes LHCI and CP29 was however retarded by a decrease in ambient light intensity.
The senescence-specific decrease in minor LHCs relative to the bulk LHCII results in a decrease in energy transfer to the reaction centers and seems to represent an additional protection mechanism against photodamage, the risk of which is increasing during senescence. In contrast, ELIP levels fluctuated independently from the stage of development, being high at days with high PAR and being low at days with low PAR, in mature as well as in senescing flag leaves.
Northern analyses using a cDNA clone specific for CP29 and another one specific for ELIP revealed that both, the senescence-specific decline in the minor complex CP29 and the light dependent fluctuations in ELIP level are regulated at the transcript level.

J. H. Argyroudi-Akoyunoglou and H. Senger (eds.),
The Chloroplast: From Molecular Biology to Biotechnology, 297–302.

2. Materials and Methods

Plant Material. Spring barley (*Hordeum vulgare* L., cv. Carina) and winter barley (*Hordeum vulgare* L., cv. Trixi) were grown in 1992 under agricultural conditions in a field.

Photosynthetic Capacity. Photosynthetic oxygen evolution was measured with a leaf-disc O_2-electrode as previously described [1]. A saturating light intensity of 2050 $\mu mol{\bullet}m^{-2}{\bullet}s^{-1}$ was used for determination of photosynthetic capacity.

Chlorophyll Fluorescence Measurements. Chlorophyll fluorescence parameters were measured with a pulse amplitude modulation fluorometer. Prior to fluorescence measurements, the leaf blades were dark adapted for 30 min. The F_v/F_m ratio which is a measure of the quantum yield of photosystem II photochemistry, was determined as $(F_m-F_0)/F_m$.

Chlorophyll Content. Chlorophylls were extracted from whole flag leaf blades with hot methanol. Their concentrations were determined spectrophotometrically.

Immunological Analysis of Protein Levels. Proteins were extracted from whole flag leaf blades and immunologically analyzed as described earlier [1]. Antibodies used were directed towards LHCI [2], towards LHCII [3] and towards barley low molecular mass ELIP [4]. The antibody specific for LHCI also detects CP29 [1].

Isolation and Analysis of RNA. Total RNA was isolated and analyzed as described[1]. The homologous cDNA insertion of pHv60, specific for low molecular mass ELIP [5], was used for the detection of ELIP mRNA. For analysis of *Lhcb4* mRNA levels , the cDNA clone HvCP29 (EMBL nucleotide sequence database accession number AJ006296) was used.

3. Results

Characterization of flag leaves. PSII-efficiency and Chlorophyll content of flag leaves of spring barley cv. Carina were high until about 15 days after they had reached their full length (Figs. 1 and 2). After that day, due to senescence, both parameters constantly decreased. Flag leaves of winter barley cv. Trixi entered the phase of senescence about 20 days after they had reached full length (Figs. 1 and 2). In this case, chlorophyll content decreased a few days earlier than PSII-efficiency.

Influence of leaf development and light intensity on the levels of LHCs. Levels of LHCII, CP29, LHC I and ELIP were analyzed in mature and senescing flag leaves of spring barley cv. Carina (Fig.3). In mature flag leaves (0,2,4 and 6 days after flag leaves had reached full length) levels of the minor complexes CP29 and LHC I and of the bulk LHCII were high. Contrarily, ELIP content fluctuated drastically being high on days with high light intensity and low on days with low light intensity. During senescence of flag leaves of spring barley (days 18, 20, 22 and 24) the minor LHCs CP29 and LHC I decreased drastically (Fig. 3).

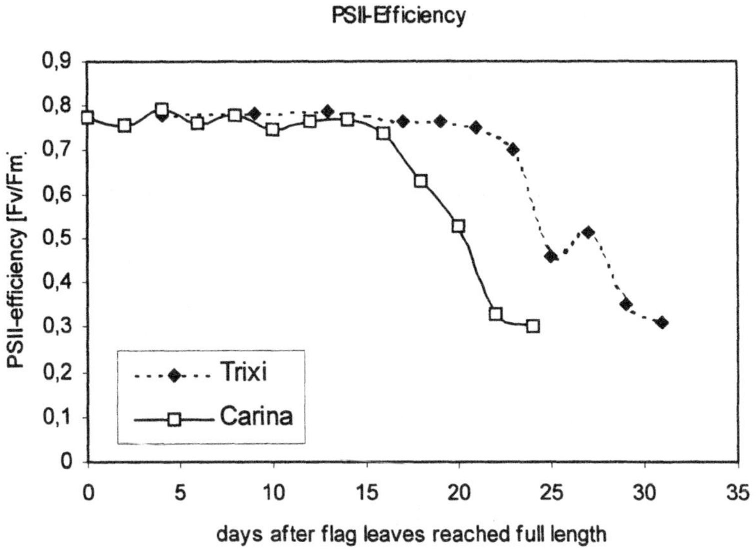

Figure 1:PSII-efficiency was determined in flag leaves of spring (cv. Carina) and winter (cv. Trixi) barley with a pulsed chlorophyll fluorometer after dark adaptation. ±SD was less than 0.03.

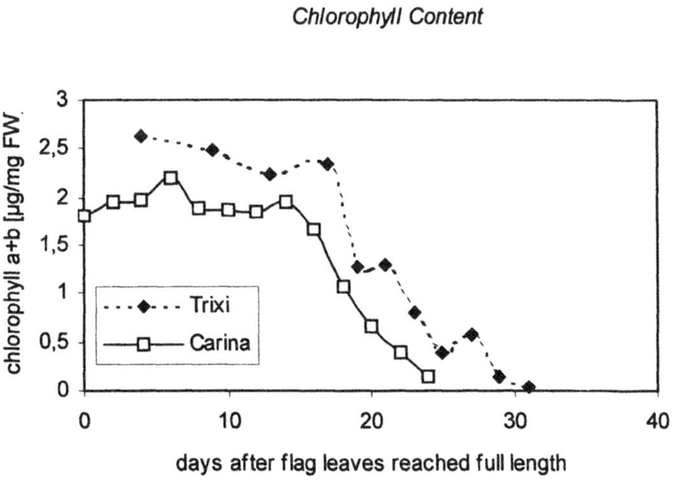

*Figure 2:*Chlorophyll content was determined in flag leaves of spring (cv. Carina) and winter (cv. Trixi) barley after extraction with hot methanol spectrophotometrically. ±SD was less than 0.18.

The levels of these proteins seem not only to depend on developmental stage but also on the environmental conditions. The senescence specific decrease was partly reversed when days with high light intensities were followed by cloudy days with low light intensities (Fig. 3, days 18 and 20). Similar results were obtained with flag leaves of winter barley cv. Trixi (Fig. 4). Here also the levels of CP29 and LHCI decreased during senescence and here also a decrease in light intensity reversed this decrease. In contrast, levels of LHCII decreased in flag leaves of spring and winter barley not until a late stage of senescence (Figs. 3 and 4). Compared to CP29 and LHCI, ELIP levels showed an inverse behaviour being high at days with high light intensities and low at days with low light intensities (Figs. 3 and 4).

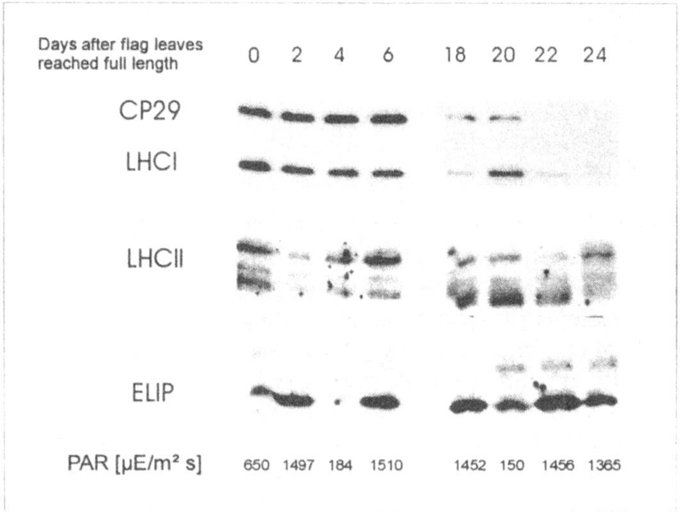

Fig. 3: Immunological analysis of levels of CP29, LHCI, LHCII and ELIP in mature and senescent flag leaves of spring barley cv. Carina. Photosynthetically active radiation (PAR) was measured with a metereological station located in the field.

Influence of leaf development and light intensity on levels of transcripts specific for CP29 and ELIP. In order to further examine the underlying regulatory mechanism of the changes in protein level, total RNA was extracted from similar flag leaves as used for protein analyses. Transcript levels specific for CP29 were then compared to those specific for ELIP (Figs. 5 and 6). In mature flag leaves of spring barley (Fig. 5) levels of *Lhcb4* mRNA specific for CP29 are high showing only minor fluctuations. Contrarily, ELIP mRNA levels drastically changed in response to the changes in light intensity as seen also on the protein level. During senescence the level of transcripts specific for CP29 drastically decreased in flag leaves of cv. Carina as well as in flag leaves of cv. Trixi (Figs. 5 and 6). In contrast, the level of mRNA specific for ELIP is only affected by light intensity and not by the developmental stage of the leaves.

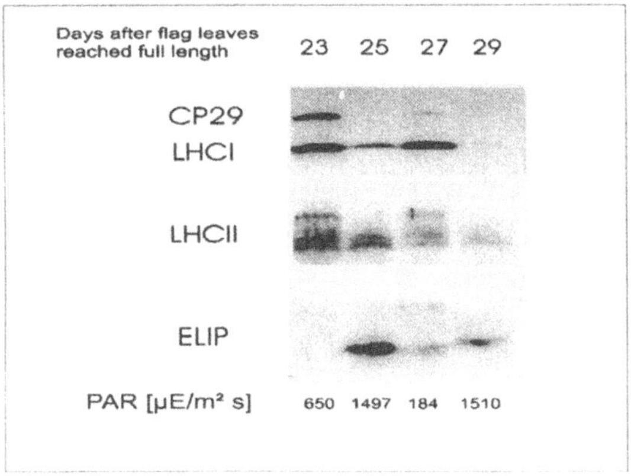

Fig. 4: Immunological analysis of levels of CP29, LHCI, LHCII and ELIP in mature and senescent flag leaves of winter barley cv. Trixi . Photosynthetically active radiation (PAR) was measured with a metereological station located in the field.

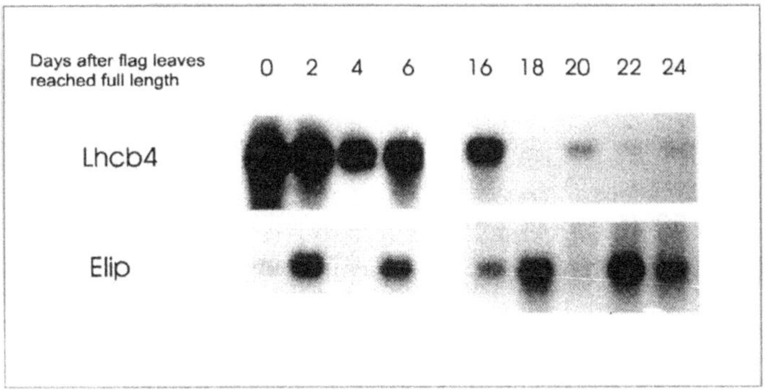

*Fig. 5:*Levels of transcripts specific for CP29 (Lhcb4) and ELIP in flag leaves of spring barley cv. Carina during maturation and senescence.

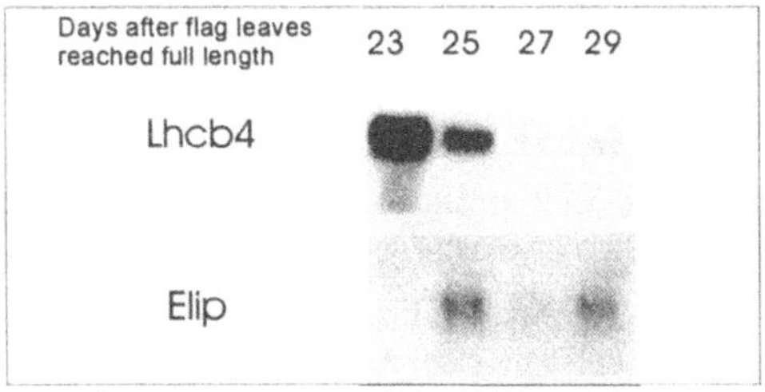

*Fig. 6:*Levels of transcripts specific for CP29 (Lhcb4) and ELIP in flag leaves of winter barley cv. Trixi during senescence.

4. Conclusions

During senescence the levels of inner light-harvesting complexes CP29 and LHCI decrease much faster than levels of the bulk LHCII. The early decrease in minor complexes correlates with a decrease in efficiency of energy transfer within PSII measured as a decrease in Fv/Fm. This might represent a protection mechanism against photodestruction during the early phase of senescence as shown in the model in Fig. 7.

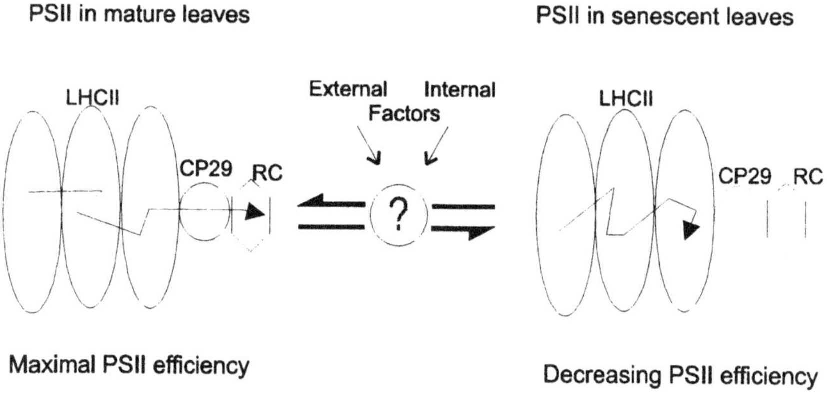

Fig. 7: *Model of changes in energy transfer within PSII during the early phase of senescence. During senescence levels of minor complex CP29 decrease much faster than LHCII. Under this condition energy transfer from the outer LHCII to the reaction center is reduced which might represent a protection mechanism against photodestruction within PSII.*

Northern analyses with a specific cDNA probe revealed that the senescence-specific decrease in CP29 is regulated at the transcript level. A change from high to low light intensities in the early phase of senescence retards or even reverts the decrease in minor LHCs. Levels of ELIP are regulated in a different way, being high at days with high light intensities and low at days with low light intensities in mature and in senescent leaves.

5. References

1. Humbeck, K., Quast S. and Krupinska K. (1996) Functional and molecular changes in the photosynthetic apparatus during senescence of flag leaves from field-grown barley plants. *Plant Cell & Environment* **19**, 337-344.
2. Hoyer-Hansen G., Bassi R., Honberg L.S. and Simpson D. (1988) Immunological characterization of chlorophyll a/b-binding proteins of barley thylakoids. *Planta* **173**, 12-21.
3. Harrison M.A. and Melis A. (1992) Organization and stability of polypeptides associated with the chlorophyll a/b light-harvesting complex of photosystem II. *Plant Cell Physiology* **33**, 627-637.
4. Pötter E. and Kloppstech K. (1993) Effects of light stress on the expression of early light-inducible proteins in barley. *European Journal of Biochemistry* **214**, 779-786.
5. Grimm B. and Kloppstech K. (1987) The early light-inducible proteins of barley. *European Journal of Biochemistry* **167**, 493-499.

BIOTECHNOLOGICAL APPROACHES

MEMBRANE PROTEINS OF PHOTOSYNTHESIS: STRUCTURE - FUNCTION - BIOTECHNOLOGY

M. RÖGNER, S. BERRY, U. BORONOWSKY, A. FLUEREN, C. JÄGER, H. KUHL, D. SCHNEIDER, A. SEIDLER, S.-O. WENK, N. KARAPETYAN* & J. KRUIP

Plant Biochemistry, Ruhr-University Bochum, 44780 Bochum, Germany

**A.N. Bakh Institute of Biochemistry, Russian Academy of Sciences, 117071 Moscow, Russia*

1. Biotechnology of Cyanobacteria: Mass culture

Prerequisite for the structural and functional investigation of membrane proteins of photosynthesis, i.e. photosystem 1 (PS1), photosystem 2 (PS2) and the cytochrom b_6f-complex (b_6f) is the mass production of such complexes in appropriate systems. Mesophilic (i.e. *Synechocystis* PCC 6803) and thermophilic (i.e. *Synechococcus elongatus*) cyanobacteria can be grown in mass culture under controlled conditions in a new developed 25 L photobioreactor especially designed for these organisms; its main features are an illumination unit on wheels and the use of transparent polyamid foil instead of glass which enables a safe *in situ* sterilisation. Under pH-, light, gas-mixing and temperature-controlled condition the yield of cells can be more than doubled, resulting in more than 100 g cells (wet weight) out of 25 L culture in about 3 days. This set-up may be interesting for various applications in biotechnology involving cyanobacteria and green algae.

J. H. Argyroudi-Akoyunoglou and H. Senger (eds.),
The Chloroplast: From Molecular Biology to Biotechnology, 305–310.

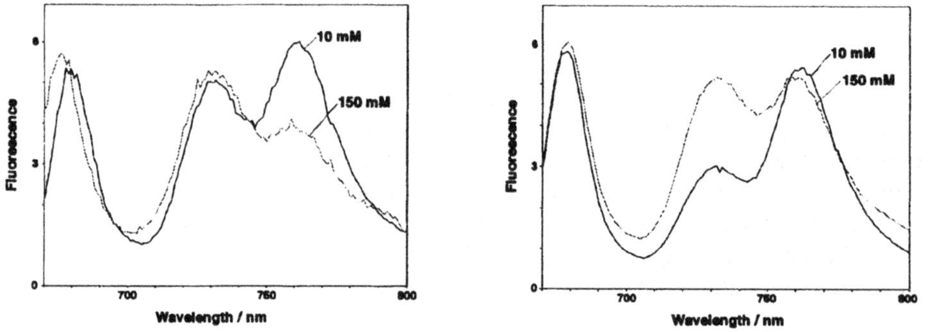

Fig.1: 77 K fluorescence emission spectra of purified PS1 monomers (left) and PS1 trimers (right) reconstituted into liposomes and exposed to low and high salt concentrations (10 mM and 150 mM $MgSO_4$, resp.)

2. Photosystem 1

We had shown before (1) especially with PS1 of *Synechocystis* 6803, that oligomeric and monomeric structures of the photosynthetic membrane proteins, i.e. PS1, PS2 and the Cyt. b6f-complex, seem to be in a dynamic equilibrium in the cyanobacterial thylakoid membrane. This equilibrium can be shifted *in vitro* by changing the surface charge density, for example by changing the salt concentration of the medium. For a direct observation of this dynamic equilibrium we developed a model system consisting of PS1 from *Spirulina platensis* reconstituted into liposomes. The advantage of this system is that trimeric PS1 - in contrast to monomeric PS1 - shows chlorophyll-fluorescence at 760 nm, which can be used as a kind of „internal marker“ for trimerization. Fluorescence shows that homogenous monomeric PS1 when reconstituted into liposomes forms about 30% trimers and homogenous trimeric PS1 yield about 40% monomers; the transformation of monomers into trimers within the liposomes can be considerably induced by a change to low-salt-conditions, while reconstituted trimers dissociate distinctly into monomers when exposed to high-salt-conditions (Fig.1), as predicited by our model (1). Such reconstitution experiments could also show that PS1 monomers lacking the stromal subunits PsaC, -D and -E do not form trimers, confirming our suggestion (2) that PsaD - in addition to PsaL - plays a central role for trimerization of PS1.

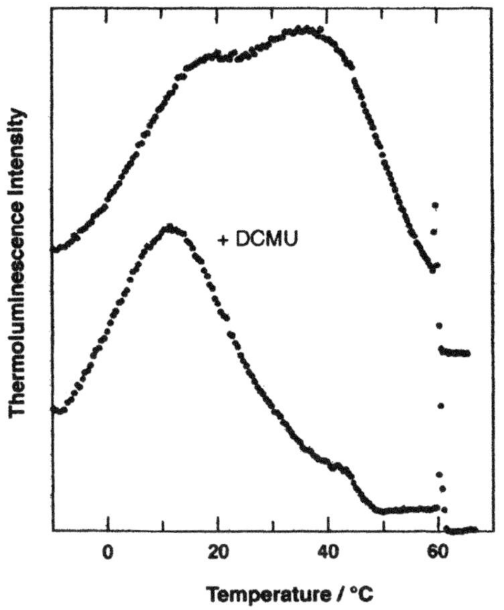

Fig. 2: Thermoluminescence of purified dimeric PS2; measurements have been performed with a heating rate of 0.4°C s^{-1} from -10 to 70°C; the Q-band is at 13.8°C and the B-band at 40.7°C (collab. with A. Krieger, Univ. Freiburg, Germany)

3. Photosystem 2

Our preparation of monomeric and dimeric PS2 from the thermophilic cyanobacterium *Synechococcus elongatus* could be further improved: Oxygen evolution measurements yielded values of 2,500 μmol O_2/ mg Chl h (±20%) for the monomer and 6,000 μmol O_2/ mg Chl h (±20%) for the dimer (25°C, DCBQ as electron acceptor). EPR-measurements with these dimers at 10 K could show a new spin state in S_2 (between g=5 and g=9) which was discovered for the first time in cyanobacteria (3). Both by EPR and by thermoluminescence measurements, also the intactness of the acceptor side could be shown (Fig. 2), yielding an estimation of at least 60% of occupied Q_B-sites. The extremely high activity and stability of this preparation - especially the PS2 dimer retains its activity if kept at 4°C for several days without substantial loss - make it an ideal candidate for the formation of 2D- and 3D-crystals.

HPLC analysis could also show the existence of a dimeric PS2 in the mesophilic cyanobacterium *Synechocystis* 6803 due to a new developed preparation procedure. This PS2 dimer had a higher oxygen-evolving activity than the monomer which is in line with the results from *Synechococcus elongatus*. However, the absolute values are at least a factor of 10 lower than with PS2 isolated from the thermophile and the preparation is much less stable.

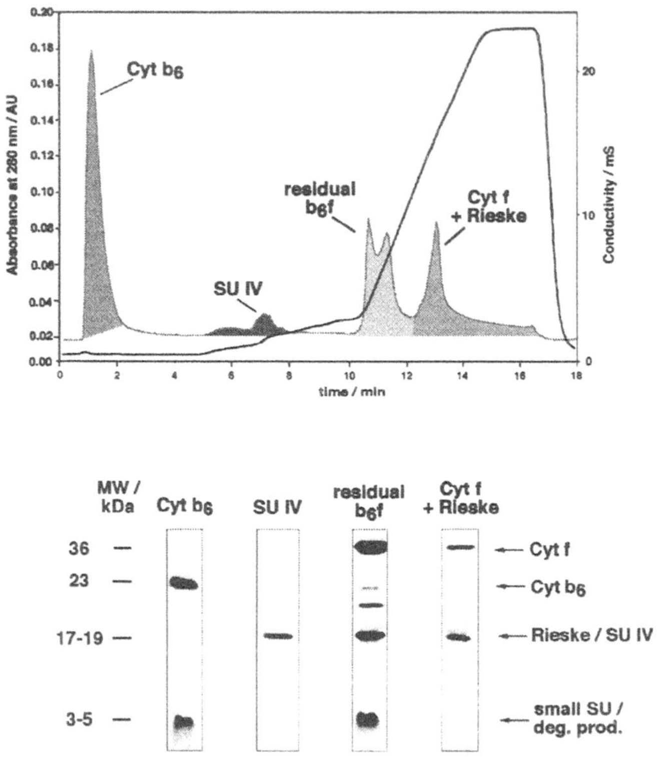

Fig. 3

Separation of native Cyt. b6f subunits by HPLC anion exchange chromatography (upper part) and SDS-PAGE of the respective fractions of this run (lower part)

4. The Cytochrome b6f-complex

4.1. LOCALIZATION OF TIGHTLY BOUND PIGMENTS: STRUCTURAL ROLE?

We succeeded in developing a new isolation procedure for the b6f-complex of the cyanobacterium *Synechocystis* 6803 which contains all the major subunits including the Rieske-protein; in contrast to the preparation from *Chlamydomonas*, our complex is monomeric, suggesting that the Rieske-subunit is not lost upon monomerization. The *Synechocystis*-complex contains one tightly bound chlorophyll molecule which has been characterized in detail in (4); in addition we now find for the first time in a cyt. b6f-

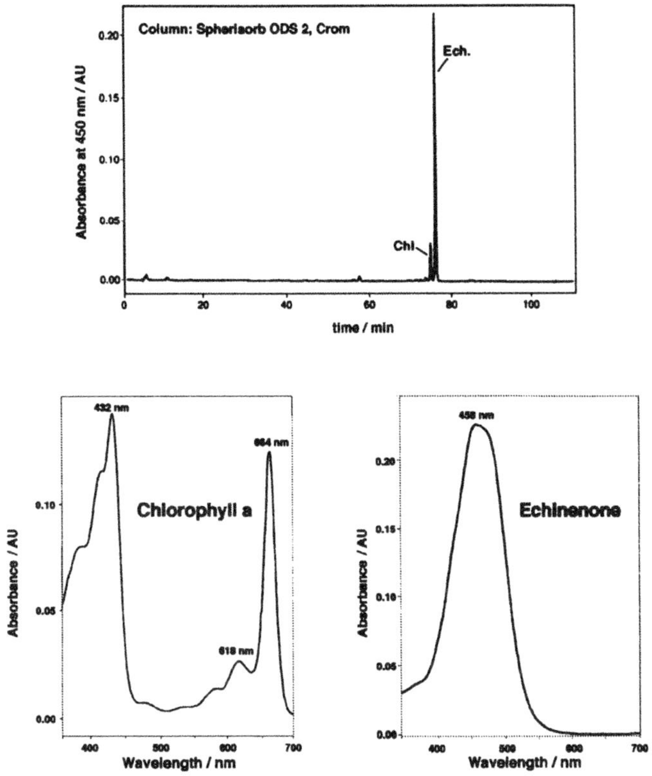

Fig. 4

Reversed phase analysis of the pigments bound to the isolated cyt. b6 subunit (upper part). The analysis was done from acetone extracts of all major subunits of the cyt. b6f complex and yielded positive results only in case of the b6-subunit.
UV/VIS spectra of the left RP-peak revealed Chl a (lower left). The right peak is the oxidized form of a carotenoid only found in cyanobacteria, echinenone.

complex a carotenoid - Echinenone - in substoichiometric amounts (<1 per complex). Dissociation of the complex under mild conditions leaving the single subunits intact and purification of the isolated subunits by HPLC (Fig. 3) can show that both pigments are bound by the isolated b6-subunit exclusively. As no functional role could be attributed to these pigments, a structural role should be considered.

4.2. OVEREXPRESSION OF RECOMBINANT CYT. B6F SUBUNITS

b6-subunit and subunit IV of *Synechocystis* have been successfully overexpressed in *E. coli* and purified to homogeneity by HPLC. Also, a fusion protein of b6 and SU IV, mimicking the b-subunit of the bc1-complex, has been overexpressed successfully. Refolding of these proteins has been achieved in the presence of detergent, but without any urea-treatment. Characterization of these recombinant proteins which include b6 subunits with site-directed mutations of the heme-binding ligands are in progress.

4.3. CHARACTERIZATION OF THE PetM$^-$-MUTANT

A PetM$^-$-mutant of *Synechocystis* has been successfully generated; comparison with WT cells shows similar growth rates and a similar PS2/PS1-ratio; also, phenotype and oxygen evolution activity are quite similar (although the mutant culture shows a more greenish colour). Pronounced differences are a reduced phycobilin/Chl-ratio in the mutant and a more reduced PQ-pool of petM$^-$ in comparison with WT as judged from PAM-measurements. This indicates a reduced activity of the cyt. b6f-complex in the mutant which has to be further investigated.

5. References

1. Kruip, J., Boekema, E.J., Bald, D. and Rögner, M. (1994) Evidence for the existence of trimeric and monomeric photosystem 1 complexes in the thylakoid membranes from cyanobacteria, *Photosynth. Res.* **40(3)**, 279-286
2. Kruip, J., Chitnis, P.R., Lagoutte, B., Rögner, M. and Boekema, E.J. (1997) Structural organization of the major subunits in cyanobacterial photosystem 1, *J. Biol. Chem.* **272**, 17061-17069
3. Boussac, A., Kuhl, H., Un, S., Rögner, M. and Rutherford, A.W. (1998) Effect of near-infrared light on the S2-state of the manganese complex of photosystem 2 from *Synechococcus elongatus*, *Biochemistry* **37**, 8995-9000
4. Peterman, E.J.G., Wenk, S.O., Pullerits, T., Palsson, L.-O., van Grondelle, R., Dekker, J.P., Rögner, M. and van Amerongen, H. (1998) Fluorescence and absorption spectroscopy of the weakly fluorescing chlorophyll a in cytochrome b6f of *Synechocystis* PCC 6803, *Biophys. J.* **75**, 389-398

tRNA^{Glu} MEDIATED δ-AMINOLEVULINATE BIOSYNTHESIS AND BIOTECHNOLOGY

C. GAMINI KANNANGARA
Department of Physiology, Carlsberg Laboratory
Gamle Carlsberg Vej 10
DK-2500 Copenhagen-Valby
Denmark

1. Abstract

δ-Aminolevulinate is the first universal committed intermediate for the biosynthesis of tetrapyrrole molecules functioning in respiration, photosynthesis and as cofactors in other important enzyme reactions. There are two ways for the biosynthesis of δ-aminolevulinate. In one, a single enzyme condenses succinyl-CoA and glycine to produce δ-aminolevulinate. In the other route glutamate is converted into δ-aminolevulinate with the help of three enzymes via glutamyl- tRNA^{Glu}. Aminoacylated tRNAs are ordinarily used for protein biosynthesis where they serve as substrates in the transfer of amino acids to develop the polypeptide chains following the codons of a given mRNA. In δ-aminolevulinate synthesis, glutamyl tRNA^{Glu} is reduced to glutamate 1-semialdehyde and this is the only known example of a reduction of an aminoacylated tRNA to an α-aminoaldehyde. The genes encoding the tRNA^{Glu} and the enzymes as well as catalytically active recombinant enzymes are now available. Several aspects of utilising the tRNA^{Glu} mediated δ-aminolevulinate biosynthetic pathway in industry and agriculture are considered.

2. Biosynthesis of δ-Aminolevulinate

Variety of tetrapyrrole molecules are encountered in nature. These are the chlorophylls, bacteriochlorophylls, hemes, siroheme, vitamin B_{12}, porphyrins, bilins, phycobilins, the phytochrome chromophore and factor F_{430} of methanogenic bacteria. Two or more of these tetrapyrroles are usually produced in a single organism either simultaneously or at different stages of development. Tetrapyrroles function as light absorbing pigments, photosynthetic reaction center components, electron transfer molecules and enzyme cofactors and are essential for all biological energy metabolism.

The pyrrole rings in all known tetrapyrroles arise by the condensation of two δ-aminolevulinate molecules. δ-Aminolevulinate is a non-protein amino acid. Animals, yeast and some bacteria such as *Rhizobium* produce their δ-aminolevulinate by condensing

J. H. Argyroudi-Akoyunoglou and H. Senger (eds.),
The Chloroplast: From Molecular Biology to Biotechnology, 311–315.

succinyl-CoA with glycine using a single enzyme, the δ-aminolevulinate synthase. In contrast to this, plants, algae, cyanobacteria and other bacteria such as *Escherichia coli* and *Bacillus subtilis* convert glutamate to δ-aminolevulinate with the help of three enzymes.

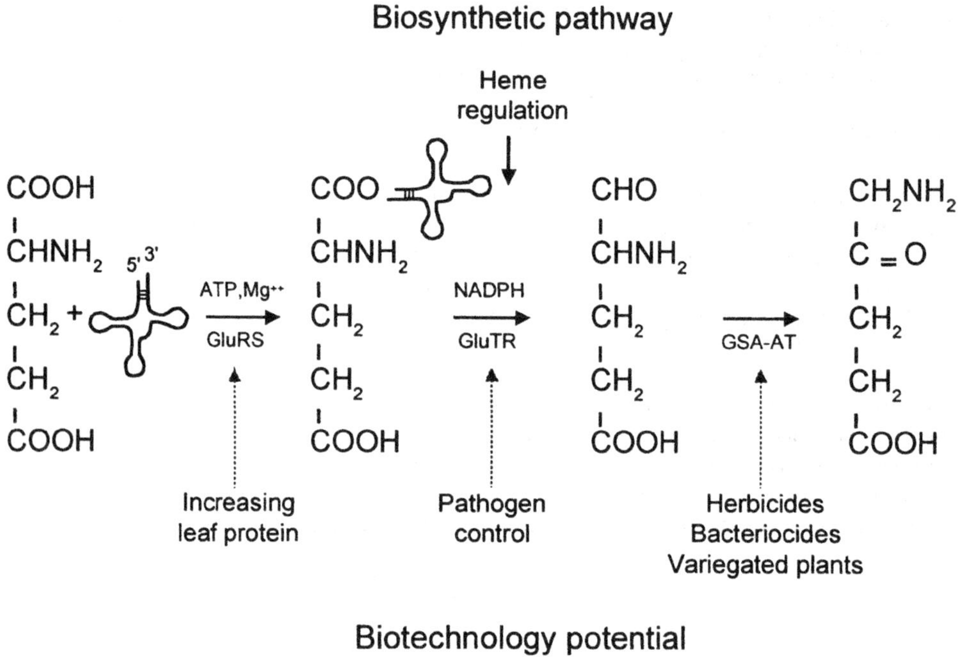

Glutamate is first activated by ligation to α-carboxyl group of $tRNA^{Glu}$ with the help of an aminoacyl tRNA synthetase [5]. An NADPH dependent reductase then converts the glutamyl $tRNA^{Glu}$ to glutamate 1-semialdehyde which is finally converted to δ-aminolevulinate by an aminotransferase. The components involved in this process are located in the chloroplasts stroma in plants and algae. Furthermore in plants and algae the $tRNA^{Glu}$ is encoded in chloroplast DNA while the enzymes are encoded in nuclear DNA. The $tRNA^{Glu}$ has a hypermodified nucleotide 5-methyl aminomethyl 2-thiouridine in the anticodon which probably plays a role in the light-dark regulation of δ-aminolevulinate synthesis. Ligation of glutamate to $tRNA^{Glu}$ requires ATP and Mg^{2+} and proceeds via a ternary intermediate. Glutamyl $tRNA^{Glu}$ reduction appears to involve complex formation.
Glutamate 1-semialdehyde aminotransferase contains vitamin b_6 [6]. This enzyme can be converted into pyridoxamine or pyridoxal 5' phosphate form (with different absorption spectra) by treatment with 4,5-diaminovalerate or 4,5-dioxovalerate. Pyridoxamine 5' phosphate form of the enzyme converts (S)-glutamate 1-semialdehyde to δ-aminolevulinate via 4,5-diaminovalerate using a Bi Bi ping pong mechanism.

The basic tools such as the genes encoding the enzymes and the $tRNA^{Glu}$ involved in δ-aminolevulinate synthesis as well as catalytically active over-produced recombinant enzymes are now available [6,8,9]. It is therefore timely to investigate ways to utilise the $tRNA^{Glu}$ mediated δ-aminolevulinate biosynthetic pathway in industry and agriculture.

3. Biotechnology potential

3.1. REGULATION OF LEAF PROTEIN

The intriguing feature of the conversion of glutamate to δ-aminolevulinate is the involvement of the $tRNA^{Glu}$. Aminoacylated tRNAs are ordinarily used for protein biosynthesis where they serve as substrates in the transfer of amino acids to develop the polypeptide chains following the codons of a given mRNA. In δ-aminolevulinate synthesis glutamyl $tRNA^{Glu}$ is reduced to glutamate 1-semialdehyde and this is the only known example of a reduction of an aminoacylated tRNA to an α-aminoaldehyde. Aminoacyl tRNA synthetases display a stringent specificity towards the tRNA substrate and the amino acid activated by it. However, the first enzyme of the δ-aminolevulinate biosynthetic pathway, glutamyl tRNA synthetase activate glutamate molecules using tRNA molecules having glutamate or glutamine anticodons producing glutamyl-$tRNA^{Glu}$ and glutaminyl-$tRNA^{Gln}$. The glutamyl-$tRNA^{Glu}$ is then used for both protein and δ-aminolevulinate synthesis. A glutaminyl $tRNA^{Gln}$ synthetase is not found in many bacteria and chloroplasts of plants and algae. They produce their glutaminyl $tRNA^{Gln}$ required for protein synthesis by the conversion of glutamyl-$tRNA^{Gln}$ to glutaminyl $tRNA^{Gln}$ with the help of an amidophosphoribosyl transferase [7]. There must be a mechanism to coordination chlorophyll and protein synthesis in leaves. It is envisaged that transforming plants with additional copies of the gene encoding glutamyl tRNA synthetase involved in δ-aminolevulinate synthesis will increase the leaf protein content.

3.2. HORTICULTURALLY ATTRACTIVE PLANTS

Antisense RNA expression of a given gene leads to elimination or reduction of its protein product. In a previous study in tobacco the cDNA sequence encoding the glutamate 1-semialdehyde aminotransferase was inserted in reverse orientation under the control of a cauliflower mosaic virus 35S promoter [4]. The transformed plants exhibited a wide variety of leaf variegations due to partial and/or complete suppression of glutamate 1-semialdehyde aminotransferase. Depending on the location of insertion of the antisense gene, the variegations ranged from entire pale green leaves to only specific regions with chlorophyll reductions. These variegations patterns were inherited in the progenies. Although these pioneering studies were performed in tobacco other plant species could be transformed to strains with attractive patterns of leaf variegation. It is thus concluded that new horticultural strains could be developed by silencing the glutamate 1-semialdehyde aminotransferase gene.

3.3. HERBICIDES AND BACTERIOCIDES

Compounds which inhibit plant growth by inhibiting the $tRNA^{Glu}$-mediated δ-aminolevulinate synthesis are potential herbicides. It has already been shown that salicylate inhibits the ligase while the glutamate analogue of puromycin and heme inhibits the reductase of barley [2,6,9]. Furthermore several compounds such as gabaculine, 6-amino-5-hydroxy-1,3-cyclohexadiene-1-carboxylic acid, glutamic-γ-monohydroxamate and 4-amino hex 5-ynoic acid, irreversibly inactivated glutamate 1-semialdehyde aminotransferase. Gabaculine, which is most potent, is a neurotoxin and is not suitable as a herbicide. However, with the elucidation of the enzyme mechanism of glutamate 1-semialdehyde aminotransferase, it is now possible to design herbicides targeted to this enzyme.

Expensive farming practices are required to reduce or eliminate the presence of *Salmonella* in chicken. In *Salmonella typhimurium*, δ-aminolevulinate is synthesised from glutamate via glutamyl $tRNA^{Glu}$ and glutamate 1-semialdehyde. In birds, δ-aminolevulinate is synthesised by condensation of glycine and succinyl-CoA. Accordingly a potential way to reduce or eliminate *Salmonella* from birds is by specifically inhibiting the $tRNA^{Glu}$-mediated δ-aminolevulinate synthesis.

3.4. PLANT PATHOGEN CONTROL USING THE *TIGRINA -D* GENE

In barley a group of mutants designated as *tigrina* mutants have been found where the capacity to repress δ-aminolevulinate biosynthesis in darkness is lost [8]. As a consequence these *tigrina* mutants overproduce protochlorophyllide to such an extent that they are pale green when grown in darkness for seven days and reach 1.5 to 15 fold higher protochlorophyllide than the wild type seedlings. When grown in day and night cycles these mutants develop a characteristic tiger's tail phenotype with green bands alternating with white or necrotic bands [4]. The green bands correspond to leaf segments elongated in the day while the pale or necrotic bands correspond to segments elongated during the night. Protochlorophyllide which is overproduced in the night is bleached by the light of the morning and causes destruction of cellular membranes and induces cell death. This destruction of cells of the leaf is localized and is confined exclusively to the areas accumulating excess protochlorophyllide. The mutation *tigrina-d*^{12} only affects the synthesis of protochlorophyllide in the darkness. *Tigrina-d*^{12} mutants develop normally into mature plants and set seed when grown in continuous light in a phytotron. However a single night period is sufficient to cause cellular destruction in the developing leaves even on a mature *tigrina-d*^{12} plant. The other *tigrina* mutants of barley are pleiotropic mutants having additional defects in the carotenoid biosynthetic pathway and chloroplast ribosome stability [1].

To curtail plant pathogens, one could introduce the *tigrina-d*^{12} gene in antisense orientation under the control of a pathogen inducible-promoter. Silencing the *tigrina-d*^{12} gene lead to de-repression of δ-aminolevulinate synthesis and to overproduction of protochlorophyllide in darkness. Induction of the antisense DNA expression by the pathogen would deregulate δ-aminolevulinate biosynthesis causing the infected leaf cells and the neighboring cells to accumulate excessive amounts of protochlorophyllide in darkness. Daylight will then selectively photodynamically destroy the pathogen and the host cells in the areas attacked by the pathogen and thereby curtail the infection. This method is only useful if the pathogen initially attacks the leaves and the spread of the pathogen is favored by darkness. An example of such a pathogens is *Phytophthora infestans*. Dark wet nights favor germination of *Phytophthora infestans* spores on potato leaves and cause severe loss of the crop.

Acknowledgements I thank Inge Sommer and Nina Rasmussen for help in composing this manuscript.

References

1. Casadoro, G., Høyer-Hansen, G., Kannangara, C.G., and Gough, S.P. (1983) Analysis of temperature and light sensitivity in *tigrina* mutants of barley, *Carlsberg Res. Commun.* **48**, 95-129.
2. Gough, S.P., Kannangara, C.G., and von Wettstein, D. (1993) Glutamate 1-semialdehyde aminotransferase as a target for herbicides, in P. Böger and G. Sandmann (eds), *Target assays for modern herbicides and related phytotoxic compounds*, Lewis Publishers, Chelsea, Mich., USA, pp. 21-27.
3. Hansson, M., Gough, S.P., Kannangara, C.G., and von Wettstein, D. (1997) Analysis of RNA and enzymes of potential importance for regulation of 5-aminolevulinic acid synthesis in the protochlorophyllide accumulating barley mutant *tigrina-d*^{12}, *Plant Physiol. Biochem.* **35**, 827-836.
4. Höfgen, R., Axelsen, K.B., Kannangara, C.G., Schüttke, I., Pohlenz, H.-D., Wilmitzer, L., Grimm, B., and von Wettstein, D. (1994) A visible marker for antisense mRNA expression in plants: Inhibition of chlorophyll synthesis with a glutamate 1-semilaldehyde aminotransferase antisense gene, *Proc. Natl. Acad. Sci.* **91**, 1726-1730.
5. Kannangara, C.G., Gough, S.P., Bruyant, P., Hoober, J.K., Kahn, A., and von Wettstein, D. (1988) $tRNA^{Glu}$ as a cofactor in δ-aminolevulinate biosynthesis: Steps that regulate chlorophyll synthesis, *TIBS* **13**, 139-143.
6. Kannangara, C.G., Andersen, R.V., Pontoppidan, B., Willows, R.D., and von Wettstein, D. (1994) Enzymic and mechanistic studies on the conversion of glutamate to 5-aminolevulinate, in D.J. Chadwick and K. Ackrill (eds), *Biosynthesis of the Tetrapyrrole Pigments*, The Ciba Foundation Symposium 180, Wiley, Chichester, England, pp. 3-20.
7. Schön, A., Kannangara, C.G., Gough, S.P., and Söll, D. (1988) Protein biosynthesis in organelles requires misacylation of transfer RNA, *Nature* (London) **331**, 187-190.
8. von Wettstein, D., Gough, S., and Kannangara, C.G. (1995) Chlorophyll Biosynthesis, *Plant Cell* **7**, 1039-1057.
9. Vothknecht, U.C., Kannangara, C.G., and von Wettstein, D. (1998) Barley glutamyl $tRNA^{Glu}$ reductase: Mutations affecting haem inhibition and enzyme activity, *Phytochem.* **47**, 513-519.

A TRANSGENIC APPROACH TO CHARACTERIZE THE PLASTID TRANSCRIPTION MACHINERY IN HIGHER PLANTS

P. MALIGA, K. LIERE, P. SRIRAMAN AND Z. SVAB
Waksman Institute, Rutgers, The State University of New Jersey,
190 Frelinghuysen Road, Piscataway, NJ, 08854-8020 USA

The plastid genes of higher plants are transcribed by two RNA polymerases: the plastid-encoded plastid RNA polymerase or PEP and the nuclear-encoded plastid RNA polymerase or NEP. Below is a brief summary of what we know about the two plastid RNA polymerases, highlighting recent data from this laboratory. Most of the data were obtained using transgenic approaches, facilitated by the development of technology for plastid transformation in tobacco [1, 2].

1. PEP, the Eubacterial-Type Plastid RNA Polymerase

The plastids of photosynthetic higher plants have evolved from eubacteria-like endosymbionts closely related to contemporary cyanobacteria. PEP is derived from the RNA polymerase of this ancestral bacterium. The plastid genome encodes proteins which are homologous to the eubacterial *(E. coli*-like*)* DNA-dependent RNA polymerase α, β and β' subunits. The subunit structure of the plastid enzyme is similar to that of the eubacterial enzyme, except that the bacterial β' subunit N- and C-terminus are encoded in separate genes. The plastid genes were named *rpoA*, *rpoB*, *rpoC1* and *rpoC2* to indicate homologies with the *E. coli* genes [3, 4]. While the genes for the PEP core enzyme have been maintained in the plastid genome, σ-like factors required for promoter recognition [5, 6] have been transferred to the nucleus [7-9]. PEP promoters are similar to the eubacterial σ^{70}-type promoters: the core promoter is comprised of two conserved blocks of hexameric sequences corresponding to the eubacterial -35 (TTGACA) and -10 (TATAAT) promoter elements. The hexamers are spaced 17-19 nucleotides apart; transcription initiates 5-7 nucleotides downstream of the -10 box sequence [3, 4].

1.1 THE *psbD* UPSTREAM LIGHT-RESPONSE ELEMENT

The *psbD* operon in tobacco plastids contains the *psbD* and *psbC* genes, encoding the D1 and CP34 photosystem II subunits. In mature leaves, transcription of most plastid genes is down-regulated. In contrast, transcription of the *psbD* operon is activated by high-irradiance blue and UV A light [10]. The σ^{70}-type core is poorly conserved, and the hexameric boxes are spaced closer than the consensus, with 15 instead of 17-19 bp between the –10 and –35

J. H. Argyroudi-Akoyunoglou and H. Senger (eds.),
The Chloroplast: From Molecular Biology to Biotechnology, 317–323.

promoter elements [11, 12]. This core promoter is faithfully recognized *in vivo* [11] and *in vitro* [12-14]. Study of transgenic tobacco plants indicates, that *psbD* transcription is 150-fold enhanced by light *via* A-rich sequences upstream of the core, the binding site for nuclear-encoded factors [11]. Transcription *in vitro* is light-responsive in extracts from wheat [14] but not from barley [13].

1.2. SPECIES-SPECIFIC ACTIVATING FACTOR FOR THE *rrn* OPERON

The plastid *rrn* operon in tobacco, maize, and pea is transcribed by the PEP from a σ^{70}-type promoter (P1 or Nt-Prrn-114 and its homologues [15, 16]). In addition, the *rrn* operon in some species has a non-essential NEP promoter (see below). The nature of the *rrn* operon promoter in spinach has been an enigma for a long time. Although *rrn* transcription in spinach initiates in the general region of P1, the transcription initiation site *in vivo* has been mapped between the –10 and –35 boxes (Pc promoter [17]). However, when the spinach Pc promoter was introduced into tobacco plastids, transcription initiated at the consensus σ^{70} site, leading to the conclusion that selection of the Pc promoter in spinach is due to a species-specific transcription factor [18]. The Pc promoter is also active in mustard and Arabidopsis [18, 19]. *In vitro* transcription data suggest, that the specificity factor interacts with the PEP [19].

1.3. RNA STABILITY IS REGULATED *VIA* THE mRNA 5' UTR

The tobacco *rbcL* gene is transcribed by the PEP from a single promoter with relatively well conserved -35 and -10 elements (TTGCGC, TACAAT), canonical spacing (18 nt), and no promoter elements outside the core [20]. Transcription of *rbcL* is developmentally regulated [21]. In chloroplasts, the *rbcL* mRNA accumulates to a high level in both the light and the dark, although the rate of *rbcL* transcription is reduced in the dark. Study of plastid transgenes indicates, that high levels of *rbcL* mRNA in the dark are due to stabilization of the mRNA *via* a segment of the 5' untranslated region [20]. Regulation of mRNA stability *via* the 5'-UTR is also known in the unicellular alga *Chlamydomonas reinhardtii* for *rbcL* [22] and *petD* [23].

2. NEP, the Phage-Type Plastid RNA Polymerase

Mitochondria, like plastids, have evolved from an eubacteria-like endosymbiont. During this evolutionary process they lost the ancestral multi-subunit RNA polymerase and acquired instead a nuclear-encoded enzyme related to the phage T3/T7 RNA polymerases. The search for the gene of the *Arabidopsis* mitochondrial RNA polymerase gene lead to the cloning of three cDNAs. One of these is targeted to plastids, and is probably the catalytic subunit of NEP, the phage-type plastid RNA polymerase. The gene is proposed to have evolved by duplication and re-targeting of the mitochondrial RNA polymerase to plastids [24].

Previously, several lines of evidence suggested the existence of a second RNA polymerase [25-28]. The existence of this second RNA polymerase was confirmed by deleting the PEP subunits from the tobacco plastid genome [16, 29] and demonstrating transcription from non-σ^{70}-type promoters in plastids lacking PEP activity [16, 30]. NEP

Plant mitochondrial consensus		YRTA	**NNNNNNNN**	[35,36]
Nt Prrn16-64	GTTATGC	CTTG	GAA**T**G**A**AAGA	[15,16]
Nt Prps2-152	AAAAGAA	TAAT	GAAT**A**GAAAA	[30]
Nt Prps16-107	CTATCTC	TATA	GTATAT**A**GAT	[30]
Nt Prpl32-1018	GATCAAG	TATC	CGAAG**T**CTAA	[32]
Nt PatpB-289	CGAATTC	CATA	GAAT**A**GATAA	[30,32]
Nt PatpI-207	ATAATAA	CATA	ATA**A**GATAAA	[30]
Nt PclpP-511	CTGTATG	CATT	ATATG**A**ACAG	[30]
Nt PclpP-173	TATTTAT	CATT	AGAAAG**A**CTT	[30]
Nt Pycf1-41	AAATCAT	CATA	TAAT**A**ATCCA	[30]
Nt Pycf2-1577	TATATAT	CGTA	GAATGA**A**GTC	[30]
Nt PaccD-129	ATATTAA	TATA	TAGTG**A**AGTG	[30]
Nt PrpoB-345	GTATTAT	CATA	ATAA**T**GGTAG	[29]
Zm PatpB-601	TTTCATT	CATA	TAT**A**ATGTGA	[34]
Zm PclpP-111	GTATTCT	TATA	GAAT**A**AGAAA	[34]
Zm PrpoB-147	TGAAATA	CATA	TATG**A**AATAC	[34]
Hv Prpl23-73	GTATATT	CATA	CC**AT**AACATA	[33]
Hv PclpP-133	GTATTCA	TATA	GAA**T**AGAATA	[33]
Plastid NEP consensus		YATA	NN**NNNNNN**	

Figure 1. The plastid Type I NEP and mitochondrial promoters have related core consensus sequence motifs. Shown is the sequence required for PrpoB-345 function, and cognate regions in other promoters. The conserved YATA motif is boxed.

promoters were also mapped in the tobacco BY4 tissue culture line [15, 31, 32], and in ribosome-less barley (*albostrians*) [33] and maize (*iojap*) [34] mutants.

Two types of NEP promoters have been identified. Type I tobacco NEP promoters characterized thus far (PaccD-129 and PrpoB-345) are contained in a small DNA fragment (-17 to +2) upstream of the transcription initiation site (+1). Most NEP promoters fall into this class (Figure 1). A salient feature of NEP promoters is a conserved YATA sequence motif centered at about -8 to -6, forming the promoter core. The YATA sequence motif is also present in the conserved YRTAT (YATAT or YGTAT) core motif of higher plant mitochondrial promoters [35, 36], supporting the common origin of the plastid NEP and the mitochondrial RNA polymerases.

Type II NEP promoters share sequences downstream (-5 to +25) of the transcription initiation site (Figure 2; ref. 37]. Thus far only one gene, *clpP*, was found to have a Type II NEP promoter. The 30-bp region shown to be important for PclpP-53 promoter function contains a 28-bp stretch which is conserved and functions in liverwort [39] and conifers [40] indicating that the phage-type plastid RNA polymerase appeared early on during the evolution of land plants. While the tobacco PclpP-53 region is well conserved in rice, it does not function as a promoter suggesting that cereals lack the Type II NEP specificity factor [37].

Spinach[38]	AACCGTAATTA**T**TACGTTTCCACATCAAAGTGAAA
Tobacco [30]	AAAAAAAATT**G**TTACGTTTCCACCTCAAAGTGAAA
Arabidopsis [37]	AAACCCCAATT**T**TACGTTTCCACATCAAAGTGAAA
Rice [37]	TCTTTCCATTCTTACGTTTCCATATTAAAGTGTAG
Marchantia [39]	GAATTTCA**T**TTTTACGTTTTTTTATTATAGAAGAG
Pinus [40]	CTTCATAT**A**CTTTACGTTCCCATATTATAGTATAG

Figure 2. Conservation of the tobacco Type II *clpP* promoter. Shown is the 35 bp sequence required for full promoter strength, and the 30 bp region sufficient to direct specific transcription initiation (underlined). Transcript 5' ends are in bold. Based on [37].

Since the plastid NEP is related to phage-type yeast and mammalian mitochondrial enzymes, it probably also associates with at least two other factors, one of which confers promoter specificity to the polymerase [41]. Given the distinct Type I and Type II recognition specificities, it is likely that the plastid NEP utilizes a similar mechanism.

3. The Role of Two RNA Polymerases in Plastid Gene Expression

The two plastid RNA polymerases transcribe distinct groups of genes in plastids: photosynthetic genes are transcribed by the PEP; most genes have promoters for both RNA polymerases, while a few housekeeping genes are transcribed exclusively by the NEP [15, 16, 30-32; reviewed in ref. 42]. Exclusively transcribed by the NEP are the *rpoB* operon encoding three of the PEP core subunits, *clpP* in monocots [34] and *accD* in dicots [30]. Subsequent to accidental targeting to plastids, transcription of plastid genes by the NEP probably occurred from spurious promoters, generating additional sets of RNAs for plastid genes. The NEP became indispensable later, when it took over transcription of essential genes, such as the *rpoB* operon encoding PEP subunits. Transcription of PEP genes by the NEP was probably a critical evolutionary step of the nucleus indirectly taking control of the transcription of plastid genes, thereby fully integrating plastids into the regulatory network of multicellular plants.

Most NEP promoters are inactive in chloroplasts, and could be identified only in non-photosynthetic plastids [30-34]. These NEP promoters should function in some non-photosynthetic plastid types. However, their plastid type and cell type specificities remain to be determined. Indeed, one of these, the Type I *rpoB* promoter, is highly active during early chloroplast development [34]. However, there are also examples for constitutive expression of Type I NEP promoters, such as the *clpP* promoter in rice and maize (PclpP-111), which actively transcribe the *clpP* protease subunit gene in both leaves and in embryogenic cells [34, 43]. The tobacco Type II *clpP* promoter (PclpP-53) is also constitutive [37]. It appears therefore, that NEP promoters are highly regulated in a promoter-specific fashion.

The existence of promoters for both RNAPs upstream of many plastid housekeeping genes is an intriguing feature. Is this accidental, or could this have evolved to ensure transcription of essential genes from alternative promoters when PEP or NEP are in limited supply? The plastid ribosomal RNA operon (*rrn*) offered a good model for testing the role of a developmentally-timed promoter switch in chloroplast differentiation. In chloroplasts, the *rrn* operon is transcribed by PEP from P1, a σ^{70}-type promoter with conserved -10/-35 hexameric sequences. The *rrn* operon has a second promoter, P2 recognized by the NEP, which is inactive in chloroplasts (Prrn-64 [15]; Prrn-62 [16]). Homology-directed DNA sequence manipulation was employed to obtain transgenic plants in which the *rrn* operon had only the P1 or P2 promoter. We have found that the P1 plants are phenotypically indistinguishable from wild-type (P1P2) plants. Accordingly, transcription of the *rrn* operon by NEP is not essential at any time during plastid development. Lack of an essential role for the *rrn* NEP promoter argues against a promoter switch for the expression of genes which have promoters for both PEP and NEP during development, and suggest a parallel, rather than a hierarchical role for the two RNA polymerases.

4. References

1. Svab, Z., and Maliga, P. (1993) High-frequency plastid transformation in tobacco by selection for a chimeric *aadA* gene, *Proc. Natl. Acad. Sci. USA* **90,** 913-917.
2. Svab, Z., Hajdukiewicz, P., and Maliga, P. (1990) Stable transformation of plastids in higher plants, *Proc. Natl. Acad. Sci. USA* **87,** 8526-8530.
3. Gruissem, W. and Tonkyn, J.C. (1993) Control mechanisms of plastid gene expression, *Crit. Rev. Plant Sci.* **12,** 19-55.
4. Link, G. (1996) Green Life: control of chloroplast gene transcription, *Bioessays* **18,** 465-471.
5. Tiller, K., Eisermann, A., and Link, G. (1991) The chloroplast transcription apparatus from mustard (*Sinapis alba* L.) Evidence for three different transcription factors which resemble bacterial s factors, *Eur. J. Biochem.* **198,** 93-99.
6. Tiller, K., and Link, G. (1993) Sigma-like transcription factors from mustard (*Sinapis alba* L.) etioplast are similar in size to, but functionally distinct from, their chloroplast counterparts, *Plant Mol. Biol.* **21,** 503-513.
7. Liu, B. and Troxler, R.F. (1996) Molecular characterization of a positively photoregulated nuclear gene for a chloroplast RNA polymerase σ factor in *Cyanidium caldarium*, *Proc. Natl. Acad. Sci. USA* **93,** 3313-3318.
8. Isono, K., Shimizu, M., Yoshimoto, K., Niwa, Y., Satoh, K., Yokota, A., and Kobayashi, H. (1997a) Leaf-specifically expressed genes for polypeptides destined for chloroplasts with domains for σ^{70} factors of bacterial RNA polymerases in *Arabidopsis thaliana*, *Proc. Natl. Acad. Sci. USA* **94,** 14948-14953.
9. Tanaka, K., Tozawa, Y., Mochizuki, N., Shinozaki, K., Nagatani, A., Wakasa, K., and Takahashi, H. (1997) Characterization of three cDNA species encoding plastid RNA polymerase sigma factors in *Arabidopsis thaliana*: evidence for the sigma factor heterogeneity in higher plant plastids, *FEBS Lett.* **413,** 309-313.
10. Christopher, D.A., Kim,M., and Mullet, J.E. (1992) A novel light-regulated promoter is conserved in cereal and dicot chloroplasts, *Plant Cell* **4,** 785-798.
11. Allison, L.A. and Maliga, P. (1995) Light-responsive and transcription-enhancing elements regulate the plastid *psbD* core promoter, *EMBO J.* **14,** 3721-3730.
12. To, K.Y., Cheng, M.C., Suen, D.F., Mon, D.P., Chen, L.F.O., and Chen, S.C.G. (1996) Characterization of the light-responsive promoter of rice chloroplast *psbD-C* operon and the sequence-specific DNA binding factor, *Plant Cell Physiol.* **37,** 660-666.
13. Kim, M. and Mullet, J.E. (1995) Identification of a sequence-specific DNA binding factor required for transcription of the barley chloroplast blue light-responsive *psbD-psbC* promoter, *Plant Cell* **7,** 1445-1457.
14. Satoh, J., Baba, K., Nakahira, Y., Shiina, T., and Toyoshima, Y. (1997) Characterization of dynamics of the *psb*D light-induced transcription in mature wheat chloroplasts, *Plant Mol. Biol.* **33,** 267-278.

15. Vera, A., and Sugiura, M. (1995). Chloroplast rRNA transcription from structurally different tandem promoters: an additional novel-type promoter, *Curr. Genet.* **27**, 280-284.
16. Allison, L.A., Simon, L.D., and Maliga, P. (1996) Deletion of *rpoB* reveals a second distinct transcription system in plastids of higher plants, *EMBO J.* **15,** 2802-2809.
17. Iratni, R., Diederich, L., Harrak, H., Bligny, M., and Lerbs-Mache, S. (1997) Organ-specific transcription of the *rrn* operon in spinach plastids, *J Biol. Chem.* **272**, 13676-13682.
18. Sriraman, P., Silhavy, D.. and Maliga, P. (1998a) Transcription from heterologous promoters rRNA operon promoters in chloroplasts reveals requirement for specific activating factors, *Plant Physiol.* **117,** 1495-1499.
19. Pfannschmidt ,T. and Link, G. (1997) The A and B forms of plastid DNA-dependent RNA polymerase from mustard (*Sinapis alba* L.) transcribe the same genes in a different developmental context, *Mol. Gen. Genet.* **257,** 35-44.
20. Shiina, T., Allison, L.A., and Maliga, P. (1998) *rbcL* transcript levels in tobacco plastids are independent of light: Reduced dark transcription rate is compensated by increased mRNA stability, *Plant Cell* **10,** in press.
21. Isono, K., Niwa, Y., Satoh, K., and Kobayashi, H. (1997) Evidence for transcriptional regulation of plastid photosynthesis genes in *Arabidopsis thaliana* roots, *Plant Physiol.* **114,** 623-630.
22. Salvador, M.L., Klein, U., and Bogorad, L. (1993). 5' sequences are important positive and negative determinants of the longevity of *Chlamydomonas* chloroplast gene transcripts, *Proc. Natl. Acad. Sci. USA* **90,** 1556-1560.
23. Sakamoto, W., Chen, X., Kindle, K.L., and Stern, D.B. (1994). Function of the *Chlamydomonas reinhardtii petD* 5' untranslated region in regulating the accumulation of subunit IV of the cytochrome b_6/f complex, *Plant J.* **6,** 503-512.
24. Hedtke, B., Börner, T., and Weihe, A. (1997) Mitochondrial and chloroplast phage-type RNA polymerases in Arabidopsis, *Science* **277,** 809-811.
25. Morden, C.W., Wolfe, K.H., dePamphilis, C.W., and Palmer, J.D. (1991) Plastid translation and transcription in a non-photosynthetic plant: intact, missing and pseudo genes, *EMBO J.* **10,** 3281-3288.
26. Hess, W.R., Prombona, A., Fieder, B., Subramanian, A.R., and Börner ,T. (1993) Chloroplast *rps15* and the *rpoB/C1/C2* gene cluster are strongly transcribed in ribosome-deficient plastids: evidence for a functioning non-chloroplast-encoded RNA polymerase, *EMBO J.* **12**, 563-571.
27. Han, C.D., Patrie, W., Polacco, M., and Coe, E.H. (1993) Aberrations in plastid transcripts and deficiency of plastid DNA in striped and albino mutants in maize, *Planta* **191,** 552-563.
28. Lerbs-Mache, S. (1993) The 110-kDa polypeptide of spinach plastid DNA-dependent RNA polymerase: Single-subunit enzyme or catalytic core of multimeric enzyme complexes? *Proc. Natl. Acad. Sci. USA* **90,** 5509-5513.
29. Serino,G., and Maliga, P. (1998) RNA polymerase subunits encoded by the plastid *rpo* genes are not shared with the nucleus-encoded plastid enzyme, *Plant Physiol.* **117,** 1165-1170.
30. Hajdukiewicz, P.T.J., Allison, L.A., and Maliga, P. (1997) The two RNA polymerases encoded by the nuclear and the plastid compartments transcribe distinct groups of genes in tobacco plastids, *EMBO J* **16**, 4041-4048.
31. Vera, A., Hirose, T., and Sugiura, M. (1996) A ribosomal protein gene (*rpl32*) from tobacco chloroplast DNA is transcribed from alternative promoters: similarities in promoter region organization in plastid housekeeping genes, *Mol. Gen. Genet.* **251**, 518-525.
32. Kapoor, S., Suzuki, J.Y., and Sugiura, M. (1997) Identification and functional significance of a new class of non-consensus-type plastid promoters, *Plant J.* **11,** 327-337.
33. Hübschmann,T. and Börner, T. (1998) Characterization of transcript initiation sites in ribosome-deficient barley plastids, *Plant Mol. Biol.* **36,** 493-496.
34. Silhavy, D., and Maliga, P. (1998) Mapping of promoters for the nucleus-encoded plastid RNA polymerase (NEP) in the *iojap* maize mutant, *Curr. Genet.* **33,** 340-344.
35. Caoile, A.G.F.S. and Stern, D. (1997) A conserved core element is functionally important for maize mitochondrial promoter activity, *Nucleic Acids Res.* **25,** 4055-4060.
36. Yan, B., and Pring, D.R. (1997) Transcriptional initiation sites in sorghum mitochondrial DNA indicate conserved and variable features, *Curr. Genet.* **32,** 287-295.
37. Sriraman, P., Silhavy, D., and Maliga, P. (1998b) The phage-type PclpP-53 promoter comprises sequences downstream of the transcription initiation site, *Nucleic Acids Res.* in press.
38. Westhoff, P. (1985) Transcription of the gene encoding the 51 kd chlorophyll *a*-apoprotein of the photosystem II reaction center from spinach, *Mol. Gen. Genet.* **201**, 115-123.

39. Kohchi, T., Ogura, Y., Umesono, K., Yamada, Y., Komano, T., Ozeki, H., and Ohyama, K. (1988) Ordered processing and splicing in a polycistronic transcript in liverwort chloroplasts, *Curr. Genet.*, **14**, 147-154.

40. Clarke, A.K., Gustafsson, P., and Lidholm, J.A. (1994) Identification and expression of the chloroplast clpP gene in the conifer *Pinus contorta, Plant Mol. Biol.* **26,** 851-862.

41. Tracy, R.L. and Stern, D.B. (1995) Mitochondrial transcription initiation: promoter structures and RNA polymerases. *Curr. Genet.*, **28**, 205-216.

42. Maliga, P. (1998) Two plastid RNA polymerases of higher plants: an evolving story, *Trends Plant Sci.* **3,** 4-6.

43. Silhavy, D., and Maliga, P. (1998). Plastid promoter utilization in rice embryogenic cell culture, *Curr. Genet.* **34,** 67-70.

STRUCTURE AND FUNCTION OF THE PHOTOSYNTHETIC APPARATUS IN TRANSGENIC TOBACCO

R. VALCKE & L. QUANTEN

Limburgs Universitair Centrum, Laboratory of Botany, Dept. S.B.G., Universitaire Campus, B-3590 Diepenbeek, Belgium.

1. Introduction

Extensive studies about the role of cytokinins on the development of higher plants has been published. Due to its pleiotropic action, the understanding of the physiological function of cytokinins is very difficult. To study the interaction between the cytokinins and the plastids, the response at the level of gene expression and photosynthetic activity was usually analysed by applying exogenous cytokinins. Using such experimental systems in which high, non-physiological concentrations of natural or synthetic cytokinins were applied at leaf discs, wound response and pharmacological effects can not be avoided. Moreover, two tools that have been valuable in determining the physiological functions of some plant hormones, namely, well-defined biosynthetic or perception mutants and chemical inhibitors of hormone action, are not available for cytokinins as stated by Kende and Zeevaart [1]. Therefore, in this work, a different model is being used. Transgenic tobacco plants in which the *ipt*-gene was introduced using the *Agrobacterium tumefaciens* transformation system, are studied. The *ipt*-gene, coding for the isopentenyl transferase, a key enzyme in the biosynthesis of cytokinin, is under the control of different promotors. Physiological effects of this gene were already extensively described by Beinsberger et al. [2], [3] and Estruch et al. [4].

The model systems used were the P*hsp70-ipt* and the P*ssu-ipt* chimeric constructs in which the *ipt*-gene is coupled to the heat-inducible promotor of the 70kDa heat shock protein of *Drosophila melanogaster* or to the light-inducible promotor of the small subunit of the RubisCO-enzyme (Ribulose-1,5-diphosphate Carboxylase/oxygenase) of *Pisum sativum* respectively.

In this paper, the effect of an increased amount of endogenous cytokinin on the morphology of the P*hsp-ipt* and P*ssu-ipt* seedlings and on the structure and function of the photosynthetic apparatus of the P*ssu-ipt* line will be described. The photosynthetic

J. H. Argyroudi-Akoyunoglou and H. Senger (eds.),
The Chloroplast: From Molecular Biology to Biotechnology, 325–330.

capacity was studied under "*in vivo*"-conditions following the oscillation patterns in oxygen evolution and chlorophyll fluorescence emission using the "Leaf Disc Electrode" according to Walker [5] and by quenching analysis, essentially as described by van Kooten and Snel [6]. The results indicate that due to an increase in the endogenous cytokinin level, the thermodynamic control mechanisms of the primary photochemical reaction and the metabolic regulation (electron transport vs Calvin cycle) are strongly affected.

2. Materials and Methods

2.1. PLANT MATERIAL

Genetically modified *Nicotiana tabacum* cv. Petit Havana SR1 containing a supplementary *ipt*-gene, was obtained using the *Agrobacterium tumefaciens* transformation system as described in Beinsberger et al. [7] and Van Loven et al. [8]. Two types of transgenic plants are used: P*ssu-ipt*, a chimeric construct with the *ipt*-gene coupled to the light inducible promotor of the small subunit of the RubisCO-enzyme and P*hsp-ipt*, the *ipt*-gene in a chimeric construction with the 70 kDa heat inducible promotor of *Drosophila melanogaster*. The plants were cultivated in a greenhouse: temperature 25°C day/18°C night, relative humidity of 60%, additional illumination was provided 16h a day with AgroSon T (400W) and HTQ (400W) lamps (photon flux density of 200μmol quanta m^{-2} s^{-1}).

2.2. OXYGEN AND FLUORESCENCE MEASUREMENTS

The measuring routine was according to the three-part procedure described by Walker [5]. Following a wake-up, first, oxygen evolution and chlorophyll fluorescence kinetics of 10 cm^2 leaf discs were recorded. Second, following a 5 min dark interval a photon flux density curve was obtained. If necessary, the oxygen evolution under low irradiance was measured more accurately in the third part. The fluorescence emission was recorded simultaneously. The non-modulated fluorescence according to Schreiber et al. [9] was used to calculate the photochemical and non-photochemical quenching.

2.3. PIGMENT AND PHYTOHORMONE ANALYSIS

Quantification of the photosynthetic pigments in 80% aceton was according to Lichtenthaler and Wellburn [10]. Phytohormones were analysed as described by Beinsberger et al. [3].

3. Results and Discussion

3.1. ENDOGENOUS CYTOKININ CONTENT AND MORPHOLOGY

Repeated heat shock treatment of the P*hsp-ipt* transgenic plants resulted in an increase of the endogenous cytokinin content (table 1). The treated plants were charaterised by an inhibition of the apical dominance, short internodes, strongly reduced root system and a change in the ultrastructure of the chloroplasts as already described by Van Loven et al. [8]. The endogenous cytokinin content of the P*ssu-ipt* line is much higher (table 1) but in contradiction to the wild type SR1, the distribution of the cytokinins throughout the leaves from top to bottom is inversed.

TABLE 1: Endogenous cytokinin content in ng g^{-1} fresh mass of SR1 type and P*hsp-ipt* and P*ssu-ipt* transgenic tobacco leaves.

Plant type	SR1 wild type	P*hsp-ipt*	P*ssu-ipt* Grafts	P*ssu-ipt* Rooted plants
Z	25.9 ± 10	-	154 ± 30	251 ± 145
ZR	28.9 ± 8	271 ± 10	467 ± 172	238 ± 64
IP	29.4 ± 6	54 ± 3	30.3 ± 5	46.8± 19
Ipa	11.9 ± 3	-	55.8 ± 14	30.9 ± 15

Compared to the P*hsp-ipt* plants, the morphology of the P*ssu-ipt* transgenic line is more pronounced with a very strong reduction in the root system of the *Pssu-ipt* seedlings and a bushy growth pattern of the grafted shoots (extreme outgrowth of the axiallary shoots). Furthermore, leaves were strongly curled and the pigment distribution is very heterogenous as was shown earlier by Beinsberger et al. [3].

3.2. PIGMENT CONTENT (SR1 vs P*ssu-ipt*)

The pigment content of the transgenic P*ssu-ipt* plants is significantly increased compared to the SR1 wild-type (table 2). The total chlorophyll amount depends on the endogenous cytokinin level and consequently on the stature of the plant material (e.g. graft or seedling).

Important to notice here is the problem of the physiological age. Due to the endogenous overproduction of cytokinins and the concomitantly change in phytohormone ratio's, the absolute age of the plant material is not any longer in relation with the physiological age.

TABLE 2: Pigment content (on fresh weight and on surface area), chl*a/b* and chl/car ratio's of mature leaves of SR1 type and of P*hsp-ipt* and P*ssu-ipt* transgenic plants.

Plant type	SR1	P*ssu-ipt* Grafts	P*ssu-ipt* rooted plants
Chl (mg kg^{-1} f.m.)	1200 ± 200	2100 ± 200	1600 ± 200
Chl (mg m^{-2})	236 ± 50	450 ± 80	310 ± 40
Chl a/b	4 ± 0.3	3.7 ± 0.4	4.0 ± 0.5
Chl/Car	6.1 ± 0.8	7.0 ± 1	6.0 ± 1

No significant changes in the Chl *a/b* ratio nor in the Chl/car ratio could be observed in transgenic tobacco compared to the wild type.

3.3. PHOTOSYNTHETIC CHARACTERISTICS

As a basis for the calculation of the apparent quantum requirement (QR), dark respiration (R_D), light compensation point, maximal photosynthetic capacity (P_{max}) and the covexity coefficient, light response curves were used (table 3). The quantum requirement calculated as the initial slope of the photon flux density curve is in the P*ssu-ipt* rooted plants significantly higher compared to the P*ssu-ipt* graft and wild type. This indicates a less efficient conversion of the absorbed light quanta in PSII photochemistry. Nevertheless, the maximal yield of photochemistry is not affected.
The higher light compensation point means that in this transgenic plants, according to Walker and Osmond [11], a higher light intensity is needed to compensate for the dark respiration which is of comparable magnitude in the three plants studied. On the other hand, the significantly higher maximal photosynthetic capacity at light saturation (800 µmol m^{-2}s^{-1}) in the P*ssu-ipt* rooted plants would suggest that due to an increased level of endogenous cytokinin the metabolic capacity has been improved. This statement is supported by the higher value of the convexity coefficient, which indicates a more efficient link between the thermodynamic and metabolic capacities as shown by Terashima and Takenaka [12]. The P*ssu-ipt* grafts, which are characterised by an even higher endogenous cytokinin content (table 1), behave in a different way.

TABLE 3: Photosynthetic characteristics: quantum requirement (Q_R), light compensation point (LCP), maximal photosynthetic capacity (P_{max}), dark respiration (D_R), convexity coefficient (Θ), maximal quantum yield for photochemistry (Φ_{PSII}), photochemical quenching (q_p), non-photochemical quenching (q_n).

	SR1	*Pssu-ipt* Grafts	*Pssu-ipt* rooted plants
Q_R	13 ± 1	12.2 ± 0.6	17 ± 3
LCP	22 ± 2	24 ± 3	30 ± 5
P_{max}	11 ± 2	15 ± 4	16 ± 2
D_R (µmol/cm²/chl)	1.7 ± 0.3	1.9 ± 0.3	1.7 ± 0.8
Θ (dark-adapted)	0.78 ± 0.03	0.6 ± 0.1	0.86 ± 0.02
Φ_{PSII}	0.75 ± 0.02	0.80 ± 0.02	0.78 ± 0.02
q_p	0.5 ± 0.2	0.8 ± 0.1	0.5 ± 0.2
q_N	0.6 ± 0.1	0.16 ± 0.03	0.4 ± 0.3

The strong reduction in the convexity coefficient but not in the other measured parameters as compared to the wild type (even a higher P_{max}), shows that a high content of cytokinin harms the conversion of the thermodynamic to the metabolic capacity. This is confirmed by the low non-photochemical quenching coefficient (energetic quenching) in the *Pssu-ipt* grafts indicative for a reduction in the membrane energisation resulting in a lower production of ATP and reducing power as suggested by Synková et al. [13]. These results must be interpreted with care due to the inaccuracy of the determination of F_M under the used experimental conditions. At the end of the wake up period, q_N-relaxation was probably not completed. The remaining q_N could have interfered in the F_M determination. A lower calculated value of q_N in the subsequent light period, as is observed in the *Pssu-ipt* grafts, can also be explained by assuming a lower decay of q_N in the dark, resulting in a lower apparent F_M. However, the lower coupling status gives evidential support for a reduced membrane energisation.

The results presented in this paper clearly shows that due to an increase in endogenous cytokinin content the photosynthetic capacity of a leaf is affected. Changes in the ultrastructure of the chloroplast, in the composition of the polypeptides of the thylakoid membrane as shown by Synková et al. [14] and in the capacity of the photosynthetic electron transport measured *in vitro* as indicated by Synková et al. [13] and Baldewijns and Valcke [15], support this statement. Moreover, the stature of the transgenic plant material as well as its general physiological respons depend not only on several threshold values of endogenous cytokinin content and subsequently on the changed endogenous phytohormone ratio's but also on the expression of the chimeric construct itself that is most probably tissue dependent (developmental regulation) and on the interrelation with environmental constraints (light, water status). Thorough analysis of the regulation of gene expression of the photosynthetic apparatus will give more insight in this processes.

4. References

1. Kende, H.and Zeevaart, J.A.D .(1997), The five "classical" plant hormones, *The Plant Cell* **9**, 1197-1210.
2. Beinsberger, S., Valcke, R., Deblaere, R., Clijsters, H., De Greef, J. and Van Onckelen, H. (1991) Effects of the introduction of *Agrobacterium tumefaciens* T-DNA *ipt* gene in *Nicotiana tabacum* L.cv. Petit Havana SR1 plant cells, *Plant Cell Physiol.* **32**, 489-496.
3. Beinsberger, S., Clijsters, H., Valcke, R. and Van Onckelen, H. (1992) Morphological characteristics and phytohormone content of *ipt*-transgenic tobacco, in C.M. Karssen, L.C. van Loon and D. Vreugdenhill (eds), *Progress in Plant Growth Regulation*, Kluwer Acaddemic Publishers, Dordrecht, pp. 738-745.
4. Estruch, J., Prinsen, E., Van Onckelen, H., Schell, J. and Spena, A. (1991) Viviparous leaves produced by somatic activation of an inactive cytokinin-synthesizing gene, *Science* **254**, 1364-1367.
5. Walker, D. (1990) *The use of oxygen electrode and fluorescence probes in simple measurements of photosynthesis*, Oxygraphics Ltd., Sheffield, UK.
6. van Kooten, O. and Snel, J. (1990) The use of chlorophyll nomenclature in plant stress physiology, *Photosynthesis Research* **25**, 147-150.
7. Beinsberger, S., Valcke, R., Clijsters, H., De Greef, J. and Van Onckelen, H. (1992) Effects of enhanced cytokinin levels in *ipt* transgenic tobacco, in M. Kaminek, D.W.S. Mok and E. Zažímalová (eds.), *Physiology and Biochemistryof cytokinins in Plants*, SPB Acad. Publ. bv., The Hague, pp. 77-82.
8. Van Loven, K., Beinsberger, S., Valcke, R., Van Onckelen, H. and Clijsters, H. (1993) Morphometric analysis of the growth of *Phsp-ipt* transgenic tobacco plants, *J. Exp. Botany* **44(268)**, 1671-1678.
9. Schreiber, U., Neubauer, C. and Klughammer, C. (1989) Devices and methods for room-temperature fluorescence analysis, *Phil. Trans. Roy. Soc. Lond.* **B323**, 241-251.
10. Lichtenthaler, H. K. and Wellburn, A.R. (1983) Determinations of total carotenoids and chlorophylls a and b of leaf extracts in different solvents, *Biochem. Soc. Trans.* **11**, 591-592.
11. Walker, D. A.and Osmond, C.B. (1986) Measurement of photosynthesis *in vivo* with a leaf disc electrode: correlations between light dependence of steady-state photosynthetic O_2 evolution and chlorophyll a fluorescence transients, *Proc. Roy. Soc. Lond.* **B227**, 267-280.
12. Terashima, I. And Takenaka, A. (1986) Organisation of photosynthetic systems of dorsiventral leaves as adapted to the irradiation from the adaxial side, in R. Marcelle, Clijsters, H. and Van Poucke, M. (eds.), *Biological Control of Photosynthesis*, Martinus Nijhoff Publ., Dordrecht, pp. 219-299.
13. Synková, H., Van Loven, K., Pospíšilová, J. and Valcke, R. (1998) Photosynthesis of transgenic *Pssu-ipt* tobacco, submitted.
14. Synková, H., Van Loven, K. and Valcke, R. (1997) Increased content of endogenous cytokinins does not delay degradation of photosynthetic apparatus in tobacco, *Photosynthetica* **33(3-4)**, 595-608.
15. Baldewijns, L. and Valcke, R. (1998) Photosystem II electron transport in transgenic tobacco, *Proc. Xith Intern. Congress on Photosynthesis*, Budapest, XXX.

LIGHT, TEMPERATURE AND REDOX CONTROL OF THE DEVELOPMENT OF THE PHOTOSYNTHETIC APPARATUS

N.P.A. HUNER, K.E.WILSON, M. KRÓL, A.G. IVANOV, E. MISKIEWICZ, R.M. MORGAN, L.V. SAVITCH
The University of Western Ontario,
Dept. of Plant Sciences, London, Canada N6A 5B7

1. Introduction

The absorption of light through the primary photochemical reactions of photosystem II (PSII) and PSI occurs at least 15 orders of magnitude faster than temperature-dependent redox reactions of intersystem electron transport and subsequent enzyme-catalyzed carbon assimilation. It is imperative that photosynthetic organisms maintain a balance between energy supplied through photochemistry and energy consumed through electron transport and metabolism [1]. According to Durnford and Falkowski [2], a balance between energy absorbed and energy utilized is attained when

$$\sigma\mathbf{PSII} \cdot \mathbf{I} = \mathbf{n} \cdot \tau^{-1} \quad (1)$$

where **σPSII** is the functional absorption cross section of PSII, **I** is the incident radiation, **n** is the number of electron sinks and τ^{-1} is the turnover rate of these sinks. Thus, an energy imbalance will occur whenever

$$\sigma\mathbf{PSII} \cdot \mathbf{I} > \mathbf{n} \cdot \tau^{-1} \quad (2)$$

This will be manifested when organisms are exposed to excessive irradiance (I) at normal temperatures or, alternatively upon exposure to low temperature and normal irradiance since τ^{-1} is temperature dependent. An energy imbalance will be reflected in a change in the redox state of PSII which reflects an increase in PSII excitation pressure [1]. This may be estimated by Chl a fluorescence as $1-q_P$, where q_P is the photochemical quenching parameter [3].

2. Mechanisms of Photosynthetic Adjustment

According to equation (2), organisms can adjust to exposure to high excitation pressure by reducing **σPSII** by either reducing the efficiency of energy transfer by increasing nonphotochemical quenching (q_N) through modulation of the xanthophyll cycle [4] and / or by reducing the amount of light harvesting complex associated with PSII [5]. Alternatively, organisms could increase **n** in response to high excitation pressure, and thus, increase their

J. H. Argyroudi-Akoyunoglou and H. Senger (eds.),
The Chloroplast: From Molecular Biology to Biotechnology, 331–336.

capacity to utilize the absorbed light energy through an increase in sink capacity ($\mathbf{n} \cdot \tau^{-1}$)[1]. The mechanisms employed to counteract an environmentally-induced energy imbalance are species dependent.

2.1 GREEN ALGAE AND CYANOBACTERIA

Maxwell *et al.* [6, 7] reported that photosynthetic adjustment of *Chlorella vulgaris* and *Dunaliella salina* to growth at low temperature and moderate irradiance is comparable to cells grown at high light with respect to the decrease in Pmax, the increase in the ratio of Chl a/b, the decrease in Chl content with a concomitant increase in zeaxanthin, resistance to photoinhibition as well as the reduction in *Lhcb* mRNA and polypeptide content coupled to the accumulation of Cbr, the carotenoid-binding protein [8]. These adjustments cannot be explained as either a simple growth temperature effect nor a simple irradiance response. Rather, cultures grown at either 5°C /150 μmol m^{-2} s^{-1} (5/150) or 27/2200 are photosynthetically adjusted to growth at high PSII excitation pressures whereas cultures grown at either 27/150 or 5/20 are photosynthetically adjusted to growth at low PSII excitation pressures. Similar conclusions regarding the role of PSII excitation pressure have been reported for thermal and photoacclimation of *Laminaria saccharina* [9]. *Chlorella* grown at either 5/150 or 27/2000 exhibit limitations with respect to Pi utilization, are feedback limited, and appear unable to adjust carbon metabolism. As a consequence, green algae appear unable to adjust sink capacity [10]. Thus, green algae appear to respond to high light- or low temperature-induced PSII excitation pressure by reducing σ**PSII** through a reduction in light harvesting capacity coupled with an increased capacity to dissipate excess light nonphotochemically [6]. However, not all green algae respond to high PSII excitation pressure in the same way. In contrast to *Chlorella*, the antarctic psychrophile, *Chlamydomonas subcaudata*, does not alter its Chl content significantly when grown at either 8/20 or 8/150. This may be due, in part, to the fact that this species exhibits a significantly altered organization of LHCII-PSII and LHCI-PSI units compared to the mesophile, *Chlamydomonas reinhardtii* [11].

When cultures of the filamentous cyanobacterium, *Plectonema boryanum*, are grown at either 29/150 or 15/10 they exhibit the expected blue-green pigmentation. However, when this cyanobacterium is grown at either 15/150 or 29/750, the culture exhibits a red to orange pigmentation due to the accumulation of the carotenoid, myxoxanthophyll, with the concomitant decreases in Chl a, phycobilisome content and energy transfer efficiency from the phycobilisomes to PSII. These adjustments result in an increased resistance to photoinhibition.The myxoxanthophyll has been localized to the outer cell membrane and acts as a sunscreen, thus, reducing **I** whereas the decreased phycobilisome content and energy transfer efficiency protect PSII from over-excitation by reducing σ**PSII**. Shifting cultures grown at 15/150 to conditions of 29/150 induces the rapid depletion of myxoxanthophyll and the concomitant accumulation of Chl a and phycobilisomes which are coupled to a decreased resistance to photoinhibition (Miskiewicz and Huner, unpublished).

It appears that these photoautotrophs detect energy imbalances induced either by excessive irradiance or by low temperature through a common sensing mechanism involving the redox state of PSII. If this is correct, we predict, first, that the Chl a/b ratio and Lhcb content should vary in a linear fashion with respect to 1-q_P [7]. Second, and more important, variations in the Chl a/b ratio and Lhcb content measured as a function of 1-q_P should be independent of the method used to modulate PSII excitation pressure. The results of Fig. 1 illustrate that, as expected, Chl a/b measured under steady state growth is positively and linearly correlated with $1\text{-}q_P$ whether one modulates PSII excitation pressure by low temperature (open symbols) or by high light (closed symbols). However, there is a clear temperature dependence since the curve obtained for the modulation of $1\text{-}q_P$ by low

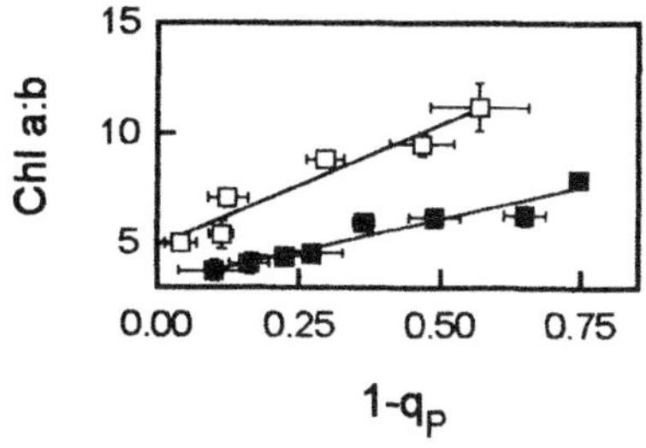

Figure 1. Effects of excitation pressure on the Chl a/b ratio in *Chlorella*. Cultures were grown at either 27° (closed squares) or 5°C (open squares).

temperature is clearly distinct from that obtained for the modulation of $1\text{-}q_P$ by high light. The same conclusion was reached when Lhcb content was monitored as a function of $1\text{-}q_P$ (data not shown).

The epoxidation state (EPS) of the xanthophyll cycle intermediates reflects the conversion of the light harvesting xanthophyll, violaxanthin, to the energy quenching xanthophyll, zeaxanthin [4]. As expected, the epoxidation state of the xanthophyll cycle intermediates decreased in a linear fashion with respect to $1\text{-}q_P$ (Fig. 2). However, in contrast to either Chl a/b or Lhcb content, the relationship between EPS and $1\text{-}q_P$ appeared to be independent of whether $1\text{-}q_P$ was modulated by low temperature or high light. Thus, it appears that during growth and development of *Chlorella*, EPS is regulated directly by the redox state of PSII whereas light harvesting content measured either as Chl a/b or Lhcb content is regulated by a combination of temperature as well as the redox state of PSII.

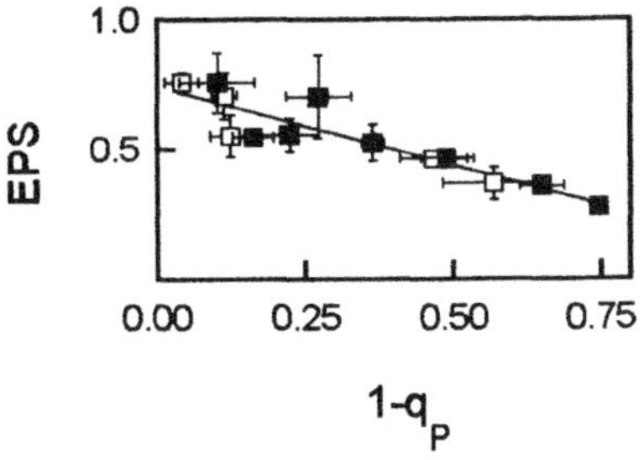

Figure 2. The effects of excitation pressure on EPS in *Chlorella*. Cultures were grown at either 27° (open squares) or 5°C (closed squares).

However, is PSII the redox sensor? To address this question, *Chlorella* was grown under high PSII excitation pressure at 5/150. When these yellow cultures are shifted to 29/150, they green within about 12 h during which time PSII excitation pressure decreases and the Chl a/b ratio decreases from 10 to about 3 whereas the Chl content increases to that of control cultures grown at 29/150 [7]. We tested the ability of DCMU and DBMIB to block this greening process. Although PSII remained totally reduced in the presence of 1 μM DCMU, the cultures still exhibited normal greening and the accumulation of Lhcb. Clearly, PSII can not be the redox sensor controlling the expression of *Lhcb*. However, 10 μM DBMIB completely blocked this greening response. Thus, we conclude that photosynthetic electron transport regulates *Lhcb* expression and that a putative redox sensor probably is localized between PSII and the Cyt *b6/f* complex. This is consistent with the reports for cyanobacteria by Fujita and co-workers [12]as well as for *D. tertiolecta* by Escoubas *et al.*[13].

2.2 CEREALS

Recently, we reported that photosynthetic acclimation to either low temperature or high light in wheat and rye can also be explained as a response to the redox state of PSII [1, 14]. Plants grown at either 5/250 or 20/800 are photosynthetically adjusted to high PSII excitation pressure whereas plants grown at either 5/50 or 20/250 are acclimated to growth and development at low PSII excitation pressure [1]. In contrast to algae, wheat and rye plants grown under high PSII excitation pressures exhibit no significant differences in either Chl a/b ratios or the contents of Lhcb1 and Lhcb2, D1, Cyt *f*, PsaA/PsaB and CF1 [15]. The increased resistance to photoinhibition observed in wheat and rye grown under high PSII excitation pressures is largely a consequence of an increased Pmax. Thus, cold tolerant cereals adjust to growth and development under high PSII excitation pressures by increasing sink capacity [14, 16].

A strong correlation has been established between the non-photochemical dissipation of excess light energy and the zeaxanthin content, thus, protecting PSII reaction centers from over-excitation [4]. Xanthophyll cycle pigments are associated with the major and minor Lhcb polypeptides of LHCII and the Lhca polypeptides of LHCI [17]. Kloppstech and co-workers were the first to report that early light-inducible proteins (ELIPs) are transiently expressed during greening of etiolated barley seedlings as well as mature barley leaves exposed to high light stress [18]. Furthermore, ELIPs and the PSII-S protein, thylakoid polypeptides induced under high light stress and related to the Lhcb family of light harvesting polypeptides, may also bind carotenoids to protect the photochemical apparatus from potential photo-oxidative damage upon exposure to excess light [19].

Chloroplast biogenesis and assembly of the photosynthetic apparatus has generally been examined in monocots by exposure of etiolated seedlings to continuous or intermittent illumination [20]. Greening of dark grown seedlings results in the conversion of etioplasts characterized by the presence of prolamellar bodies to mature chloroplasts exhibiting typical granal stacks with intervening stromal thylakoids. Clearly, photoautotrophs should be especially sensitive to potential energy imbalances during the biosynthesis and assembly of the photochemical apparatus which is itself light dependent.

The results illustrated in Fig. 3 indicate that greening of etiolated wild-type (WT) barley plants under continuous illumination at either 20/250 (circles) or 20/800 (squares) results in transient fluctuations in EPS as a function of greening time. The extent of the

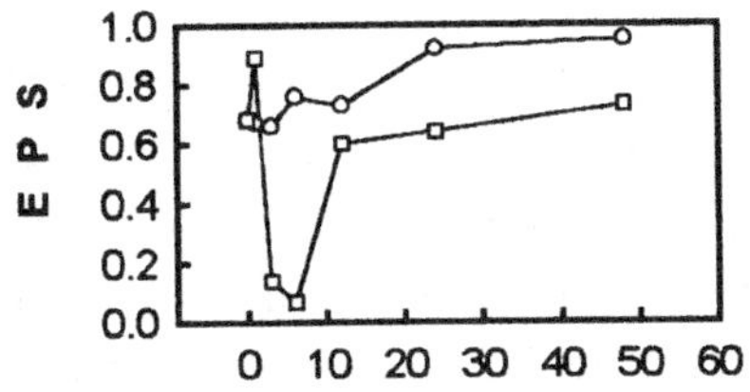

Figure 3. Changes in EPS as function of greening time for WT barley. Greening occurred at either 20/250 (circles) or 20/800 (squares).

transient fluctuations was light dependent. Similar trends were observed during the greening of the *chlorina f2* mutant. Furthermore, these trends were mimicked during greening of both the WT and the *f2* mutant at either 5/50 or 5/250 except that the greening time at which the minimum EPS was observed was shifted to 72h (data not shown). These transients in EPS reflected transient accumulations of zeaxanthin accompanied by concomitant transient decreases in the levels of violaxanthin (data not shown). The kinetics for the transients in EPS correlated with the kinetics for the transient accumulation of the 14 kDa ELIP in both the WT and *f2* mutant during greening at either 20°or 5°C. As expected, Lhcb5 did not exhibit transient kinetics but accumulated gradually during the course of greening at either 20°or 5°C in both the WT and *f2* mutant. Since Lhcb2 is not present in the *f2* mutant of barley, we conclude that this LHCII polypeptide is not essential for the regulation of the xanthophyll cycle in barley (data not shown). These patterns of accumulation for zeaxanthin and the 14 KDa ELIP are consistent with the notion that ELIPs may be zeaxanthin-binding proteins which protect the developing photosynthetic apparatus from over-excitation [19].

Greening under either high light (20/800) or low temperature (5/250, 5/50) causes an increase in the lag time prior to maximum Chl a and Chl b accumulation relative to greening in controls (20/250). Furthermore, a decrease in irradiance during greening at low temperature (5/50) resulted in a doubling in the rate of Chl accumulation in both the WT and the *f2* mutant (data not shown). These results indicate that, although light is required for Chl synthesis in angiosperms, light can also inhibit Chl accumulation and the development of the photochemical apparatus. More important, the optimal irradiance for greening appears to be temperature dependent indicating an important interaction between light and temperature during chloroplast biogenesis. Assuming that the transient fluctuations in EPS during greening are due to the modulation of the xanthophyll cycle primarily, the signal which induces the observed fluctuation in EPS cannot be due to either a light or temperature *per se*. We suggest that the fluctuations in EPS and ELIP levels reflect a response to transient imbalances between energy absorbed by the photochemical apparatus versus energy utilized by metabolism due to limitations in photosynthetic capacity during various stages of chloroplast biogenesis [1]. This is supported by the fact that development at either 20/800 or 5/250 resulted in a higher reduction state of PSII than development at either 20/250 or 5/50. This conclusion is consistent with the recent report of Montane *et al.* [21] regarding the regulation of ELIP accumulation by PSII excitation pressure rather than light or temperature.

3. References

1. Huner, N.P.A., Öquist, G. and Sarhan, F. (1998) Energy balance and acclimation to light and cold, *Trends Plant Sci.* **3**, 224-230.
2. Durnford, D.G. Falkowski, P.G. (1997) Chloroplast redox regulation of nuclear gene transcription during photoacclimation, *Photosynth. Res.* **53**, 229-241.
3. Schreiber, U., Schliwa, U., and Bilger, W., (1986) Continuous recording of photochemical and non-photochemical chlorophyll fluorescence quenching with a new type of modulation fluorometer, *Photosynth.. Res.* **10**, 63-73.
4. Demmig-Adams, B. and Adams III, W.W. (1996) The role of xanthophyll cycle carotenoids in the protection of photosynthesis, *Trends Plant Sci.* **1**, 21-26.
5. Tanaka, A. and Melis, A. (1997) Irradiance dependent changes in the size and composition of the chlorophyll *a-b* light harvesting complex in the green alga *Dunaliella salina, Plant Cell Physiol.* **38**, 17-24.
6. Maxwell, D.P., Falk, S. and Huner, N.P.A. (1995) Photosystem II excitation pressure and development of resistance to photoinhibition I. LHCII abundance and zeaxanthin content in *Chlorella vulgaris, Plant Physiol*. **107**, 687-694.
7. Maxwell, D.P., Laudenbach, D.E. and Huner, N.P.A. (1995) Redox regulation of light-harvesting complex II and *cab* mRNA abundance in *Dunaliella salina* , *Plant Physiol.* **109**, 787-795.
8. Król, M., Maxwell, D.P. and Huner, N.P.A. (1997) Exposure of *Dunaliella salina* to low temperature mimics the high light-induced accumulation of carotenoids and the carotenoid binding protein (Cbr), *Plant Cell Physiol.* **38**, 213-216.
9. Machalek, K.M., Davison, I.R. and Falkowski, P.G. (1996) Thermal acclimation and photoacclimation of photosynthesis in the brown alga *Laminaria saccharina, Plant Cell Environ.* **19**, 1005-1016.
10. Savitch, L.V., Maxwell, D.P. and Huner, N.P.A. (1996) Photosystem II excitation pressure and photosynthetic carbon metabolism in *Chlorella vulgaris, Plant Physiol.* **111**, 127-136.
11. Morgan, R.M., Ivanov, A.G., Priscu, J.C., Maxwell, D.P. and Huner, N.P.A. (1998) Structure and composition of the photochemical apparatus of the antarctic green alga, *Chlamydomonas subcaudata, Photosynth. Res.*, in press.
12. Fujita, Y., Murakami, A., Aizawa, K. and Ohki, K. (1994) Short-term and long-term adaptation of the photosynthetic apparatus: homeostatic properties of thylakoids, in Bryant, D.A. (ed.) *Advances in Photosynthesis, Vol.* **1** ,*The molecular biology of cyanobacteria*, Kluwer Academic, Dordrecht, pp. 677-692.
13. Escoubas, J-M., Lomas, M., LaRoche, J. and Falkowski, P.G. (1995) Light intensity regulates cab gene transcription via the redox state of the plastoquinone pool in the green alga, *Dunaliella tertiolecta, Proc. Natl. Acad. Sci. U.S.A.* **92**, 10237-10241.
14. Gray, G.R., Savitch, L.V., Ivanov, A.G. and Huner, N.P.A. (1996) Photosystem II excitation pressure and development of resistance to photoinhibition. II. Adjustment of photosynthetic capacity in winter wheat and winter rye, *Plant Physiol.* **110**, 61-71.
15. Gray, G.R., Ivanov, A.G., Krol, M. and Huner, N.P.A. (1998) Adjustment of thylakoid plastoquinone content and photosystem I electron donor pool size in response to growth temperature and growth irradiance in winter rye (*Secale cereale* L.), *Photosynth.. Res.* **56**, 209-221.
16. Savitch, L.V., Gray, G.R. and Huner, N.P.A. (1997) Feedback-limited photosynthesis and regulation of sucrose-starch accumulation during cold acclimation and low-temperature stress in a spring and winter wheat, *Planta* **201**, 18-26.
17. Bassi, R., Pineau, B., Dainese, P. and Marquardt, J. (1993) Carotenoid-binding proteins of photosystem II, *Eur. J. Biochem.* **121**, 297-303.
18. Meyer, G. and Kloppstech, K. (1984) A rapidly light-induced chloroplast protein with high turnover coded for by pea nuclear DNA, *Eur. J. Biochem.* **138**, 201-207.
19. Adamska, I. (1997) ELIPs-light-induced stress proteins, *Physiol. Plant.* **100**, 794-805.
20. Akoyunoglou, G. (1984) Thylakoid biogenesis in higher plants: assembly and reorganization, in C. Sybesma (ed.) *Advances in Photosynthesis Research, Vol.* **4**, *Martinus Nijhoff/Dr W Junk Publishers, The Hague*, pp. 595-602.
21. Montane, M.H., Dreyer, S., Triantaphylides, C., and Kloppstech, K. (1997) Early light-inducible proteins during long-term acclimation of barley to photooxidative stress caused by light and cold - high level of accumulation by posttranscriptional regulation, *Planta* **202**, 293-302.

INFLUENCE OF A CO_2-PARTIAL PRESSURE OF 700 PPM ON THE LIPID AND FATTY ACID COMPOSITION OF HIGHER PLANTS.

G. H. Schmid[a], A. Radunz[a], P. He[b] and Ch. Kedeinis[a]
[a]Lehrstuhl für Zellphysiologie, Fakultät für Biologie, Universität Bielefeld, Germany
[b]Central South Forestry University, Zhuzhou/Hunan 41206, People's Republic of China

Introduction

The CO_2-content of the atmosphere is steadily increasing (1,2). It is generally assumed that the CO_2-content of air will increase within the next 50 years from now 350 ppm to 700 ppm. In context with this expected atmospheric change it is of interest to know what impact the CO_2-increase will have on growth of higher plants. With *Nicotiana tabacum* and the Chinese "Tung Oil Tree" *Aleurites montana* we were able to show that under the influence of the increased CO_2-content 30-40% more biomass is formed (3,4,5). Differences in dependence on the CO_2-content are observed for the chlorophyll and sugar content on the one hand but also with respect to the content of the biomass forming enzyme ribulose 1.5-bisphosphate carboxylase and the energy equivalent-forming coupling factor of photophosphorylation CF_1. Molecular changes are observed in the composition of peptides of photosystem II and photosystem I. Thus, the light-harvesting complex of photosystem II (3,4) is reduced and the same is true for photosystem I (5). The oxygen-evolving-complex (OEC)-peptides are reduced in plants cultivated under increased CO_2 (3,4). In addition the electron transport components (plastoquinone, cytochrome f, plastocyanin, ferredoxin, ferredoxin-NADP-reductase) are reduced under these conditions as shown by rocket-immune-electrophoresis (6).

Lipids not only form the bimolecular lipid membrane, but are also adsorbed as prosthetic groups onto peptides and peptide complexes such as the light harvesting complex in photosystem I and photosystem II (7-9) and the CF_1 complex of the ATPase (10). Lipids are important for the anchoring of these peptides within the membrane and the holding together of the peptides within the complex and also play an important role for the conformational state of these functional peptides. Therefore, it was to be expected that the lipids with their fatty acid components which decisively influence the fluidity of a membrane would suffer changes under the influence of an increased CO_2-partial pressure. In order to obtain an information on the extent of the lipid changes, in orienting studies the leaf lipids of *Aleurites montana,* as a representative of a multiannual plant (4), and those of the wild-type of *Nicotiana tabacum* var. JWB as a representative of an annual plant, were analyzed. In addition, in comparison the chlorophyll deficient tobacco mutant *N. tabacum* Su/su was analyzed. The chloroplasts of wild type tobacco and of the Su/su mutant differ in their chloroplast structure and their photosynthetic performance (11-15).

J. H. Argyroudi-Akoyunoglou and H. Senger (eds.),
The Chloroplast: From Molecular Biology to Biotechnology, 337–342.

Material and Methods

1. Cultivation of Plants

Plants were cultivated in growth chambers at 12.000 lux and a light/dark cycle of 14/10 hours, a CO_2-partial pressure of 350 ppm and the relative humidity of 60%. *CO_2-Plants* of *Nicotiana tabacum* var. John William's Broadleaf and of the Su/su-mutant, as well as of *Aleurites* were cultivated in the same growth chambers in separated glass-compartments under a CO_2-partial pressure of 700 ppm. Harvesting of tobacco plants was done 4 weeks after transplantation of the cotyledons. *Aleurites* being a multiannual plant was cultivated for 7 months in the growth chamber, *CO_2-Plants* of *Aleurites* were cultivated for 5 months in the glass compartments under the increased CO_2-concentration (after an initial 2 months adaptation under normal CO_2 in the same growth chamber, hence, 7 months in total).

2. Lipid and Fatty Acid Analyses

The harvested leaves were freeze-dried, ground to powder and the lipids extracted on a glass sinter funnel with methanol, acetone and then with diethylether. The lipids were taken up in diethylether and separated from insoluble components by centrifugations at 5.000 xg. Identification was achieved by thin layer chromatography on silica gel G with the solvent chloroform/methanol/acetic acid/water (85:15:10:3.5, v/v/v/v). For quantitative determination the lipids were first separated by thin layer chromatography and the glycolipids analyzed with the anthrone-sulfuric acid reagent. The phospholipids were analyzed photometrically with the Fiske-Subbarow reagent (4). Fatty acids were analyzed by gas chromatography on a capillary column with the stationary phase being ethylene glycol-succinate. The column temperature was 190°C and the detector and injection temperature 300°C. The gas chromatograph was from Hewlett Packard, type 5890, series II plus.

Results

Ether soluble lipids make up for 11.5% of leaf dry weight in the wild-type tobacco and for 10.8% in the chlorophyll-deficient tobacco mutant Su/su. The lipid composition of the tobacco mutant is characterized in comparison to that of the wild type, by the fact that due to the reduced size of the light harvesting complex and the thereby reduced grana stacks (11-14), the chlorophyll content is only half as high as in the wild type with carotenoids being substantially reduced. Due to this, also the phospholipids and among the glycolipids the monogalctolipid appear in higher concentration (Table I).

Under the influence of the increased CO_2-partial pressure in the atmosphere, the lipid composition in the wild type as well as in the mutant undergoes a quantitative change. As can be seen from Table I in the wild-type as well as in the mutant, a decrease in the monogalactosyldiglyceride(MGDG)-content is observed together with a simultaneous increase in digalactosyldiglyceride (DGDG) and sulfoquinovosyldiglyceride (SQDG), which means that the total amount of glycolipids stays more or less constant. However, with phospholipids differences between the two plants are observed. Whereas in the wild-type total phospholipids, with the exception of the extraplastidic phosphatidylethanolamine, increase, the phospholipids seem to decrease in the mutant. Chlorophylls and carotenoids behave in the wild-type and the mutant under the influence of the increased CO_2-content differently. Whereas the chlorophylls in the wild-

Table I. Composition of the purified ether soluble leaf lipids of *Nicotiana tabacum* var. JWB and of the mutant Su/su grown for 3 weeks under 700 ppm CO_2 in air.

Plants	Chlorophyll			Glycolipids				Phospholipids						Carotenoids					Chlorophyll / Carotenoid	Glyco-/ Phospholipids
	a	b	a/b	MGDG	DGDG	SQDG	total	CA	PE	PG	PC	PI	total	β-Ca	Lu	Vio	Neo	total		
JWB																				
Control plants	10.7	4.1	2.6	39.2	19.8	0.7	**59.7**	0.8	4.0	3.3	8.5	1.3	**17.9**	1.5	4.7	0.69	0.69	**7.6**	2.1	3.3
CO_2-plants	10.3	4.3	2.4	35.3	22.7	1.5	**59.5**	0.9	3.8	3.7	9.1	1.8	**19.3**	1.2	4.1	0.56	0.74	**6.6**	3.9	3.1
Su/su																				
Control plants	4.9	1.6	3.0	41.7	19.9	0.9	**62.5**	0.9	6.4	4.7	12.2	2.3	**26.5**	1.2	2.4	0.60	0.27	**4.5**	1.4	2.4
CO_2-plants	6.0	2.5	2.4	38.4	23.3	1.1	**62.9**	0.6	5.5	4.0	11.7	1.6	**23.4**	1.6	3.1	0.29	0.25	**5.2**	1.6	2.7

Lipid values are given in per cent of total lipids which were put 100%. Not included in this calculation are plastoquinones, tocopherols and waxes. Values are averages of 6 to 8 individual determinations. The deviation of the values given is between 2 to 3 percent.

MGDG = Monogalactosyldiglyceride, DGDG = Digalactosyldiglyceride, SQDG = Sulfoquinovosyldiglyceride, CA = Cardiolipin, PE = Phosphatidylethanolamine, PG = Phosphatidylglycerol, PC = Phosphatidylcholine, PI = Phosphatidylinositol, β-Ca = β-Carotene, Lu = Lutein, Vio = Violaxanthin, Neo = Neoxanthin.

type besides a slight increase in chlorophyll b stay constant, the carotenoids decrease. In the mutant, however, chlorophylls and carotenoids are increased. Here, it should be noted that chlorophyll b increases stronger than chlorophyll a, which could mean that in the *CO_2-plants* a stronger expression of the light-harvesting complex occurs.

Corresponding to the changes in the lipid composition in the wild-type, namely the decrease in MGDG and the increase of the phospholipids, an increase of the saturation degree of the fatty acids under the increased CO_2-content was expected. This is indeed the case as seen from Table II. Corresponding to the decomposition of MGDG the two main components of this membrane lipid, namely linolenic and the hexadecatrienoic acids, decrease. In total, the unsaturated fatty acids decrease from 84% in control plants to 75% in *CO_2-plants* and saturated fatty acids increase from 16% to 25% of total fatty acids. Also unusually high is the increase of the unsaturated fatty acids with 12 carbon atoms from 4 to 8% in *CO_2-plants*. Besides the changes in the saturation degree under the influence of the increased CO_2-partial pressure also changes in the chain length are observed. Thus, fatty acids with 18 carbon atoms are decreased at the expense of fatty acids with 12 carbon atoms. Apparently, a compensation of the increase of saturated fatty acids is achieved by an increase of fatty aids with shorter chain lengths in order to obtain a defined degree of flexibility of the membrane. The lipids of the mutant are characterized by the fact that MGDG as well as the phospholipids are reduced. This observation lets assume, since MGDG contains the trienoic fatty acids and since the phospholipids contain only by 50% unsaturated fatty acids, that the fatty acid pattern would stay constant. This , however, is not the case. Due to the reduction of the phospholipids the unsaturated fatty acids stay constant or increase slightly. The decrease of the MGDG must have obviously been compensated by the increase in linolenic acid. The 2nd fatty acid typical for MGDG, namely hexadecatrienoic acid remains constant (Table II).

Discussion

Quantitative comparative analyses by thin layer chromatography of lipids and gas chromatographic analyses of fatty acids have led to the result, that with *N. tabacum* under the influence of a CO_2-partial pressure of 700 ppm in comparison to the normal pressure of 350 ppm quantitative modifications occur in the lipid and fatty acid composition. Phospholipids as components of mitochondria and cell membranes increase in the wild type and decrease in the mutant. The monogalactolipid as a component of the thylakoid membrane and of the envelope suffers a reduction. As the monogalactolipid contains up to 90% polyenoic fatty acids and as phospholipids are built up by approx. one half saturated and by one half of unsaturated fatty acids, in parallel with these lipid shifts, the polyenoic fatty acids decreased and the saturated fatty acids increase. In addition a shortening of the chain length of the fatty acids occurs (Table II).

Lipid analyses in leaves of the Chinese "tung oil tree" *Aleurites montana*, taken as an example of a multiannual plant, which has been grown for 5 months under identical conditions of an increased CO_2-partial pressure of 700 ppm, led in their lipid and fatty acid composition to the same modifications. The difference to tobacco was that not only monogalactosyldiglyceride but also the digalactolipid and the sulfolipid suffered a reduction under these conditions (4).

From these quantitative changes observed for glycolipids and phospholipids as well as from the shift of the saturation degree of the fatty acids we can conclude that

Table II. Comparison of the fatty acid composition of the leaf lipids of *Nicotiana tabacum* var. JWB and of the mutant Su/su grown for 3 weeks under 700 ppm CO_2 in air. Fatty acids are given in % of total fatty acids.

Plants	$C_{12:0}$	$C_{12:3/4}$	$C_{14:0}$	$C_{16:0}$	$C_{16:1}$ cis	$C_{16:1}$ trans	$C_{16:2}$	$C_{16:3}$	$C_{18:0}$	$C_{18:1}$	$C_{18:2}$	$C_{18:3}$	C_{16}	C_{18}	C_{16}/C_{18}	$C_{:0}$	$C_{:1-4}$	$C_{:0}/C_{:1-4}$
JWB																		
Control plants	0.2	3.6	1.3	13.1	0.6	2.6	0.8	7.2	1.5	2.3	15.0	51.8	24.3	70.6	0.34	16.1	83.9	0.19
CO_2-plants	0.4	8.2	5.7	15.7	0.7	2.4	0.7	4.8	3.1	2.4	13.6	42.4	24.3	61.4	0.40	24.8	75.2	0.27
Su/su																		
Control plants	0.1	1.1	0.4	13.5	0.4	2.1	0.4	9.1	1.4	1.3	12.6	57.8	25.4	73.0	0.34	15.4	84,6	0.18
CO_2-plants	0.2	1.0	1.1	10.6	0.3	1.9	0.5	8.0	1.3	1.2	12.9	60.9	21.3	76.4	0.28	13.2	86.8	0.15

C_{16}: fatty acids with 16 carbon atoms; C_{18}: fatty acids with 18 carbon atoms; $C_{:0}$: saturated fatty acids; $C_{:1-4}$: unsaturated fatty acids.

under the influence of the increased CO_2-partial pressure the fluidity of the membranes has been changed, but in addition also the molar content/ratio of the lipid-protein-complexes of the thylakoid membrane as well as of cell membranes has been changed.

Moreover, in studies of the photosynthetic activity in the region of Photosystem I when analyzing the reduction of methylviologen with the electron donor couple dichlorphenolindophenol (DCPiP), a shift of the pH-optimum to the acid region is observed. Earlier analyses of photosystem I and II have shown that above all monogalactolipid with its highly unsaturated fatty acids is an essential component of photosystem I (16). Whether the observed decrease of the monogalactolipid content has to be seen in context/relation with the functional diminution of photosystem I will be elucidated in future studies.

Literature

1. Houghton J.T., Jenkins G.J., Ephraums J.J. and Woodwell G.M. (1990), Climate change: the IPCC scientific assessment. Cambridge University Press, Cambridge, U.K.
2. Lindzen R. (1993), Res. Explor. 9. 191-200.
3. He P., Radunz A., Bader K.P. and Schmid G.H. (1996a), Z. Naturforsch. 51c, 441-453.
4. He P., Radunz A., Bader K.P. and Schmid G.H. (1996b); Z. Naturforsch. 51c, 833-840.
5. Makewicz A., Radunz A. and Schmid G.H. (1995), Z. Naturforsch. 50c, 511-520.
6. Alfermann K., Radunz A. and Schmid G.H., unpublished.
7. Voß R., Radunz A. and Schmid G.H. (1992), Z. Naturforsch. 47c, 46-415.
8. Kruse O., Radunz A. and Schmid G.H. (1994), Z. Naturforsch. 49c, 115-124.
9. Makewicz A., Radunz A. and Schmid G.H. (1996), Z. Naturforsch. 51c, 319-328.
10. Haase R., Unthan M., Couturier A., Radunz A. and Schmid G.H. (1993), Z. Naturforsch. 48c, 623-631.
11. Schmid G.H., Price J.M. and Gaffron H. (1966), J. de Microscopie 5, 205-216.
12. Schmid G.H. (1967), J. de Microscopie 6, 485-497.
13. Schmid G.H. and Gaffron H. (1967), J. Gen. Physiol. 50, 563-582.
14. Homann P.H.and Schmid G.H. (1967); Plant Physiol. 42, 1619-1632.
15. Radunz A. and Schmid G.H. (1989), Z. Naturforsch. 44c, 689-697.
16. Bednarz J., Radunz A. and Schmid G.H. (1988), Z. Naturforsch. 43c, 423-430.

THE STABILIZING EFFECT OF GLYCINE BETAINE ON THE MODIFIED PHOTOSYSTEM II PARTICLES UNDER HEAT STRESS

Y. M. ALLAKHVERDIEVA , N. N. FERIMAZOVA &
R. A. GASANOV
Department of Biophysics, Institute of Botany, Azerbaijan Academy of Sciences, Matbuat ave. 2, Baku 370073, Azerbaijan

1. Introduction

The compatible solute glycine betaine is strongly implicated as an osmoprotectant in cyanobacteria, green algae and higher plants. It was shown that glycine betaine protects photosystem II (PSII) complex against the high salt induced dissociation of the extrinsic proteins ant the manganese cluster (1). A similar type of protective effect of glycine betaine has been observed in the oxygen evolution by cyanobacterial thylakoid membranes against heat inactivation (2). By contrast, glycine betaine as assumed has no stabilizing effect on processes not connected with the oxygen evolving mechanisms of the PSII complex.

The data recently obtained suggest that, in addition to water oxidation, partial electron transfer reactions more intimate to the reaction center PSII complex are stabilized by glycine betaine (3).

In the present study we examined systematically the effects of glycine betaine on heat stability of photosynthetic electron transport reactions in various type PSII particles (with selectively removed peripheral proteins).

2. Materials and methods

BBY-type PSII particles were prepared from spinach leaves *(Spinacia oleracea)* as described earlier (4) and kept at liquid nitrogen until use. Oxygen evolving (OE) PSII particles depleted of 17- and 23-kDa extrinsic polypeptides (d PSII) were obtained as described by (5). PSII particles lacking all three extrinsic polypeptides (dd PSII) were obtained as described by (6). The samples were incubated for 10 min at designated temperatures in darkness in a medium containing 25 mM MES-NaOH (pH 6.5), 10 mM NaCl, 0.4 mM sucrose in the absence or presence of 0.5 M glycine betaine.The suspension was then cooled down to 20°C and after the addition of various electron donors and acceptors photosynthetic activities were measured. The rate of oxygen evolution was recorded using Clark-type closed Pt electrode at saturating intensity of white light in the presence of 0.4 mM PBQ as an exogenous electron acceptor. The photoreduction of DCIP by dd PSII particles was measured spectrophotometrically by monitoring the absorption changes at

J. H. Argyroudi-Akoyunoglou and H. Senger (eds.),
The Chloroplast: From Molecular Biology to Biotechnology, 343–346.

600 nm, with DPC as artificial electron donor. In all experiments, samples were added to a final concentration that corresponded to 10 μg Chl/ml in the oxymetric measurements and 5 μg Chl/ml in the spectrophotometric measurements.

3. Results

The effect of high temperature on the transport of electrons from H_2O to PBQ in the PSII particles is shown in Figure 1. When the samples were heated without glycine betaine for 10 min at various temperatures the extent of inactivation was considerable at a temperature above 30°C. In the presence of glycine betaine, by contrast, the rate of oxygen evolution in PSII particles increased up to 35°C and then declined very slowly in samples treated at more elevated temperatures. Temperatures for 50% inactivation were found to be about 38°C in the absence and 47°C in the presence of glycine betaine. This observation is consistent with previous data derived from thylakoid membranes from *Synechocystis sp.* PCC6803 (2).

Figure 2 shows the effect of high temperature on the oxygen-evolving activity of the dPSII particles. In the course of rise of incubation temperature the control dPSII particles demonstrated very quick fall in the rate of oxygen evolving activity. The OE activity of dPSII particles in the presence of glycine betaine remained practically equal to that in untreated PSII in the presence of glycine betaine (Fig. 1). The results presented here (Fig. 2) strongly suggest the protective role of glycine betaine against heat destabilization of dPSII particles. At the same time, in the presence of glycine betaine, the temperature for 50% inactivation, was the same as in the case of dPSII particles and PSII particles. This indicates that heat stress results in dissociation of 33 kDa protein from the OE complex and glycine betaine prevents this process.

For conviction, we analyzed the effect of glycine betaine on the DCIP photoreduction with dd PSII particles. Figure 3 shows that there is a stabilizing effect of glycine betaine on electron transport reaction with dd PSII particles which have lost all peripheral proteins that maintain oxygen evolution of the complex. Moreover, 50% inactivation of electron transport with dd PSII particles took place at higher temperatures both in the presence and in the absence of glycine betaine. This result could be explained by greater thermostability of RC PSII than the oxygen-evolving complex in general. Glycine betaine shifts the temperature for 50% inactivation of the DCIP photoreduction from 47°C to 55°C. The result obtained with dd PSII particles confirms the stabilizing effects of glycine betaine on the reactions near RC PSII including probably the acceptor side of this photosystem.

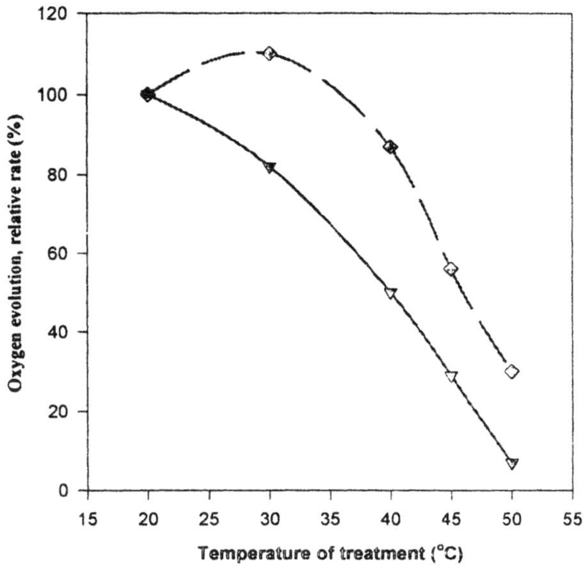

Figure 1. Dependence on incubation temperature of the rate of oxygen evolution in **PSII** particles. Samples were incubated for 10 min at designated temperatures in 25 mM Mes-NaOH (pH 6.5), 10 mM NaCl, 0,4 M sucrose and then cooled to 20°C. After the addition of 0,4 mM PBQ the electron-transport activity was measured by monitoring the oxygen evolved. ▼ - in the absence of glycine betaine ■ - in the presence of 0,5 M glycine betaine

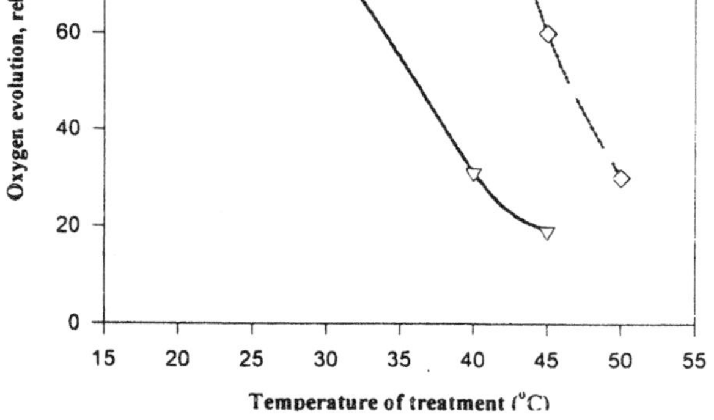

Figure 2. Dependence on incubation temperature of the rate of oxygen evolution in **dPSII** particles. Samples were incubated for10 min at designated temperatures in 25 mM Mes -NaOH (pH 6.5), 10 mM NaCl, 0,4 M sucrose and then cooled to 20°C. After the addition of 0,4 mM PBQ the electron-transport activity was measured by monitoring theoxygen evolved. ▼ - in the absence of glycine betaine ■ - in the presence of 0,5 M glycine betaine

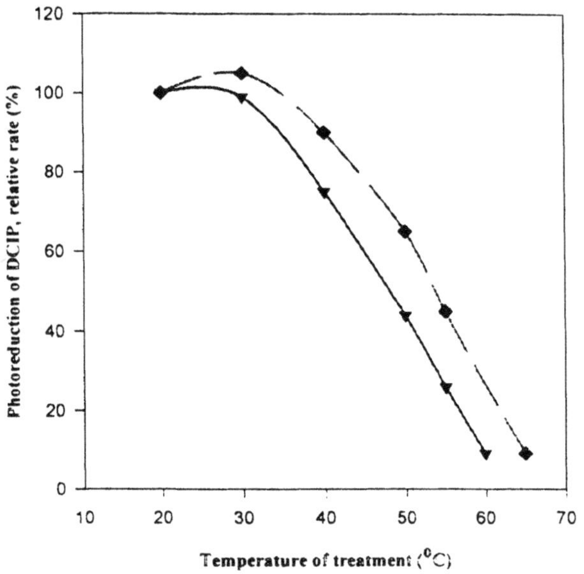

Figure 3. Dependence on incubation temperature of the rate of electron-transport from DPC to DCIP in ddPSII particles. The samples were incubated for 10 min at designated temperatures in 25 mM Mes-NaOH (pH 6.5), 10 mM NaCl, 0,4 M sucrose and then cooled to 20°C. After the addition of 0,05 mM DCIP, 0,5 mM DPC the electron-transport activity was measured by monitoring the photoreduction of DCIP ▼ - in the absence of glycine betaine ■ - in the presence of 0,5 M glycine betaine

4. References

1. Papageorgiou, G.C. and Murata, N. (1995) *Photos. Res.* 44, 243 - 252
2. Mamedov, M.D., Hayashi, H., Murata, N. (1993) *Biochim. Biophys. Acta* 1142, 1-5
3. Allakhverdiev, S.I., Feyziev, Ya.M., Ahmed, A., Hayashi, H., Aliev, Ja.A., Klimov, V.V., Murata, N., Carpantier, R. (1996) *J. Photochem. Photobiol.* 34, 149-157
4. Bertold, D.A., Babcock, G.T. and Yocum, C.F. (1981) *FEBS Lett* 134: 231-234
5. Kuwabara, T. and Murata, N. (1983) *Plant Cell Physiol.* 24, 741-747
6. Ono, T. and Inoue, Y. (1984) *FEBS Lett.* 168, 281-286

APPLICATION OF THYLAKOIDS IN HERBICIDE-DETECTING SYSTEM

E.V. PILETSKAYA[a], S.A. PILETSKY[b], T.A. SERGEYEVA[b],
A.V. EL'SKAYA[b], A.A. SOZINOV[a], J.-L. MARTY[c], R. ROUILLON[c]

[a]*Institute of Agroecology and Biotechnology, Academy Agrarian Sciences of Ukraine, Metrologicheskaya, 12 - 252143 Kiev, Ukraine*
[b]*Institute of Molecular Biology and Genetics, Academy of Sciences of Ukraine, Zabolotnogo 150 - 252143 Kiev, Ukraine*
[c]*Universite de Perpignan, Centre de Phytopharmacie, URA CNRS 461 Avenue de Villeneuve 52 - 66860 Perpignan Cedex, France*

1. Abstract

A simple method of the photosynthesis inhibiting herbicides detecting is proposed as practical approach for environmental monitoring. Method combines the microtiter-plate assay based on the inhibition of the thylakoid membrane activity with the sample preconcentrating on the solid-phase cartridges. The C_{18} cartridge provides acceptable recovery (>85%) of atrazine, simazine and cyanazine in broad range of concentrations. The method proposed allows detecting the photosynthesis inhibiting herbicides in the concentration range of 0.025–25 μg/l. To improve system stability thylakoid immobilization in poly(vinylalcohol) bearing styrylpyridinium groups (PVA-SbQ) on microtiter plates is performed.

2. Introduction

Almost uncontrolled utilization of the triazine herbicides results in pollution of the agricultural areas and, subsequently, drinking water and food. The detection of minor quantities of herbicides becomes now the barest necessity for both health care and economic demands. The most common methods of herbicide detection using now are gas chromatography with mass spectrometry detection (GC/MS) and high performance liquid chromatography (HPLC) [1]. Last years a new kinds of biosensors for herbicide detection, based on usage of antibodies were developed [2, 3].
Our method is based on the binding of the certain classes of herbicides to the photosynthetic reaction center. The application of the receptor properties of thylakoid membranes seems the most direct scheme of biorecognition assay suitable to detect all possible herbicides, which inhibit photosystem II. The assay explores the catalytic photoreduction of redox dye by the photosystem II reaction centers in accordance with Hill equation:

J. H. Argyroudi-Akoyunoglou and H. Senger (eds.),
The Chloroplast: From Molecular Biology to Biotechnology, 347–350.

$$2H_20 \quad + \quad 2A \xrightarrow[\text{thylakoid membranes}]{\text{light}} 2AH_2 \quad + \quad O_2$$

where A - 2,6-dichlorphenolindophenol (DPIP) is electron and proton acceptor, AH_2 - its reduced form. Under illumination thylakoids reduces the blue DPIP to the colorless leuco form in minutes, while herbicide inhibited thylakoid suspension remains blue.

3. Experimental

3.1. MATERIALS

All chemicals were obtained from commercial sources and were of analytical or HPLC grade: DPIP (Sigma), herbicides (Erhenstorfer, Germany), PVA-SbQ with degree of polymerization 3500, degree of saponification 88, SbQ content 0.83 mol %, was a gift from Toyo Gosei Kogyo, Japan.

3.2. THYLAKOID MEMBRANES PREPARATION

Thylakoid membranes were isolated from spinach leaves (*Spinacea oleracea L.*) accordingly by [4].

3.3. DETERMINATION OF THYLAKOID MEMBRANE ACTIVITY AND HERBICIDE ASSAY

30 μl of thylakoid membranes suspension (0.9 μg chlorophyll in the probe) was mixed with 20 μl distilled water (control) or herbicide solution, 100 μl 0.3 mM DPIP in 0.1 M 4- morpholinepropanesulfonic acid (MOPS), adjusted to pH 6.5, in the microtiter plate wells. The samples were illuminated for 10 min with 100-W lamp and absorbance was measured at 530 nm using reader Titertec Uniscan II (France). The activity of thylakoid membranes preparations was calculated from the data on the amount of reduced DPIP, concentration of chlorophyll and the time of illumination. The triazine herbicides (atrazine, simazine, and cyanazine), triazinone herbicide (metribuzin) and phenylurea (diuron) were used in herbicide assay.

3.4. HERBICIDE SOLITIONS PRECONCENTRATING

All herbicides were dissolved in the distilled water at concentrations 0.1-10 nM. Solid-phase extraction cartridges containing the solid-phase silicon particles with attached 18-carbon chains (C_{18}) were used. One liter of herbicide solution was passed through the cartridge with a flow rate 5 ml min^{-1}. After the cartridge drying in air stream the herbicide extraction was performed with 1 ml of ethyl acetate.

3.5. THYLAKOID MEMBRANES IMMOBILIZATION WITH PVA-SbQ

1.5 g PVA-SbQ was diluted by 5 time with buffer containing 0.35 M saccharose, 10 mM sodium chloride, 10 mM MOPS, pH 7.9, 1% BSA and mixed with 250 μl of thylakoid membranes suspension (2 mg chlorophyll·ml^{-1}). The 150 μl aliquots of this mixture were placed in the microtiter plate wells. To remove access of the water microtiter plates were dried during 24 hours at +4°C in the dark until 30 μl of solution left. Resulting mixture was exposed to UV radiation for 2 min, incubated at –20°C for 2 days and stored in a dry state at +4°C or at –20°C in the dark.

4. Results and discussion

Application of Hill reaction for the herbicides detection is very attractive, because of the possibility to detect simultaneously and specifically all photosynthesis inhibiting herbicides, presented in the sample. However, it has limitation, connected with extremely short life of isolated thylakoid membranes. Also it is found that atrazine, cyanazine and simazine at low concentrations (less than $1\cdot10^{-7}$ M) have small inhibitory effect on the native thylakoids. The detection limit of a 10% thylakoid membranes inhibition by sample containing a herbicide is determined as $5\cdot10^{-7}$ M for atrazine, cyanazine and simazine. Because of stronger inhibition it is possible to detect metribuzin and diuron at lower concentrations ($3\cdot10^{-8}$ M and $1.5\cdot10^{-8}$ M correspondingly). All photosynthesis inhibiting herbicides can be determined simultaneously at concentration range $5\cdot10^{-7}$ - $1\cdot10^{-5}$ M (see Fig.1).

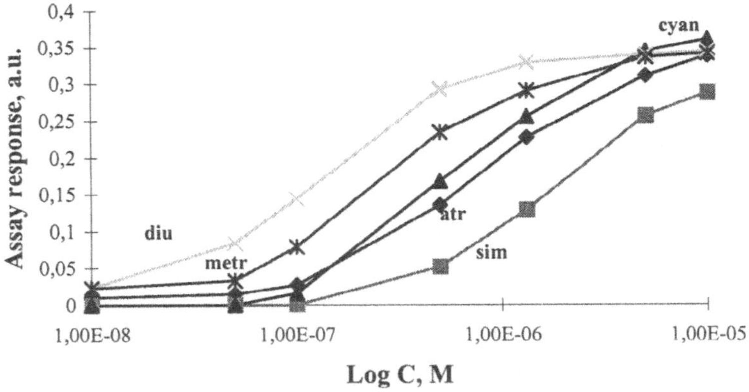

Figure 1. Assay response for herbicide concentrations using native thylakoid membranes. Herbicides are atrazine (atr), simazine (sim), cyanazine (cyan), metribuzin (metr), diuron (diu).

To perform a sample preconcentrating solid-phase C_{18} cartridges were used for herbicides extraction giving 1000 enrichment. C_{18} cartridges are suitable for the samples preconcentration in concentrations 0.1 nM - 10 nM (see Table 1).

TABLE 1. Atrazine recovery on C_{18} cartridges for different herbicide concentrations

Atrazine concentration	Water	0.1 nM	0.5 nM	1 nM	5 nM	10 nM
Recovery	0	110%	93%	85%	97%	87%

Immobilization in PVA-SbQ was used in order to improve stability of thylakoid membranes. The PVA-SbQ with entrapped thylakoids is homogeneous and transparent film covering the microtiter plate wells. The advantage of this method is phenomenal stability of the prepared polymers which makes possible to discuss idea of the commercial test-kits production. The results of these studies are presented in Table 2.

TABLE 2. A comparison of stability and sensitivity of native and PVA-SbQ-entrapped thylakoid membranes

Thylakoid preparations	Activity, μmol DCPIP/h·mg chl	Detection limit	Stability(-20°C)	Reproducibility
Native	50	0.5 μM	7 h	excellent
PVA-SbQ	25.2	0.5 μM	1 year – 25% [5]	good

5. Conclusions

The method, based on the combination of preconcentration step on C_{18} cartridges and Hill reaction in the microplate format permits to detect trace quantities of photosynthesis inhibiting herbicides in water at the level requested by EC regulations. The advantage of thylakoids immobilization in PVA-SbQ is high stability, which permit to store and use prepared test-system during 1 year at –20°C or 2 months at +4°C. Easy preparation of the test-kits and inexpensive reagents utilization permit to consider them as practical approach to detect herbicides in drinking water.

6. Acknowledgement

The authors are thankful to INTAS UA-0161 for financial support of the project.

7. References

1. Kisinski, H.G. (1991) Multi-residue analysis of plant protectives and metabolites in water with HPLC and diode-array detection, *CLB Chem. Labor Biotech.* **42**, 487-488, 491-493.
2. Minunni, M.and Mascini, M. (1993) Detection of pesticide in drinking water using real-time biospecific interaction analysis, *Anal.Lett.* **26**, 1441-1460.
3. Dzantiev, B. B., Zherdev, A. V., Romanenko, O. G., Sapegova, L. A. (1996) Development and comparative study of different immunoenzime techniques for pesticides detection, *Intern. J. Environ. Anal. Chem.* **65**, 95-111.
4. Rouillon, R., Mestres, J.-J., Marty, J.-L. Entrapment of chloroplasts and thylakoids in polyvinylalcohol-SbQ. Optimization of membrane preparation and storage conditions, *Anal.Chem.Acta* **311**, 437-442 (1995).
5. Rouillon, R., Sole, M., Carpentier, R., Marty, J.-L. (1995) Immobilization of thylakoids in polyvinylalcohol for the detection of herbicides, *Sensors and Actuators B* **26-27**, 477-479.

AUTHOR INDEX

SUBJECT INDEX

GPSR Compliance

The European Union's (EU) General Product Safety Regulation (GPSR) is a set of rules that requires consumer products to be safe and our obligations to ensure this.

If you have any concerns about our products, you can contact us on ProductSafety@springernature.com

In case Publisher is established outside the EU, the EU authorized representative is:

Springer Nature Customer Service Center GmbH
Europaplatz 3
69115 Heidelberg, Germany

Batch number: 10799857

Printed by Printforce, the Netherlands